Advanced Structured Materials

Volume 139

Common engineering materials reach in many applications their limits and new developments are required to fulfil increasing demands on engineering materials. The performance of materials can be increased by combining different materials to achieve better properties than a single constituent or by shaping the material or constituents in a specific structure. The interaction between material and structure may arise on different length scales, such as micro-, meso- or macroscale, and offers possible applications in quite diverse fields.

This book series addresses the fundamental relationship between materials and their structure on the overall properties (e.g. mechanical, thermal, chemical or magnetic etc.) and applications.

The topics of *Advanced Structured Materials* include but are not limited to

- classical fibre-reinforced composites (e.g. glass, carbon or Aramid reinforced plastics)
- metal matrix composites (MMCs)
- micro porous composites
- micro channel materials
- multilayered materials
- cellular materials (e.g., metallic or polymer foams, sponges, hollow sphere structures)
- porous materials
- truss structures
- nanocomposite materials
- biomaterials
- nanoporous metals
- concrete
- coated materials
- smart materials

Advanced Structured Materials is indexed in Google Scholar and Scopus.

More information about this series at http://www.springer.com/series/8611

Andrei K. Abramian · Igor V. Andrianov ·
Valery A. Gaiko
Editors

Nonlinear Dynamics of Discrete and Continuous Systems

Springer

Editors
Andrei K. Abramian
Institute of Problems in Mechanical Engineering
Russian Academy of Sciences
St. Petersburg, Russia

Igor V. Andrianov
Institute of General Mechanics
RWTH Aachen University
Aachen, Nordrhein-Westfalen, Germany

Valery A. Gaiko
United Institute of Informatics Problems
National Academy of Sciences of Belarus
Minsk, Belarus

ISSN 1869-8433 ISSN 1869-8441 (electronic)
Advanced Structured Materials
ISBN 978-3-030-53005-1 ISBN 978-3-030-53006-8 (eBook)
https://doi.org/10.1007/978-3-030-53006-8

This Springer imprint is published by the registered company Springer Nature Switzerland AG
The registered company address is: Gewerbestrasse 11, 6330 Cham, Switzerland

This book is dedicated to
Dr. Wim van Horssen
on the occasion of his 60th birthday

Dr. Wim van Horssen

Preface

This Commemorative volume is a collection of papers contributed by colleagues and disciples in tribute to eminent scientist, Dr. Wim van Horssen on the occasion of his 60th birthday. The researches of Dr. Van Horssen cover various fields of applied mechanics, in particular, ordinary and partial differential equations, difference equations, delay equations, asymptotic theory, theory of dynamical systems and bifurcation theory. Working at Delft University of Technology (The Netherlands), he brought up many pupils and created well known and highly branched scientific school that made a significant contribution to the aforementioned fields of science.

This book contains articles by well-known scientists who actively work in the fields where Dr. Van Horssen was very active over 35 years. Geographically, this volume covers researches from Belarus, P.R. China, Germany, Indonesia, The Netherlands, Russia, Serbia, Ukraine, UK and USA. It includes 16 articles

We believe that this Commemorative volume will be of great interest to researchers and practitioners in the fields of applied and pure mathematics.

St. Petersburg, Russia — Andrei K. Abramian
Köln, Germany — Igor V. Andrianov
Minsk, Belarus — Valery A. Gaiko

Contents

Contributors

Andrei K. Abramian Institute of Problems in Mechanical Engineering, Russian Academy of Sciences, St. Petersburg, Russia

Igor V. Andrianov Institute of General Mechanics, RWTH Aachen University, Aachen, Germany

Andrey V. Bochkarev Yuri Gagarin State Technical University of Saratov, Saratov, Russia

Sudipto Choudhury University of Central Florida, Orlando, Fl, USA

Yulia Danik Federal Research Center "Computer Science and Control" of Russian Academy of Sciences (FRC CSC RAS), Moscow, Russia

T. G. de Jong Media Analytics and Computing Laboratory, School of Information Science and Engineering, Xiamen University, Xiamen, China

Mikhail Dmitriev Federal Research Center "Computer Science and Control" of Russian Academy of Sciences (FRC CSC RAS), Moscow, Russia

Johan L. A. Dubbeldam Delft Institute of Applied Mathematics, Delft University of Technology, Delft, The Netherlands

I. Elishakoff Department of Ocean and Mechanical Engineering, Florida Atlantic University, Boca Raton, FL, USA

Andrei B. Fărăgău Faculty Civil Engineering & Geosciences, Tu Delft, Delft, The Netherlands

Valery A. Gaiko United Institute of Informatics Problems, National Academy of Sciences of Belarus, Minsk, Belarus

Feng Gao School of Electrical Engineering, Shandong University, Jinan, Shandong, China

Stefan Kaczmarczyk The University of Northampton, Northampton, UK

E. Kaplunov Whizz-Kidz Charity, London, UK

J. Kaplunov Department of Computer Science and Mathematics, Keele University, Keele, UK

Ivana Kovacic Faculty of Technical Sciences, Centre of Excellence for Vibro-Acoustic Systems and Signal Processing, University of Novi Sad, Novi Sad, Serbia

Hai Xiang Lin Delft Institute of Applied Mathematics, Delft University of Technology, Delft, The Netherlands

Victor I. Malyi HSE Tikhonov Moscow Institute of Electronics and Mathematics, Moscow, Russia

Mario Martinelli Faculty Civil Engineering & Geosciences, Tu Delft, Delft, The Netherlands

Andrei V. Metrikine Faculty Civil Engineering & Geosciences, Tu Delft, Delft, The Netherlands

Anna A. Orlova Saratov Social and Economic Institute, Plekhanov Russian University of Economics, Saratov, Russia

Federico Pisanò Faculty Civil Engineering & Geosciences, Tu Delft, Delft, The Netherlands

Aleksandr V. Ratushny Saratov State University, Saratov, Russia

Huibert Reijm Delft University of Technology, Delft, The Netherlands

A. E. Sterk Bernoulli Institute, University of Groningen, Groningen, The Netherlands

Enxhi Sulollari Faculty Civil Engineering & Geosciences, Tu Delft, Delft, The Netherlands

Johan M. Tuwankotta Analysis and Geometry Group, Faculty of Mathematics and Natural Sciences, Institut Teknologi Bandung, Bandung, Indonesia

Sergei A. Vakulenko Institute of Problems in Mechanical Engineering, Russian Academy of Sciences, St. Petersburg, Russia

Karel N. van Dalen Faculty Civil Engineering & Geosciences, Tu Delft, Delft, The Netherlands

Jan H. van Schuppen Delft Institute of Applied Mathematics, Delft University of Technology, Delft, The Netherlands

Pim van der Male Faculty Civil Engineering & Geosciences, Tu Delft, Delft, The Netherlands

Ferdinand Verhulst Mathematisch Instituut, Utrecht, Netherlands

Cornelis Vuik Delft University of Technology, Delft, The Netherlands

Fei Wang School of Mathematics and Statistics, Beijing Institute of Technology, Beijing, People's Republic of China

Jun-Min Wang School of Mathematics and Statistics, Beijing Institute of Technology, Beijing, People's Republic of China

Xuefeng Wang University of Alabama, Tuscaloosa, AL, USA

Kaihua Xi School of Mathematics, Shandong University, Jinan, Shandong, China

Aleksandr I. Zemlyanukhin Yuri Gagarin State Technical University of Saratov, Saratov, Russia

Jiangyi Zhang Faculty Civil Engineering & Geosciences, Tu Delft, Delft, The Netherlands

Weidong Zhu University of Maryland, Baltimore County, Baltimore, MD, USA

Localized Waves in a Damaged Film Foundation Subjected to Periodic Impacts

Andrei K. Abramian and Sergei A. Vakulenko

Abstract Asymptotic solutions for two cases of the problem on finite numbers of impacts on a membrane are obtained: 1. the case of a small damage function values of the membrane elastic foundation, and 2. the case of significant damage function values of its elastic foundation. A condition of resonance initiation in the membrane with a small damage function was obtained. A possibility that a localized wave is a determining factor of the delaminating process was revealed. At the final stage of the damage growth, when it reaches the critical value, no localized mode and resonance are observed in the membrane, but only traveling waves. A solution of the problem of initiation and propagation of localized waves through the damaged area in the membrane was obtained. The form of the wave amplitude solution indicates that amplitude exponential reduction depends on the difference between the value of elastic foundation coefficient in a moment when the wave gets into the damaged area and the initial coefficient value.

Keywords Film delamination · Damage function · Wave localization · Membrane · Elastic foundation

1.1 Introduction

There have been a lot of researches into a possible waves and energy localization in elastic bodies with heterogeneity and inclusions in the field of deformable body mechanics in the last 30 years. Main results were obtained in [1, 2]. Those studies considered forced oscillations and found a series of resonances in the infinite zone. The proof of resonance frequency existence behind the first boundary frequency was given in [2]. It is precisely those studies that impelled the authors of this article to research possibilities of initiation of localized waves and wave localization in

A. K. Abramian (✉) · S. A. Vakulenko
Institute of Problems in Mechanical Engineering, Russian Academy of Sciences, V.O., Bol'shoy pr., 61,199178 St. Petersburg, Russia
e-mail: andabr55@gmail.com

A. K. Abramian et al. (eds.), *Nonlinear Dynamics of Discrete and Continuous Systems*, Advanced Structured Materials 139,
https://doi.org/10.1007/978-3-030-53006-8_1

elastic structures with non-homogeneous mass-elastic parameters and inclusions. In particular, works [3–5] studied localized waves initiation on a foundation—film boundary for the case of damaged material zones in the foundation adhesion layer resulted from abrupt vibration loads and a single impact. Those works revealed that even small varying forces can result in frequency localization in the vicinity of the damaged material zones and be accompanied by the adhesion layer damage growth, which leads to the partial film delaminating. A thin-film coating interlinked with a main structure through a thin intermediate adhesive layer is used in modern structures as protective or reinforcing elements. In the deformation process of such a multi-layer structure significant stresses resulting in the coating damage or delamination can occur on the foundation-coating boundary due to the difference in their physical-mechanical properties. A vast bibliography on problems of delamination in multi-layer structures subjected to static and dynamic loads is known. The in-depth review of the problem can be found in [6, 7]. Static and vibration loads effect on damage initiation and growth in an adhesive layer of a multi-layer structure are studied rather thoroughly, but similar processes caused by non-stationary loading have been studied less. Nevertheless, they are interesting because the small amplitude shock impacts can lead to oscillation localizations in the vicinity of heterogeneities and be accompanied by the growth of the adhesive layer damage,which leads to partial delamination of the film. The present article studies two aspects that have not been studied before. The first aspect is a possibility to initiate a membrane resonance on a damaged foundation by a finite number of impacts at an arbitrary velocity of the damage growth (in works [3–5] the velocity had some contingencies). The second aspect is an assessment of amplitude of localized wave passing through the zone in which the elastic foundation has damage function value more than 0.

1.2 Statement of Problem

A statement of the problem of a dynamic of a film-membrane with a zero damage function zone that is subjected to periodical impacts is the following. Taking into account the fact that a thickness of the film is much less than a characteristic dimensions of the foundation, the first approximation is to replace the film fixed to the foundation to the film on an elastic foundation. A case when the problem can be reduced to a model of an infinite length and finite width membrane on an elastic foundation with a coefficient depending on its adhesive layer is considered. The elastic foundation substitutes the substrate and the main material effect on the film. The membrane elastic foundation coefficient is equal to the total rigidity of sequential rigid springs of adhesive layer and the main material. The dynamic equation of the membrane subjected to periodic impacts has the form:

$$\gamma \Delta u - K(n)u - \rho_0 u_{tt} = Q(t, x, y), \quad x \in \Omega, \quad t > 0 \tag{1.1}$$

$$Q(x, y, t, \epsilon) = \delta_\epsilon(x - x_0)\delta_\epsilon(y - y_0)\sum_{j=0}^{M}\delta(t - j\Delta t), \quad (x, y) \in \Omega, \quad t > 0. \tag{1.2}$$

In the Eq. (1.1) $u = u(x, y, t)$ is a membrane displacement (spatial variables x, y are in the Ω area, and width zone h: $\Omega = (-\infty, \infty) \times [0, h]$, x_0, y_0-force point), t is time, M is a number of impacts, δ_ϵ is a smoothed delta function, δ is the Dirac delta function, Q is an external force, γ is a membrane uniform tension force, ρ_0 is a membrane material density, Δt is a time between two subsequent impacts, $K(n)$ is an elastic foundation rigidity depending on adhesive layer damage function n, initial rigidity of the layer G_0, and the main material rigidity k_0. The K value is found from the relation:

$$K = \mu(n)G(n), \quad \mu(n) = \frac{k_0}{k_0 + G(n)}$$

here $k_0 > 0$ is a constant,

$$G(n) = G_0(1 - n), \quad 0 \leq n \leq 1,$$

$$G(n) = 0, \quad n > 1.$$

The kinetic equation for the damage function has the form:

$$\frac{\partial n}{\partial t} = \beta H(\mu(n)|u| - \Delta)(1 - n), \tag{1.3}$$

where Δ is a critical deformation value when the damage function starts growing, β is a velocity of the damage function growth (β, $\Delta > 0$), and H is the Heaviside function, δ_ϵ is a smooth delta function:

$$\delta_\epsilon(x) = \epsilon^{-1}(2\pi)^{-1/2}\exp(-\frac{x^2}{2\epsilon^2}). \tag{1.4}$$

The boundary and initial conditions are the following:

$$u(x, y, t) \to 0 \quad (|x| \to \infty) \tag{1.5}$$

$$u(x, y, 0) = 0, \quad u_t(x, y, 0) = 0, \quad (x, y) \in \Omega \tag{1.6}$$

$$u(x, 0, t) = u(x, h, t) = 0, \quad \forall x \in (-\infty, +\infty), t > 0 \tag{1.7}$$

$$n(x, y, 0) = n_0(x, y), \quad (x, y) \in \Omega. \tag{1.8}$$

It is assumed that full destruction of the adhesive layer material resulting in delamination of some film area occurs in those x points and in that t moment when the damage function n reaches some critical value n_*. That critical value can be determined with the help of incubation time criterion proposed for a material dynamic loading [8].

1.3 Periodical Impacts and Resonance Caused by Them

1.3.1 Resonance Conditions

It is considered a case when a membrane on an elastic foundation with a damage function zone is subjected to external forces in the form of finite number of periodical localized impacts. The problem in this case is described by Eqs. (1.1–1.8) and their solutions depend on ϵ parameter characterizing localization of the load applied and the number of impacts M. Let us find a particular solution of equation (1.1) for which the resonance can occur. A limiting case when $\epsilon \to 0$ and $M >> 1$ is considered. The Fourier transformation is used for finding solution of this initial value problem:

$$u(x, y, t, \epsilon) = 2\pi^{-1/2} \int_{-\infty}^{\infty} \exp(i\omega t)\hat{u}(x, y, \omega, \epsilon) d\omega,$$

$$Q(x, t) = 2\pi^{-1/2} \int_{-\infty}^{\infty} \exp(i\omega t)\hat{Q}(x, y, \omega) d\omega, \quad i = \sqrt{-1}.$$

The Fourier coefficients $\hat{Q}(x, y, \omega)$ are found and they have the form:

$$\hat{Q}(x, y, \omega) = \delta_\epsilon(x - x_0)\delta_\epsilon(y - y_0)2\pi^{-1/2}\hat{S}(\omega),$$

$$\hat{S}(\omega) = 2\pi^{-1/2} \sum_{j=0}^{M} \exp(-ij\omega\Delta t).$$

It should be noted that for large M values $\hat{Q}(x, y, \omega)$ has the M order only if $\Delta t\omega \approx 2m_0\pi$, where m_0 is a positive integer number. If$|\Delta t\omega - 2m_0\pi| >> M^{-1}$, then $\hat{S}(\omega)$ is limited for large M:

$$\sum_{j=0}^{M} \exp(-ij\Delta t\omega) = \frac{1 - \exp(-(M+1)i\omega\Delta t)}{1 - \exp(i\omega\Delta t)} = O(1),$$

and it is larger if $|\omega\Delta t - 2m_0\pi| = 0$ for some integer m_0. Then

$$\sum_{j=0}^{M} \exp(-ji\omega\Delta t) = M.$$

It follows that the resonance condition is met for ω close to $\omega(m_0) = 2m_0\pi(\Delta t)^{-1}$. For $\hat{u}(x, y, \omega, \epsilon)$ the following equation is obtained

$$\gamma \Delta\hat{u}(x, y, \omega, \epsilon) - K(n)\hat{u}(x, y, \omega, \epsilon) + \rho_0\omega^2\hat{u}(x, y, \omega, \epsilon) = F_\epsilon(x, y), \qquad (1.9)$$

where

$$F_\epsilon(x, y) = \delta_\epsilon(x - x_0)\delta_\epsilon(y - y_0)\sqrt{2\pi}^{-1/2}\hat{S}(\omega).$$

The left part of Eq. (1.9) is a Schrödinger operator that has a localized eigenfunctions $\Psi_j(x, y)$ with E_j eigenvalues, and non-localized eigenfunctions $\Psi(x, y, k)$ corresponding to eigenvalues $E(k)$, where k is a wave number. The localized functions have square integrability, i.e they are in the space $L_2(\Omega)$, and are corresponding to the discrete spectrum, while non-localized functions do not get in that $L_2(\Omega)$ space and correspond to the continuous spectrum. The Schrödinger general theorems leads to an interesting corollary: if a damage continuous function n is positive at least in some points of the zone, then the localized eigenfunctions always exit. To obtain an analytical solution, the case when n depends only on x is considered in what follows. When such a solution for the two-dimensional case is obtained, an assessment of the qualitative membrane behavior is proposed. When n depends only on x, the separation of variables can be used and Eq. (1.9) is considerably simplified.

1.3.1.1 Asymptotic Describing Resonance

Let us consider the following cases: (**a**) the initial stage of the damage process, the damage function n is small; (**b**) the final stage, the damage function $n \approx 1$. At first, the one-dimensional case (the initial damage function n_0 and all initial data ϕ_i for the displacement depend only on x) is considered. Let us remind some results relating to the Schrödinger operator on the axis $(-\infty, +\infty)$ with a shallow potential well. Such an operator appears after the aforementioned variable separation and has the form:

$$H\Psi = (-\Delta + \rho\Phi(x))\Psi$$

in $\mathbf{R}^1$, where $\Phi(x)$ is a smooth function rapidly diminishing at $|x| \to \infty$. Then, at the conditions

$$I_\Phi = \int_{-\infty}^{+\infty} \Phi(x)dx < 0. \qquad (1.10)$$

For rather small $\rho > 0$ only one eigenfunction $\Psi_1(x)$ exists , and it corresponds to eigenvalue $E_1(\rho)$, dependent on

$$E_1 = -\eta^2, \quad \eta = \frac{\rho I_\Phi}{2}. \qquad (1.11)$$

There are no eigenvalues for the case when $I_\Phi > 0$. For Ψ_1 following asymptotic exists:

$$\Psi_1 = \eta^{3/2} \int_{-\infty}^{+\infty} \exp(ipx) a(p)(p^2 + \eta^2) dp, \quad a(p) = \frac{\tilde{\Phi}(p)}{\tilde{\Phi}(0)}, \tag{1.12}$$

where

$$\tilde{\Phi}(p) = (2\pi)^{-1/2} \int_{-\infty}^{+\infty} \Phi(x) \exp(-ipx) dx$$

is the Fourier potential form Φ.

Let us study the obtained results for the following cases. When $n << 1$, the $K(n)$ expression has the form $a_0^2 - b_0 n + O(n^2)$, where $a_0^2 = k_0 G_0/(k_0 + G_0)$ $b_0 = k_0/(k_0 + G_0) > 0$. For this case we get a Schrödinger operator for shallow potential well. Condition (1.10) is met and the only localized mode exists. The corresponding to it E_1 energy allows the following assessment $E_1 = -a_0^2 + O(n)$. For $n << 1$ the resonance condition has the form

$$k_0 G_0/(k_0 + G_0) + O(||n||) = \rho_0(\omega(m_0))^2 \tag{1.13}$$

for some m_0 integer values. To obtain this condition in a concrete form, let us consider a case when potential Φ has a form of a well of rectangular shape. Introduce detuning parameter κ

$$\kappa^2 = \gamma^{-1}(a^2 - \rho_0(\omega(m_0))^2), \quad a^2 = k_0 G_0/(k_0 + G_0). \tag{1.14}$$

Consider the resonance effect more detailed. Assume that $n << 1$, then the following asymptotic form can be used

$$K(n) = a^2 - \frac{k_0 G_0}{(k_0 + G_0)^2} n(x) + O(n^2).$$

For potential $\Phi(x)$ with a rectangular well

$$\Phi(x) = -\bar{E}, \quad |x| < l, \quad \Phi = 0, \quad |x| > l. \tag{1.15}$$

The equation (1.9) for a one-dimensional case has the form

$$\hat{u}(x, \omega)_{xx} - \bar{k}_0 \hat{u}(x, \omega) - \Phi(x)\hat{u}(x, \omega) = \gamma^{-1} \delta_\epsilon(x - x_0) \sqrt{2\pi}^{-1/2} \hat{S}(\omega). \tag{1.16}$$

Here $\bar{k}_0 = \frac{K(n)}{\gamma}$. In compliance with the aforementioned a Schrödinger operator features, the localized function exists if

$$\bar{E} - \kappa^2 = \beta^2 > 0, \quad \kappa > 0.$$

Assume that all these conditions are met. In the case when condition (1.15) is fulfilled and $\epsilon \to 0$, the solution of equation (1.16) can be found. Solution $U = \lim_{\epsilon \to 0} \hat{u}(x, \omega, \epsilon)$ has the following form for the case when $x < x_0$ (it is assumed that $|x_0| < l$, i.e. the external force is applied to the membrane in the damaged region, where l is half of the damage function zone length):

$$U = C_-(\frac{\kappa}{\beta} \sin(\beta(x+l)) + \kappa \cos(\beta(x+l)), \quad x \in (-l, x_0), \tag{1.17}$$

$$U = C_- \exp(\kappa(x+l)), \quad x < -l, \tag{1.18}$$

and for $x < x_0$ one has

$$U = C_+(-\frac{\kappa}{\beta} \sin(\beta(x-l)) + \cos(\beta(x-l)), \quad x \in (x_0, l), \tag{1.19}$$

$$U = C_+ \exp(-\kappa(x-l)), \quad x > l. \tag{1.20}$$

Let $x_0 = 0$. Then constants $C_\pm$ can be found from the following equation:

$$C_+ = C_- = \frac{1}{2(\kappa \sin(\beta l) - \beta \cos(\beta l))}. \tag{1.21}$$

Thus, the resonance condition has a form:

$$\kappa \sin(\beta l) - \beta \cos(\beta l) = 0. \tag{1.22}$$

The left part of this equation is the eigenvalue E_1 of the corresponding Schrödinger operator. This condition does not depend on x_0, but U solution depends on x_0. It is obvious that the general resonance condition has the form:

$$k_0 G_0/(k_0 + G_0) + E_j = \rho\omega(m_0)^2 \quad j = 1, \ldots \tag{1.23}$$

where E_j—is the eigenfunction of the corresponding Schrödinger operator. The accurate expression for E_j can be obtained only for some cases. Thus, using asymptotic form (1.11), for small damage function values $n_0(x)$ it can be obtained.

$$E_1 = \frac{k_0 G_0}{(k_0 + G_0)^2} \int_{-l}^{+l} n_0(x) dx. \tag{1.24}$$

For the two-dimensional case eigenvalues are presented in the form: $\Psi = \cos(m\pi y/h)\hat{\Psi}_m(x, \omega)$. Respectively, the discrete spectrum is obtained in the form $E_{j,m} = E_j + m^2\pi^2/h^2$, where m—is any positive integer number and E_j is an eigen energy obtained above. Now, we have infinite number of localized modes with energies exceeding the energies in the one-dimensional case. They are the greater, the narrower the zone is. When the zone width tends to zero ($h \to 0$), those energies tend to infinity. Therefore, we do not have a formal membrane-spring passage to the limit (it results from the condition of the membrane fixation on the edges of the zone). The most important mechanical corollary of the above mathematical reasoning is as follows. Localized modes contribute a lot into a membrane reaction effect on an impact if the resonance condition is fulfilled. To understand that fact, let us consider equation (1.9) and expand its solution in eigenfunctions in $[-L, L] \times [-h, h]$, domain , where L—is the strip length and $L >> 1$.

Then, the eigenfunction contribution to the solution is in proportion to $S(\omega)\Psi_j(x_0, y_0)$ value for the localized functions and $S(\omega)\Psi_k(x_0, y_0)$, where k is the wave vector for the non-localized functions. i.e. is proportional to the function amplitudes in an impact point. The sum contribution of the none-localized functions is in proportion to k integral that can be regularized as an principal value integral even if it contains singularity relating to the resonance. A significant resonance effect occurs at localized modes. Their amplitudes are in proportion to $h^{-1/2}$. Therefore, the narrower the zone, the stronger the resonance effect is (in case it exists). On the other hand, since the oscillation frequencies are of large values for narrow zones, the impact frequency should be of large values to cause the resonance. Below, in the next section, the other difference of a wave propagation in a membrane and a string is considered.

1.3.2 Wave Regime

The wave regime relates to a case when an external force is specified by relation (1.2) and $n \approx 1$. In this case the expression for $K(n)$ has the form $G_0(1-n) + (1-n)^2$, and the Schrödinger operator has again a small potential well. However, condition (1.10) is not fulfilled and, therefore, a localized mode does not exist, neither does the resonance. Thus, at the last stage of destruction the localized mode and the resonance do not exist . Taking into account that $K(n) \approx 0$, Eq. (1.1) has the form

$$\gamma \Delta u - \rho_0 u_{tt} = Q(x, y, t, \epsilon), \quad x \in (-\infty, +\infty), \quad t > 0. \tag{1.25}$$

Expand the displacement into a Fourier series by transversal modes:

$$u(x, y, t) = \sum_{m=1}^{+\infty} u_m(x, t) \sin(\pi m y/h).$$

For u_m one has the equation

$$\gamma u_{mxx} - \rho_0 u_{mtt} - \pi^2 m^2/h^2 u = Q_m(x, t, \epsilon), \quad x \in (-\infty, +\infty), \quad t > 0, \tag{1.26}$$

where Q_m—is the Fourier coefficients that can be found from

$$Q_m(x,t,\epsilon) = 2h^{-1} \int_{-h}^{h} Q(x,y,t,\epsilon)\sin(\pi my/h)dy. \tag{1.27}$$

Using expression (1.2) in the limit of $\epsilon \to 0$ and $h \to +\infty$ the following solution of Eq. (1.26) can be found (for $u(x,t,\epsilon) \to u(x,t)$ when $\epsilon \to 0$):

$$u_m = (2c)^{-1} \sum_{j=1}^{M} (H(x - c(t - j\Delta) - (H(x - c(t + j\Delta))H(t - j\Delta t), \quad t > 0, \tag{1.28}$$

where H is a Heaviside function . This solution is representing M traveling waves of a constant shape that is not changed during the propagation. Such a solution relates to the string case. At finite but small thicknesses h a solution of Eq. (1.25) is well known and can be obtained by the Fourier method. The waves are propagating, but their shape varies due to dispersion effects. The waves fronts are spread and oscillations on x axis occur at large t. This effect increases for large numbers of modes m. For small thicknesses h the waves do not practically propagate.

1.4 Propagation of Localized Wave Through a Zone with Damaged Foundation

It was found in [3–5] that an impact on a string lying on a damaged foundation can cause a localized wave in the damaged zone or propagating localized waves. In case of several damaged zones a question arise on how an amplitude of a localized propagation wave changes when the wave passes through a damaged foundation (substrate) zone where the foundation's rigidity differs from the rigidity outside the zone. The localized wave varies its amplitude when it is propagating through the damage foundation zone and the damage function distribution along the coordinate is not uniform. An asymptotic solution is obtained that is true for any damage function growth velocity β and for not small damage function values. Thus, this solution is free from contingencies introduced in [3–5] works. At first, let us consider a case of large width h, the fixation boundary conditions can be neglected at $y = \pm h$. Then, the displacement localized in the damage function zone can be described as:

$$u(x,y,t,\epsilon) = a(x,t)\exp(-\epsilon^{-1}S(x,t)), \tag{1.29}$$

where $S(x,t) \geq 0$ is a smooth function, such as $S(x,t) = 0$ in $x = X(t)$ point, and $\epsilon > 0$ is a small parameter. As it is explained below, these asymptotic solutions can describe waves propagation along x axis in the membrane. Inserting the displacement expression into the membrane dynamic equation, one can get the following equation for $S(x,t)$:

$$\epsilon^{-2}a(\gamma S_x^2 - \rho_0 S_t^2) + \epsilon^{-1}(\gamma(-2a_x S_x - aS_{xx}) + \rho_0(2a_t S_t + aS_{tt})) + K(n)a + \gamma a_{xx} - \rho_0 a_{tt} = 0. \tag{1.30}$$

Let us consider the terms of the equation of ϵ^{-2} order. Then, one has the following eikonal equation:

$$\gamma S_x^2 - \rho_0 S_t^2 = 0. \tag{1.31}$$

For the amplitude $a(x, t)$ one has

$$\gamma(-2a_x S_x - aS_{xx}) + \rho_0(2a_t S_t + aS_{tt}) + \epsilon(K(n)a + \gamma a_{xx} - \rho_0 a_{tt}) = 0. \tag{1.32}$$

One of the simplest solutions of Eq. (1.31) is a solution in the form $S = \frac{(x-ct)^2}{2}$. Then, $X(t) = ct$. The solution expression for $a(x, t)$ (which is very complicated for a general case) can be simplified. Introduce the functions

$$\bar{a}(t) = a(ct, t), \quad \bar{n}(t) = n(ct, t).$$

The physical interpretation of function $\bar{n}(t)$ is very simple. It is a damage function on the wave front.

Function $\bar{a}$ determines the amplitude of the wave solution. Indeed, in our case one has assessments $a(x, t) \approx a(ct, t) = \bar{a}(t) + O(\epsilon)$ and $n(x, t) \approx n(ct, t) + O(\epsilon) = \bar{n}(t) + O(\epsilon)$ because for $|x - ct| \gg \epsilon$ solution u is exponentially small. It is not easy to find $a(x, t)$ from Eq. (1.32). Therefore, we propose to use an energy balance equation here. This equation can be obtained as follows. Multiply both parts of Eq. (1.1) by u_t and integrate both parts of the obtained expression over all $x \in [-L, L]$, where L—is the membrane length in a horizontal direction (for the infinite in this direction zone we go to the limit $L \to +\infty$. Thus, one has

$$\frac{dE}{dt} = D[u], \tag{1.33}$$

where the functional

$$E[u, u_t] = \frac{1}{2}\int_{-L}^{L}(\gamma u_x^2 + \rho_0 u_t^2 + K(n)u^2)dx \tag{1.34}$$

has the meaning of the system energy and the functional

$$D = \frac{1}{2}\int_{-L}^{L} K'(n(x, t))n_t(x, t)u(x, t)^2 dx \tag{1.35}$$

determines the velocity of the energy dissipation in the system. Let us calculate a contribution of the localized solution to the system energy. At first, two auxiliary expressions are introduced:.

$$R(\epsilon, t) \approx \int_{-L}^{L} \exp(-2S/\epsilon)(\epsilon^{-2}(\gamma S_x^2 + \rho_0 S_t^2) + K(\bar{n}(t)))dx, \tag{1.36}$$

$$M_0(\epsilon) = \int_{-L}^{L} \exp(-2S/\epsilon)dx.$$

Taking into account a smallness of ϵ, the asymptotic of these expressions can be found using the Laplace method. Standard calculations give the following:

$$M_0(\epsilon) \approx \sqrt{\epsilon} \int_{-\infty}^{\infty} \exp(-w^2)dw.$$

Take into consideration (1.31) for R one has:

$$R(\epsilon, t) \approx \int_{-L}^{L} \exp(-(x - ct)^2)/\epsilon)(2\epsilon\gamma(x - ct)^2 + K(\bar{n}(t)))dx = R_0(\epsilon).$$

The last expression can be estimate by the Laplace method. In the right part of the expression $K(\bar{n}(t)$ term can be left out, because the contributions of the first and the second terms have the order of $O(\epsilon^{-1}M_0)$; the contribution of the third term of the order of $O(M_0)$. Then, the expression $R_0(\epsilon)$ can be considered as practically independent on t and vanishing with high accuracy when $\epsilon \to 0$. Finally we have:

$$R(\epsilon, t) \approx R_0 = \epsilon^{-1/2} \int_{-\infty}^{\infty} \exp(-w^2)w^2dw.$$

Using the introduced energy functions (1.35), one has the asymptotic

$$E[u, u_t] \approx R_0 \frac{\bar{a}^2(t)}{2}. \tag{1.37}$$

For dissipative functional D the asymptotic expression has the form:

$$D[u] \approx \frac{M_0}{2} \bar{a}^2(t) \frac{dK}{dn}(\bar{n}(t))\bar{n}_t. \tag{1.38}$$

Assume that a force applied to the system and its impulse is enough to start the destruction process in a point where $x = ct$, then , using expressions (1.38), (1.37) and (1.33) obtained above one has the following equation for the localized amplitude solution $\bar{a}$:

$$\frac{d\bar{a}}{dt} = \frac{dK(\bar{n}(t))}{dt} \bar{a}. \tag{1.39}$$

Thus, the localized solution amplitude varies in relation to the following formula:

$$\bar{a}(t) = \Theta(t)\bar{a}(0), \quad \Theta(t) = \exp(K(\bar{n}(t)) - K(\bar{n}(0)). \tag{1.40}$$

Since $\bar{n}(t) = n(ct, t)$, it is seem that two effects determine the amplitude variation: the wave propagation and the local growth of the damage function. The second effect always reduces the amplitude. The first effect can either increases or decreases it. Let us describe modifications that should be introduced into the solution when two-dimensionality of the problem and the boundary conditions along y are taking into account. For this purpose, again the Fourier method is applied and the solution has the form

$$u(x, y, t, \epsilon) = \sum_{m=1}^{\infty} u_m(x, t)\sin(\pi my/h)a_m(x, t)\exp(-\epsilon^{-1}S(x, t)), \tag{1.41}$$

where

$$u_m = a_m(x, t)\exp(-\epsilon^{-1}S_m(x, t)).$$

For coefficients a_m, S_m we have an equation similar to the one obtained above. However, it is not necessary to consider it again. It is easy to notice that for u_m we have the same equations as for the one-dimensional case, but instead of K we have $K_m(n) = K(n) + \pi^2 m^2/h^2$. Therefore, we have the same expression (1.40) as the above one. It should be noted that a wave propagating in a string may not be propagating in a membrane. This is evident from the fact that in the two-dimensional case a cut-off frequency increases due to growth of the elastic foundation coefficient $K_m(n)$.

1.5 Conclusion

Thus, we have obtained the following results: 1. Asymptotic solutions of the problem on a finite number of periodical impacts on a membrane with a damage function for two cases relating to small and large damage function of the elastic foundation are obtained. A condition of the membrane resonance initiation in the case of small damage functions was found. The obtained results allow us to assume that the initial stage of the damage growth is dangerous for the system because the membrane oscillation amplitude can grow and, therefore, the membrane delamination from the elastic foundation is possible. Thus, a localized wave can be a determinative factor of the delamination process. It is shown that at the final stage of the damage function growth (when its value is close to the critical) the localized mode and the resonance do not exist in a membrane on an elastic foundation; only traveling waves exist there. The main difference from the string case studied before is that those wave are dispersible and their shape varies in time and their front spread in the membrane case. 2. The obtained solution of a localized wave propagating through an elastic foundation zone shows that exponential reduction of the localized wave amplitude depends on the difference between elastic foundation coefficients at the moment when the wave has passed the damage zone and the coefficients at the initial moment of time.

References

1. Babeshko, V.A., Buzhan, V.V., Williams, R.: Localization of a vibrational process in an elastic solid by an array of rigid planar inclusions. Dokl. Phys. **47**, 156–158 (2002)
2. Babeshko, V.A., Pryakhina, O.D., Smirnova, A.V.: Dynamic problems for discontinuous media. Int. Appl. Mech. **40**, 241–245 (2004)
3. Abramyan, A.K., Vakulenko, S.A., Indeitsev, D.A., Semenov, B.N.: Influence of dynamic processes in a film on damage development in an adhesive base. Mech. Sol. **47**(5), 498–504 (2012)
4. Indeitsev, D.A., Abramyan, A.K., Bessonov, N.M.: Motion of the exfoliation boundary during localization of wave processes. Dokl. Phys. **57**(4), 179–182 (2012)
5. Abramyan, A.K., Bessonov, N.M., Indeitsev, D.A., Mochalova, Yu.A., Semenov, B.N.: Influence of oscillation localization on film detachment from a substrate. Vest. St .Petersb. Univ. Math. **44**, 1, 5–12 (2011)
6. Tran, P., Kandula, S.V., Geubelle, P.H., et al.: Dynamic delamination of patterned thin films: a numerical study. Int. J. Fract. **162**, 77–90 (2010)
7. Maeva, E., Severina, I., Bondarenko, S., Chapman, G., O'Neill, B., Severin, F., Maev, R.G.: Acoustical methods for the investigation of adhesively bonded structures. A review. Can. J. Phys. **82**, 981–1025 (2004)
8. Petrov, YuV., Smirnov, I.V., Volkov, G.A., Abramian, A.K., Bragov, A.M., Verichev, S.N.: Dynamic failure of dry and fully saturated limestone samples based on incubation time concept. J. Rock Mech. Geotech. Eng. **9**(1), 125–134 (2016)

Mathematical Models in Pure and Applied Mathematics

Igor V. Andrianov

It is better to do the right problem the wrong way than the wrong problem the right way.

Attributed to R. W. Hamming [32, p. 435]

Abstract The concepts of the mathematical model in pure and applied mathematics are analyzed, as well as the concepts of rigor proof and degree of confidence. The role of asymptotic estimations as basis for mathematical modeling is analyzed.

2.1 Introduction

August 1999, Equadiff conference at Freie Universität Berlin. A tall man in shorts comes up to me (it was very hot in Berlin!) and says: "*You have a report tomorrow. Everything that you will tell is incorrect.*" Naturally, I was delighted: a man read my abstract, became interested, there is someone to talk to! Immediately Wim (it was him!) invited me to come to TU Delft to discuss issues that were interesting to both of us. Over the years, I was a Ph.D. committee member for 12 of Wim's doctoral students, I have repeatedly enjoyed the hospitality of the Delft Institute of Applied Mathematics, and I have written several papers with Wim. In addition to mathematical problems, we often discussed the relationship between Pure Mathematics (PM) and Applied Mathematics (AM). Of course, we did not come to a single opinion, and there is no single point of view on this subject at all.

I. V. Andrianov (✉)
Institute of General Mechanics, RWTH Aachen University, Templergraben 64, 52056 Aachen, Germany
e-mail: igor.andrianov@gmail.com

A. K. Abramian et al. (eds.), *Nonlinear Dynamics of Discrete and Continuous Systems*, Advanced Structured Materials 139,
https://doi.org/10.1007/978-3-030-53006-8_2

Below is my subjective point of view, formed as a result of many years of activity in the field of AM, reading a huge number of papers and books, listening to presentations at conferences, often unimaginably boring, and, most importantly, as a result of many discussions with smart people.

2.2 Pure and Applied Mathematics—Definitions

Let us leave aside the well-known maxim: "Pure mathematicians solve only those problems that can be mathematically rigorously solved, and applied mathematicians solve the necessary problems, but without mathematical justification of the obtained results". Pure mathematicians, to be honest, often relate to the work of applied mathematicians in a snobbish way. Applied mathematicians often doubt the value of the solutions of practical problems that have been obtained by pure mathematicians (I recall last sentence of the well-known anecdote concerning riding in a hot air balloon: "... his answer was 100% correct and it was totally useless").

PM and AM are not antagonistically disconnected, of course. Poincaré [33] once said: "I hope I have said enough to show that pure analysis and mathematical physics may be reciprocally helpful without either entailing sacrifice upon the other, and that each should rejoice in whatever exalts its associate".

However, there are fundamental differences, and they are the subject of this article.

First, let's look at the definitions of PM and AM in Wikipedia [41, 43]: "Pure mathematics is the study of mathematical concepts independently of any application outside mathematics. *These concepts may originate in real-world concerns, and the results obtained may later turn out to be useful for practical applications* (marked by I.A.), but pure mathematicians are not primarily motivated by such applications. Instead, the appeal is attributed to the intellectual challenge and aesthetic beauty of working out the logical consequences of basic principles".

"Applied mathematics is the application of mathematical methods by different fields such as science, engineering, business, computer science, and industry. Thus, *applied mathematics is a combination of mathematical science and specialized knowledge*" (marked by I.A.). In other words, for the solution of quite specific problems quite specific knowledge is needed.

Arnol'd wrote in the chapter "An apologia for Applied Mathematics" of his paper [3]: "The difference between pure and applied mathematics is not scientific but only social. A pure mathematician is paid for uncovering new mathematical facts. An applied mathematician is paid for the solution of quite specific problems".

However, all these definitions do not mark an important fact: applying the PM results to the real world is possible only through an intermediate stage—the construction of a mathematical model (MM). To construct and to analyze of MM, a combination of mathematical and specialized knowledge is needed. This is the key difference between applied and pure mathematicians.

2.3 Poincaré, Lyapunov and Lord Rayleigh

It is interesting to recall a discussion between Poincaré and Lyapunov regarding figures of equilibrium for rotating, gravitating liquids [24]. Poincaré, who obtained his results using not so rigorous reasoning and often by simple analogy, wrote [34], "It is possible to make many objections, but the same rigor as in pure analysis is not demanded in mechanics". Lyapunov had entirely different position [23]: "It is impermissible to use doubtful reasonings when solving a problem in mechanics or physics (it is the same) *if it is formulated quite definitely as a mathematical one* (marked by I.A.). In that case, it becomes a problem of pure analysis and must be treated as such".

Thus, the discussion of Poincaré and Lyapunov comes down to a discussion of the level of rigor of the mathematical methods used for some particular problem and does not concern mathematical modeling itself.

Lord Rayleigh discussed the problem of the PM and AM approaches as follows [36]: "In the mathematical investigations I have usually employed such methods as present themselves naturally to a physicist. The pure mathematician will complain, and (it must be confessed) sometimes with justice, of deficient rigor. But to this question there are two sides. For, however important it may be to maintain a uniformly high standard in pure mathematics, the physicist may occasionally do well to rest content with arguments which are fairly satisfactory and conclusive from his point of view. To his mind, exercised in a different order of ideas, the more severe procedure of the pure mathematician may appear not more but less demonstrative. *And further, in many cases of difficulty to insist upon the highest standard would mean the exclusion of the subject altogether in view of the space that would be required*" (marked by I.A.).

In other words, Rayleigh assumed that AM and PM have areas of intersection, but each of these sciences has its own specific features.

2.4 Mathematical Model—Definitions

Let's turn to Wikipedia again [42]: "A model is a system, the study of which serves as a means to obtain information about another system. A model is an abstract representation of reality in some form (for example, in mathematical, physical, symbolic, graphic or descriptive) (I'd like to add 'algorithmic'—I.A.), designed to represent certain aspects of this reality and to get answers to the questions being studied' (translated from Russian by I.A.).

In a more lapidary form [5, p. 128]: "Object A is a model of object B (here the term "object" is understood in the broadest sense: any situation, phenomenon, process, etc.), if A is selected to imitate B with respect to a certain system of characteristics" (translated from Russian by I.A.). In other words, with the help of a MM we go from the real world into the world of AM, using mathematical concepts and language, after which we return to the real objects. Naturally, this is naive point of view. I did not plan to solve "the great problem of the relations between the empirical word and the mathematical word" [8]. But even Bourbaki mentioned [8]: "That there is an intimate connection between experimental phenomena and mathematical structures, seems to be fully confirmed in the most unexpected manner by the recent discoveries of contemporary physics". However, one further reads: "But we are completely ignorant as to the underlying reasons for this fact (supposing that one could indeed attribute a meaning to this words) and we shall perhaps always remain ignorant of them" [8]. Let us leave the explanation of this mysticism to philosophers.

"When studying science, the examples are more useful than the rules" (attributed to Newton). Follow the advice of a great scientist and consider a few examples from different areas of AM.

2.5 Simple Example

Transverse vibrations of stretched strings—a problem that can be found in any mathematical textbook devoted to Mathematical Physics, PDE, or physical textbooks devoted to Theory of Oscillations. However, the notion of "string" itself is interpreted differently in PM and AM. In PM the term "string" refers to an object that supports tension in its axial direction, but which does not resist bending. This is an *axiomatic definition* of an ideal mathematical object for which bending stiffness is exactly equals zero:

$$EI = 0,$$

where E is the Young's modulus and I is the static moment of the transversal string cross section.

In this case the governing initial boundary value problem is:

$$Tu_{xx} = \rho F u_{tt}, \tag{2.1a}$$

$$u = 0 \quad \text{at} \quad x = 0, L, \tag{2.1b}$$

$$u = f_1(x), \quad u_t = f_2(x) \quad \text{at} \quad t = 0, \tag{2.1c}$$

where u is the displacement in a direction perpendicular to the x axis, T is the force of tension of the string, ρ is the density, F is the area of the transversal string cross section, L is the length of the string, t is the time, and x stands for the spatial coordinate.

The solution of the initial-boundary value problem (2.1a)–(2.1c) is:

$$u = \sum_{j=1}^{\infty} \sin \frac{j\pi x}{L} \left(C_{1j} \cos \omega_j t + C_{2j} \sin \omega_j t \right), \qquad \omega_j = \sqrt{\frac{T}{\rho F}} \frac{j\pi}{L}, \tag{2.2}$$

where ω_j is the frequency of the vibration and C_{ij}, $i = 1, 2$ are constants that are determined after the expansion of the right-hand sides of the initial conditions (ICs) (2.1c) into the Fourier series.

In textbooks on the Theory of Oscillations one reads different definition: a string is a thin, tightly stretched elastic thread. In addition tension is supposed so large that bending resistance can be neglected. This is an *asymptotic definition*, in this case $\varepsilon = EI/(TL^2) \ll 1$, and we are dealing with an asymptotic model.

The equation of the free oscillations of a string that takes the bending stiffness into account is:

$$- EIu_{xxxx} + Tu_{xx} = \rho F u_{tt}. \tag{2.3}$$

The ICs have the form (2.1c), and the BCs can be defined, e.g., as follows:

$$u = u_{xx} = 0 \quad \text{at} \quad x = 0, L. \tag{2.4}$$

Equation (2.1a) can be treated as string approximation for the more exact model (2.3). The small parameter here is ε.

The solution to the initial-boundary value problem (2.1c), (2.3), (2.4) for

$$j \ll j^*, \quad j^* = \frac{1}{\varepsilon\pi^2} \tag{2.5}$$

can be approximately (with accuracy up to ε) written as follows:

$$u = \sum_{j=1}^{j^*} \sin\frac{j\pi x}{L}\left(C_{1j}\cos\omega_j t + C_{2j}\sin\omega_j t\right). \tag{2.6}$$

The restriction (2.5) for the solution can be rewritten as an asymptotic restriction for the governing problem:

$$0 \le \alpha \le 2, \qquad Lu_x \approx \varepsilon^{-\alpha} u. \tag{2.7}$$

For $j \approx j^*$ Eq. (2.3) should be used, for $j \gg j^*$ one can simplify it:

$$- EIu_{xxxx} = \rho F u_{tt}. \tag{2.8}$$

Equation (2.8) can be used for $j^* \ll j \ll (j^*)^2$, then equations of the 3D elasticity theory are needed, etc.

The results of the analysis of the described toy problem can be formulated as follows:

1. The MM of string oscillations is <u>not only</u> the problem (2.1a)–(2.1c), but also the problem (2.1a)–(2.1c) <u>and</u> the restriction (2.5) (or (2.7)), if we deal with low part of spectrum.

2. Corresponding restrictions can appear not only a priori, but also a posteriori, in the process of solving the problem.
3. One can't say that PM is more accurate than AM—they deal with different objects: PM—with axiomatically defined, AM—with asymptotically defined. PM objects are exactly defined, AM objects are fuzzy.
4. It makes sense to emphasize what objects we are dealing with in each specific problem.

2.6 Problem of Truncation

Variational approaches (Bubnov-Galerkin, Kantorovich, Trefftz, etc.) are widely used for solution of PDE or ODE. As a result, the original problem is reduced to the infinite systems of coupled ODEs or algebraic equations.

Further, as a rule, a truncation procedure is used. Usually the first few equations are used, and the accuracy estimate is based on the Runge rule of the errors estimating, i.e., if the solutions of the systems of N, $2N$, $3N$ equations are close to each other's according to some norm, then the truncation is incorrect.

However, Wim, a faithful follower of the Lyapunov line, strongly rejects such a logic and insists: if the infinite ODE system is obtained, the possibility of its truncation must be rigorously proved. Moreover, in many of his papers, e.g., in [37], it was shown that the truncation procedure for infinite systems of ODEs is incorrect.

However, let's look at the untruncated infinite system from the asymptotic standpoint. Suppose that as a result of the asymptotic procedure based on small parameter δ an ODE or PDE is obtained, which is further solved by the Bubnov-Galerkin method. How many terms of the truncated series (N) should be left in the basis functions expansion?

We deal with double limit process with small parameters δ and N^{-1}. Van Dyke mentioned: "One often encounters a double or multiple limit process, in which two or more perturbation quantities approach their limits simultaneously. Because the order of carrying out several limits cannot in general be interchanged, one must frequently specify the relative rates of approach" [39, p. 21].

In reality, in the general case one obtains

$$\lim_{\delta \to 0} \lim_{N \to \infty} \neq \lim_{N \to \infty} \lim_{\delta \to 0} . \tag{2.9}$$

Ignoring this fact can lead to erroneous conclusions.

For practice one can use the "Kruskal's principle of maximum simplification" [19]: the physically realizable states of the system are characterized by the simplest relations between small parameters. In our case one can suppose $N \sim \delta^{-1}$.

To transfer the untruncated infinite system to truncated one some regularization procedure is usually used. Often it has a rather artificial character (e.g., introduction of artificial viscosity in hydrodynamics [27]). My opinion is: a correct MM should lead to a system that allows truncation. Otherwise - look for what physical parameters you unreasonably neglected [2].

An interesting example of problems under consideration is a paradox in the problem of membrane flutter [7, ch. 4.11]. The Bubnov-Galerkin approach for plate and membrane flutter problems leads to infinite systems of algebraic equations. The infinite determinant of such a system for plate is normal and can be reduced, for membrane this is not correct (I thank Prof. I. Elishakoff who drew my attention to this results).

2.7 Navier-Stokes Equations

As it is known, the Clay Mathematics Institute proposed 7 Millenium Prize Problems. One of this problem is formulated as follows:

> Prove or give a counter-example of the following statement: In three space dimensions and time, given an initial velocity field, there exists a vector velocity and a scalar pressure field, which are both smooth and globally defined, that solve the Navier-Stokes equations (NSEs) (for more detail see [10]).

Unlike other purely mathematical Millenium Prize Problems, e.g. the Riemann hypothesis, NSEs describe a physical process and they are nothing else but an asymptotic model of reality. And, maybe, they are not the most adequate model? In reality, one has a great problem for the mathematical justification of NSEs. However, as Yudovich [45, p. 527] noted: “Doubts are often expressed about the Euler and Navier-Stokes equations themselves. If these equations are slightly altered, all problems would be immediately solved.” Yudovich himself was, as well as many pure mathematicians, strongly objected to such a change of the NSEs.

Ladyzhenskaya [22] emphasized the need for a deeper understanding of the NSEs: “After almost fifty years experience in studying the NSEs and more than a half-century of experience in studying boundary value problems and initial-boundary value problems for linear and non-linear PDE of diverse types, I would formulate the main problem concerning the NSEs in a quite different way, namely, as follows:

Do the NSEs together with the ICs and BCs conditions actually give a deterministic description of the dynamics of an incompressible fluid?”

Ladyzhenskaya [21] (also see the paper by Golovkin [13]) proposed modified NSEs with some regularizing terms. These terms represent large gradients of the velocities multiplied by some unknown parameters. Ladyzhenskaya proved the global solvability of the modified NSEs and their uniqueness under very general assumptions. The problem, however, is that no reasonable procedure has yet been proposed to determine the regularizing parameters.

Interesting attempts to construct modified NSEs using asymptotic expansions are described in [14, 35, 38].

In general, in the described situation, there is a difference in the approaches of pure and applied mathematicians to the modeling of fluid motion: for the first one the NSEs are an untouchable icon, for the second ones the NSEs are a working tool that can be modified.

2.8 Splashes

The correlation between discrete and continuous models is very nontrivial. Consider an example of continualization that leads to unexpected results [11, 12, 20, 26], i.e. the problem of the vibrations of a one-dimensional lattice of linearly connected masses. It is assumed that all the points have the same mass m and that the connections are linearly elastic with the same stiffness coefficient c. Denote the displacement of the jth point ($j = 0, 1, \ldots, n$) from its initial, unloaded equilibrium position by $y_j(t)$ and the elastic force between the j-th and $(j-1)$-th points by $\sigma_j(t) = c[y_j(t) - y_{j-1}(t)]$. The functions σ_j are the solutions of the system of ODEs

$$m\sigma_{j,tt}(t) = c(\sigma_{j+1} - 2\sigma_j + \sigma_{j-1}), \qquad j = 1, \ldots, n. \tag{2.10}$$

Let us now study the problem of lattice movement under action of a unit constant force on the point number zero. Motion of this system is governed by ODEs (2.10) with the following BCs and ICs:

$$\sigma_0(t) = 1, \qquad \sigma_{n+1}(t) = 0, \tag{2.11a}$$

$$\sigma_j(t) = \sigma_{j,t}(t) = 0 \quad \text{at} \quad t = 0. \tag{2.11b}$$

The initial boundary value problem (2.10)–(2.11b) can be used, e.g., for modeling of stresses in the chain couplers of the rolling stocks [20]. The exact solution to this problem is [20]:

$$\sigma_j = \frac{1}{1+n}\sum_{k=1}^{n} \sin\frac{\pi kj}{n+1}\cot\frac{\pi k}{2(n+1)}\left[1 - \cos(\omega_k t)\right], \quad j = 1, 2, \ldots, n. \tag{2.12}$$

For large values of n usually a continuous approximation of the discrete problem is applied. In our case it takes the form:

$$m\sigma_{tt}(x,t) = ch^2\sigma_{xx}(x,t), \tag{2.13a}$$

$$\sigma(0,t) = 1, \quad \sigma(l,t) = 0, \tag{2.13b}$$

$$\sigma(x,0) = \sigma_t(x,0) = 0, \tag{2.13c}$$

where $l = (n+1)h$, h is the distance between masses.

Table 2.1 Splashes

n	8	16	32	64	128	256	$n \to \infty$
P_n	1.7561	2.0645	2.347	2.6271	2.9078	3.1887	$P_n \to \infty$

An exact solution to the initial-boundary value problem (2.13a)–(2.13c) is [20]:

$$\sigma(x,t) = H\left(nh \arcsin\left|\sin\left(\frac{\pi}{2n}\sqrt{\frac{c}{m}}\,t\right)\right| - x\right), \tag{2.14}$$

where $H(\ldots)$ is the Heaviside function.

From equation (2.14) one obtains the following estimation:

$$|\sigma(x,t)| \le 1. \tag{2.15}$$

It was believed that the estimation (2.15) is also correct for a discrete system. However, analytical as well as numerical investigations indicate a need to distinguish between global and local characteristics of a discrete system. In other words, during the investigation of lower frequencies a transition into continuous model is allowed. However, in the case of forced oscillations the solutions for a discrete system may not be continuously transited into solution of a wave equation for $h \to 0$. Numerical investigations show that for given masses in a discrete chain quantity $P_j = \max\left|\sigma_j(t)\right|$ may exceed the initial value 1 (Table 2.1 reported in [11]).

Observe that the splash amplitude does not depend on the parameter m/c. In addition, the amplitude of the chain vibrations increases with increase of n, whereas its total energy does not depend on n. That is why the amplitude of the vibrations has an order of the sum of the quantities $\sigma_j(t)$, whereas its potential energy order is represented by a sum of squares of the quantities mentioned [26].

The explanation of the "splash paradox" is quite simple. The exact solution to the discrete problem (2.12) contains both slow and fast changes in the spatial coordinate harmonics. The continuous system (2.13a)–(2.13c) describes satisfactory only the slow components of a solution. Thus, the continuous model (2.13a)–(2.13c) of the governing discrete system must be supplemented by the restriction

$$k \ll n. \tag{2.16}$$

Continualization procedures that allow taking into account splash effect are described in [1, 12].

2.9 Correct Nonlinear Dynamic Equations of Buckled Beam

Duffing and van der Pol equations and finite systems of them are the test MMs in the Nonlinear Dynamics of systems with finite number dof.

Kirchhoff integro-differential equations of nonlinear beam dynamics play the same role in the Nonlinear Dynamics of continuous systems.

The temptation to reduce any nonlinear dynamics problem to one of these MMs is great, but sometimes it may lead to incorrect results. For further explanations, I need some well-known relations.

Consider the governing equations for nonlinear beam vibrations

$$\rho F \frac{\partial^2 W}{\partial t^2} + \frac{\partial^2 M}{\partial x^2} - \frac{\partial}{\partial x}\left(T \frac{\partial W}{\partial x}\right) = 0, \tag{2.17a}$$

$$\rho F \frac{\partial^2 U}{\partial t^2} - \frac{\partial T}{\partial x} = 0, \tag{2.17b}$$

where $M = EI\kappa, T = EF\varepsilon, \kappa = \frac{\partial^2 W}{\partial x^2}, \varepsilon = \frac{\partial U}{\partial x} + 0.5\left(\frac{\partial W}{\partial x}\right)$, κ is the curvature, which is assumed to be linear here for simplicity, U, W are the longitudinal and normal beam displacements.

Below, we consider two variants of BCs in the axial direction, namely prescribed end shortening and dead (forced) loading

$$U = U_b \quad \text{at} \quad x = 0, \quad x = L, \tag{2.18a}$$

$$T = T_b \quad \text{at} \quad x = 0, \quad x = L. \tag{2.18b}$$

The Kirchhoff model for BCs (2.18a) is [44]:

$$\rho F \frac{\partial^2 W}{\partial t^2} + EI \frac{\partial^4 W}{\partial x^4} + \frac{EF}{L}\left[U_b - \frac{1}{2}\left(\int_0^L \left(\frac{\partial W}{\partial x}\right)^2 \mathrm{d}x\right)\right]\frac{\partial^2 W}{\partial x^2} = 0. \tag{2.19}$$

Equation (2.19) can be obtained in two ways. Following Kirchhoff, one can neglect the axial inertia in Eq. (2.17b). On the other hand, Eq. (2.19) is the first approximation of an asymptotic process using as a small parameter quantity $I/\left(FL^2\right) \ll 1$.

One of the restrictions on the applicability of the simplified Eq. (2.19) is the inequality

$$\frac{EF}{L} U_b \ll T_{cr}, \tag{2.20}$$

where T_{cr} is the Euler critical load, $T_{cr} = \pi^2 EI/L^2$.

For the BCs (2.18b) very often the Eq. (2.19) is used, where $(EF/L)U_b$ is changed to T_b [17]. However, this is not correct. As Bolotin mentioned [6], in postbuckled state linear terms almost compensate each other and for a correct description of motion nonlinear terms of the first approximation must be taken into account. From the asymptotical standpoint, the situation is clear. In the original Kirchhoff equation linear terms were assumed to be of order of unity. If the linear components become small (and this is what happens in the postbuckled state), then the previously neglected higher-order nonlinear terms become significant.

Omitting details of calculations, we write out the final asymptotically correct equation for BCs (2.18b) as

$$\rho F \frac{\partial^2}{\partial t^2}\left[W - \frac{\partial}{\partial x}\left(\Psi \frac{\partial W}{\partial x}\right)\right] + EI\frac{\partial^4 W}{\partial x^4} + T_b \frac{\partial^2 W}{\partial x^2} = 0, \tag{2.21a}$$

$$\Psi = \frac{1}{2}\left[\int_0^x \left(\int_0^x \left(\frac{\partial W}{\partial x}\right)^2 \mathrm{d}x\right) \mathrm{d}x + \frac{x}{L}\int_0^L \left(\int_0^x \left(\frac{\partial W}{\partial x}\right)^2 \mathrm{d}x\right) \mathrm{d}x\right]. \tag{2.21b}$$

The general conclusion can be formulated as follows. Not taking into account the asymptotic nature of an MM, it is impossible to change it correctly, for example, to arbitrarily assume in the equations some terms small or large.

2.10 Highly Likely: Rigor of PM?

A few words about the "mathematical accuracy" of PM.

"Pure mathematics will remain more reliable than most other forms of knowledge, but its claim to a unique status will no longer be sustainable. It will be seen as the creation of finite human beings, liable to error in the same way as all other activities in which we indulge. Just as in engineering, mathematicians will have to declare their degree of confidence that certain results are reliable, rather than being able to declare flatly that the proofs are correct. Hilbert's goal of achieving perfect certainty by the laying of firm foundations died with Gödel's work, but the problem of complexity would have killed his dreams with equal finality fifty years later." [9].

The proof of theorems using a computer, when formal reasoning and numerical calculations are combined, does not differ much from the requirements for verification of results in AM. Therefore, the concept of "reasonable degree of scientific certainty", which is one of the main ones in AM [5], will sooner or later take its rightful place in PM.

"Formal verifications of complex proofs will be commonplace, but there will also be many results whose acceptance will owe as much to social consensus as to rigorous proof" [40].

Was Pythagoras theorem proven? Definitely! Was Fermat theorem proven? Highly likely! However, even if the proof itself is incorrect, which is not excluded, the statement is proved for a large number of special cases and is confirmed by so many calculations that the presence of a formal proof does not change much.

2.11 Conclusion

The main conclusion from the analysis of the above examples can be formulated as follows: any MM is asymptotic. It represents not only some ODEs, or PDEs with BCs and ICs. The MMs also include the asymptotic estimates and constraints on which they are based as an unseparable part. Only in this case a MM *is formulated quite definitely as a mathematical one* (sentence by Lyapunov).

A similar concept belongs to Mandelshtam [25, pp. 326–327]: “It can be said, a little schematically, that any physical theory consists of two complementary parts. These are the equations of the theory and the connection of the symbols (quantities), which are included in the equations with physical objects, and the connection made according *to specific recipes*” (translated from Russian by I.A.).

The practical recommendation is as follows. Be careful: if the change of your original MM is large enough, then you must also check the validity of the asymptotic estimates on the basis of which this MM was constructed. Maybe you must significantly modify your original MM.

A lot of excellent books [4, 28–31] are devoted to the analysis of paradoxes and fallacies arising in a consequence of incorrect MMs. If you do not pay enough attention to the original MM, your mathematically perfect results will be useless, if not harmful. Let’s not forget that: “Mathematics may be compared to a mill of exquisite workmanship, which grinds you stuff of any degree of fineness; but, nevertheless, what you get out depends upon what you put in; and as the grandest mill in the world will not extract wheat-flour from peascod, so pages of formulae will not get a definite result out of loose data” [18].

So, we must try to put a good grain in our mathematical mills and avoid GIGO (Garbage In–Garbage Out).

In this connection one can once again complain about the weak interaction of PM and AM communities. The real and not declarative interaction of scientists from these communities is difficult - the styles of thinking and even the languages are very different, which often leads to misunderstanding.

It seems to me that over 20 years of communication and cooperation, Wim and I have learned to better understand each other. In any case, I am very glad that in 1999 Wim read my abstract.

Dear Wim, Many Happy Returns of the Day!

Acknowledgements I owe a lot to Dr. H. Topol, who read the manuscript of the article and made a number of valuable comments and clarifications. I am grateful to Professors I. Elishakoff, A. Yu. Evkin, W. T. van Horssen, V. I. Malyi, L. I. Manevitch. Yu. V. Mikhlin, V. I. Olevskyi, A. V. Pichugin, A. A. Snarskii, F. Verhulst, D. Weichert for valuable discussions. These discussions were all the more valuable to me because, as a rule, opinions of my colleagues were directly opposite to mine. Discussio mater veritas est. I also thank IAM TU Delft for hosting my research visits in Septembers 2017, 2018, 2019.

Appendix

"I asked a few friends to read it (draft of the paper—I.A.) and to criticize it. The criticisms were excellent; they were sharp, honest, and constructive; and they were contradictory" [15].

Readers of the draft of my paper unanimously noted the most successful and undoubtedly correct phrase: "Of course, we did not come to a single opinion, and there is no single point of view on this subject at all." As for the rest of the text, the opinions were contradictory.

"Physics, watch out for metaphysics!" (attributed to Newton). On the other hand, "The purpose of computing is insight, not numbers" [16, p. 276], and here metaphysics cannot be dispensed with. BTW, Hamming emphasizes that the original MM may be changed during the solution and on the result of the solution.

The most difficult question that asked readers of my paper: "Who is this paper for? This is too early for students and useless for experienced researchers". My answer: subject of paper is very interesting for me; and I wrote it for better understanding of this important subject by myself. In this regard, I note, slightly changing the text by Halmos [15]: This is a subjective essay, and its title is misleading; a more honest title might be "How I Understand Mathematical Modeling".

The most flattering opinion: "You wrote a provocative paper!" (I would like to hope!). In the comments were questions about the relationship of hypotheses and asymptotics, the role of physical and numerical experiments, Big Data, AI, etc. I console myself with the hope that this speaks of interest, if not to my paper itself, then to its subject.

References

1. Andrianov, I.V., Awrejcewicz, J., Weichert, D.: Improved continuous models for discrete media. Math. Pr. Eng. (2010)
2. Andrianov, I.V., van Horssen, W.T.: On the transversal vibrations of a conveyor belt: applicability of simplified models. J. Sound Vibr. **313**, 822–829 (2008)
3. Arnol'd, V.I.: Topological problems of the theory of wave propagation. Russ. Math. Surv. **51**(1), 1–47 (1996)
4. Birkhoff, G.: Hydrodynamics: A Study in Logic, Fact, and Similitude, 2nd edn. Princeton U. P. (1960)
5. Blekhman, I.I., Myshkis, A.D., Panovko, Y.G.: Applied Mathematics: Subject, Logic, Peculiarities of Approaches. With Examples from Mechanics (in Russian). URSS (2007)
6. Bolotin, V.V.: The Dynamic Stability of Elastic Systems. Holden-Day (1963)
7. Bolotin, V.V.: Nonconservative Problems of Theory of Elastic Stability. Pergamon Press (1963)
8. Bourbaki, N.: The architecture of mathematics. Am. Math. Mon. **57**(4), 221–232 (1950)
9. Davies, B.: Whither mathematics? Notices AMS **52**(11), 1350–1356 (2005)
10. Fefferman, C.: Existence and Smoothness of the Navier-Stokes Equation, pp. 1–5. Clay Mathematics Institute, Cambridge, MA (2000). http://claymath.org/millenium-prize-problems/navier-stokes-equations
11. Filimonov, A.M.: Some unexpected results on the classical problem of the string with N beads. The case of multiple frequencies. C. R. Acad. Sci. Paris **315**, 957–961 (1992)

12. Filimonov, A.M.: Continuous approximations of difference operators. J. Diff. Eq. Appl. **2**(4), 411–422 (1996)
13. Golovkin, K.K.: New model equations of motion of a viscous fluid and their unique solvability. Proc. Steklov Inst. Math. **102**, 29–54 (1967). (in Russian)
14. Gorban, A.N., Karlin, I.: Hilbert's 6th problem: exact and approximate hydrodynamic manifolds for kinetic equations. Bull. AMS **51**, 186–246 (2014)
15. Halmos, P.R.: How to write mathematics. L'Enseignement mathématique **16**, 123–152 (1970)
16. Hamming, R.W.: Numerical Methods for Scientists and Engineers. McGraw-Hill (1962)
17. Holmes, P.: A nonlinear oscillator with a strange attractor. Phil. Trans. R. Soc. A **292**, 419–448 (1979)
18. Huxley, T.H.: Geological reform. Quart. J. Geol. Soc. London **25** (1869); reprinted in: Huxley, T.H.: Discourses, Biological and Geological Essays, 335–336 (2009)
19. Kruskal, M.D.: Asymptotology. In: Drobot, S., Viebrock, P.A. (eds.) Mathematical Models in Physical Sciences (Proceedings of Conference with the Same Name at University of Notre Dame, Apr. 15–17, 1962), pp. 17–48. Prentice-Hall (1963)
20. Kurchanov, P.F., Myshkis, A.D., Filimonov, A.M.: Vibrations of rolling stock and a theorem of Kronecker. J. Appl. Math. Mech. **55**(6), 870–876 (1991)
21. Ladyzhenskaya, O.A.: Modifications of the Navier-Stokes equations for large gradients of the velocities. Zap. Nauchn. Sem. Leningrad. Otd. Mat. Inst. Steklov (LOMI) **7**, 126–154 (1968) (in Russian)
22. Ladyzhenskaya, O.A.: The sixth problem of the millennium: The Navier-Stokes equations, existence and smoothness. Russ. Math. Surv. **58**(2), 251–286 (2003)
23. Liapounoff, A.: [Lyapunov A.M.]. Sur une classe de figures d'équilibre d'un liquide en rotation. Annales scientifiques de l'École Normale Supérieure **26**, 473–483 (1909)
24. Lucertini, M., Gasca, A.M., Nicolò, F. (eds.): Technological Concepts and Mathematical Models in the Evolution of Modern Engineering Systems: Controlling. Managing, Organizing. Birkhäuser (2004)
25. Mandelshtam, L.I.: Lectures in Optics, Relativity Theory and Quantum Mechanics. Nauka (1972). (in Russian)
26. Myshkis, A.D.: Mixed functional differential equations. J. Math. Sci. **129**, 4111–4226 (2005)
27. Neumann, J., Richtmyer, R.D.: A method for calculation of hydrodynamic shocks. J. Appl. Phys. **21**, 232–237 (1950)
28. Panovko, Y.G., Gubanova, I.I.: Stability and Oscillations of Elastic Systems. Fallacies, Paradoxes and New Concepts. Consultants Bureau, (1965)
29. Peierls, R.: Surprises in Theoretical Physics. Princeton U. P. (1979)
30. Peierls, R.: More Surprises in Theoretical Physics. Princeton U. P. (1991)
31. Perelmuter, A.V., Slivker, V.I.: Numerical Structural Analysis: Methods, Models, and Pitfalls. Springer (2003)
32. Petersen, J.K.: Fiber Optics Illustrated Dictionary. CRC Press (2003)
33. Poincaré, H.: The relations of analysis and mathematical physics. Bull. AMS **4**(6), 247–255 (1898)
34. Poincaré, H.: Sur la stabilité de l'équilibre des figures pyriformes affectées par une masse fluide en rotation. Phil. Trans. R. Soc. Lond. A **198**, 333–373 (1902)
35. Saint-Raymond, L.: A mathematical PDE perspective on the Chapman-Enskog expansion. Bull. AMS **51**, 247–275 (2014)
36. Strutt, J.W.: (Lord Rayleigh). The Theory of Sound, vol. 1. Preface to the 2nd edn, revised and enlarged, 1894. Here cited after First American Edition revised and enlarged, 1894. Here cited after First American Edition (Two volumes bound as one), p. XXXY. Dover Publications (1945)
37. Suweken, G., van Horssen, W.T.: On the transversal vibrations of a conveyor belt with a low and time-varying velocity. Part I: the string-like case. J. Sound Vibr. **264**(1), 117–133 (2003)
38. Theme Issue. Hilbert's Sixth Problem. Phil. Trans. R. Soc. A **376** (2018)
39. Van Dyke, M.: Perturbation Methods in Fluid Mechanics. The Parabolic Press (1975)

40. Voevodsky, V.: The Origins and Motivations of Univalent Foundations. A Personal Mission to Develop Computer Proof Verification to Avoid Mathematical Mistakes. IDEAS, The Institute Letter, Summer 2014. Institute of Advanced Study (2014). https://www.ias.edu/ideas/2014/voevodsky-origins
41. Wikipedia Contributors. Applied Mathematics—Wikipedia, the free encyclopedia (2020). [Online; accessed January 2020]
42. Wikipedia Contributors. Model—Wikipedia, the free encyclopedia (2020). [Online; accessed January 2020 (in Russian)]
43. Wikipedia contributors. Pure Mathematics—Wikipedia, the free encyclopedia (2020). [Online; accessed January 2020]
44. Yamaki, N., Mori, A.: Non-linear vibrations of a clamped beam with initial defection and initial axial displacement. J. Sound Vibr. **71**, 333–346 (1980)
45. Yudovich, V.I.: Global solvability versus collapse in the dynamics of an incompressible fluid. In Bolibruch, A.A., Osipov, Y.S., Sinai, Y.G. (eds.) Mathematical Events of the Twentieth Century, pp. 501–528. Springer (2006)

Expanding the Applicability of the Competitive Modes Conjecture

Sudipto Choudhury, Huibert Reijm, and Cornelis Vuik

Abstract The Competitive Modes Conjecture is a relatively new approach in the field of Dynamical Systems, aiming to understand chaos in strange attractors using Resonance Theory. Up till now, the Conjecture has only been used to study multipolynomial systems because of their simplicity. As such, the study of non-multipolynomial systems is sparse, filled with ambiguity, and lacks mathematical structure. This paper strives to rectify this dilemma, providing the mathematical background needed to rigorously apply the Competitive Modes Conjecture to a certain set of non-multipolynomial systems. Afterwards, we provide an example of this new theory in the non-multipolynomial Wimol-Banlue Attractor, something that up to this point has not been possible as far as the authors know.

3.1 The Competitive Modes Conjecture

This section is to serve as background knowledge, all obtained from sources [1–6].

We take a general n-dimensional autonomous system of differential equations $\dot{x}_i = F_i(\mathbf{x})$ with $\mathbf{x} \in \mathbb{R}^n$ and $i \in \{1, 2, \ldots, n\}$. We can easily transform this system into a system of interconnected oscillators as follows

S. Choudhury
University of Central Florida, 4000 Central Florida Blvd., Orlando, Fl 32816, USA
e-mail: sudipto.choudhury@ucf.edu

H. Reijm · C. Vuik (✉)
Delft University of Technology, Mekelweg 5, 2628CC Delft, The Netherlands
e-mail: c.vuik@tudelft.nl

H. Reijm
e-mail: h.a.j.reijm@student.tudelft.nl

A. K. Abramian et al. (eds.), *Nonlinear Dynamics of Discrete and Continuous Systems*,
Advanced Structured Materials 139,
https://doi.org/10.1007/978-3-030-53006-8_3

$$
\begin{aligned}
\ddot{x}_i &= \dot{F}_i(\mathbf{x}) \\
&= \sum_{j=1}^{n} \frac{\partial F_i}{\partial x_j}(\mathbf{x})\dot{x}_j \\
&= \sum_{j=1}^{n} \frac{\partial F_i}{\partial x_j}(\mathbf{x})F_j(\mathbf{x}) \\
&\equiv f_i(\mathbf{x})
\end{aligned}
\tag{3.1}
$$

This of course only works if F_i is x_j-differentiable for all $i, j \in \{1, 2, \ldots, n\}$.

Definition 3.1 (*Splitting of a Function*) In previous literature, function $f_i : \mathbb{R}^n \to \mathbb{R}$ can be split with respect to x_i if it can be rewritten as

$$f_i(\mathbf{x}) = h_i(x_1, \ldots, x_{i-1}, x_{i+1}, \ldots, x_n) - x_i g_i(\mathbf{x}) \; \forall i \in \{1, 2, \ldots, n\} \tag{3.2}$$

We name function $h_i : \mathbb{R}^{n-1} \to \mathbb{R}$ the ith forcing function. We name function $g_i : \mathbb{R}^n \to \mathbb{R}$ the ith squared frequency function.

For simplicity, let us define $\mathbf{x}_i^* = [x_1, \ldots, x_{i-1}, x_{i+1}, \ldots, x_n]^T \in \mathbb{R}^{n-1}$. If Eq. (3.1) holds and the resulting functions f_i can be split, then we can rewrite our original system of differential equations into the form given below.

$$
\begin{cases}
\ddot{x}_1 + g_1(\mathbf{x})x_1 = h_1(\mathbf{x}_1^*) \\
\ddot{x}_2 + g_2(\mathbf{x})x_2 = h_2(\mathbf{x}_2^*) \\
\cdots \\
\ddot{x}_i + g_i(\mathbf{x})x_i = h_i(\mathbf{x}_i^*) \\
\cdots \\
\ddot{x}_n + g_n(\mathbf{x})x_n = h_n(\mathbf{x}_n^*)
\end{cases}
\tag{3.3}
$$

In a sense, we have turned our system into a system of interconnected, nonlinear oscillators.

Definition 3.2 (*Competitive Modes*) Say we have the n-dimensional autonomous system of differential equations $\mathbf{x} = \mathbf{F}(\mathbf{x})$. If Eq. (3.1) holds for this system and the resulting functions f_i can be split, then the system can be transformed as shown in Eq. (3.3). The solutions x_i for Eq. (3.3) are then known as the competitive modes of the system, with g_i and h_i being the corresponding squared frequency functions and forcing functions, respectively.

Currently, there is an open conjecture connecting chaos and competitive modes together, and it is presented as follows.

Conjecture 3.1 (Competitive Modes Conjecture) *The conditions for a dynamical system to be chaotic are given below* (*assuming* Eq. (3.1) *holds and the resulting function* f_i's *can be split:*)

- *the dimension n of the dynamical system is greater than 2;*
- *at least two distinct squared frequency functions g_i and g_j are competitive or nearly competitive; that is, there exists $t \in \mathbb{R}$ so that $g_i(t) \approx g_j(t)$ and $g_i(t), g_j(t) > 0$;*
- *at least squared frequency function g_i is not constant with respect to time;*
- *at least one forcing function h_i is not constant with respect to some system variable x_j.*

3.2 Proper Splittings

Notice that the process of splitting as defined in Definition 3.1 is rather ambiguous. Therefore, we now provide a new definition for the splitting of a function. Throughout this paper, we refer to domain D, which is a uncountably infinite, open set in $\mathbb{R}^n$.

Definition 3.3 (*Splitting of a Function*) We now say that function $f : D \to \mathbb{R}$ can be split with respect to $x_i \in \mathbb{R}$ and $\mathbf{c} \in D$ if over D, it can be rewritten as

$$f(\mathbf{x}) = h(\mathbf{x}_i^*) - (x_i - c_i)g(\mathbf{x}) \tag{3.4}$$

where $\mathbf{x}_i^* = [x_1, \ldots, x_{i-1}, x_{i+1}, \ldots, x_n]^T$ and

- f is continuous in x_i for all $\mathbf{x} \in D$;
- the subset $D_i^*(\mathbf{c}) = \{\mathbf{x} \in D : x_i = c_i\}$ is not empty;
- h is constant and finite in x_i, given $x_1, \ldots, x_{i-1}, x_{i+1}, \ldots, x_n$;
- g is continuous with respect to x_i, given $x_1, \ldots, x_{i-1}, x_{i+1}, \ldots, x_n$

Here, h is again called the forcing function and g is the squared frequency function.

We then have the following results, lemmas, and theorems.

Lemma 3.1 *Say function $f : D \to \mathbb{R}$ can be split with respect to $x_i \in \mathbb{R}$ and $\mathbf{c} \in D$ into forcing function h and squared frequency function g. Then $h(\mathbf{x}_i^*) = f(\mathbf{x})|_{x_i=c_i}$.*

Proof Say function $f : D \to \mathbb{R}$ can be split with respect to $x_i \in \mathbb{R}$ into forcing function h and squared frequency function g. Then for all $\mathbf{x} \in D$, since g is continuous in x_i,

$$g(\mathbf{x})|_{x_i=\alpha} = \lim_{x_i \to \alpha} \left(\frac{h(\mathbf{x}_i^*) - f(\mathbf{x})}{x_i - c_i} \right)$$

Thus, we can conclude that

$$g(\mathbf{x})|_{x_i=c_i} = \lim_{x_i \to c_i} \left(\frac{h(\mathbf{x}_i^*) - f(\mathbf{x})}{x_i - c_i} \right) \in \mathbb{R}$$

Because of this, $\lim_{x_i \to c_i} \left(h(\mathbf{x}_i^*) - f(\mathbf{x})\right) = 0$. Otherwise, $\lim_{x_i \to c_i} g(\mathbf{x})$ would surely be infinite or undefined. Thus, we can conclude that, since f is continuous in x_i,

$$0 = \lim_{x_i \to c_i} \left(h(\mathbf{x}_i^*) - f(\mathbf{x})\right) = h(\mathbf{x}_i^*) - \lim_{x_i \to c_i} f(\mathbf{x}) = h(\mathbf{x}_i^*) - f(\mathbf{x})|_{x_i=c_i}$$

This lemma is important, as it symbolizes the ideology behind Definition 3.3. Our research started by trying to rigorously define the forcing function h, and then defining the squared frequency function g as a direct result. We noticed that in multipolynomial systems, Lemma 3.1 was always true. In fact, it seemed that previous literature had specifically defined h so that the lemma would always hold when $\mathbf{c} = \mathbf{0}$ [1–6]. We decided to expand this idea to Taylor Series, Laurent Series, and finally to general continuous functions. It is on this idea that we can build the rest of our theory.

Lemma 3.2 (Uniqueness Lemma) *Say function $f : D \to \mathbb{R}$ can be split with respect to $x_i \in \mathbb{R}$ and $\mathbf{c} \in D$ into forcing function h and squared frequency function g. Then h and g are uniquely defined.*

Proof Say function $f : D \to \mathbb{R}$ can be split with respect to $x_i \in \mathbb{R}$ and $\mathbf{c} \in D$ into forcing function h_1 and squared frequency function g_1, and also into forcing function h_2 and squared frequency function g_2. Then for all $\mathbf{x} \in D$,

$$f(\mathbf{x}) = h_1(\mathbf{x}_i^*) - (x_i - c_i)g_1(\mathbf{x}) = h_2(\mathbf{x}_i^*) - (x_i - c_i)g_2(\mathbf{x})$$

Recall that $D_i^*(\mathbf{c}) = \{\mathbf{x} \in D : x_i = c_i\}$.
Since we know from Lemma 3.1 that $h_1(\mathbf{x}_i^*) = h_2(\mathbf{x}_i^*) = f(\mathbf{x})|_{x_i=c_i}$, we can immediately conclude that $h_1 = h_2$.
As a result, for all $\mathbf{x} \in D$,

$$(x_i - c_i)(g_1(\mathbf{x}) - g_2(\mathbf{x})) = h_1(\mathbf{x}_i^*) - h_2(\mathbf{x}_i^*) = 0$$

For all $\mathbf{x} \in D \backslash D_i^*(\mathbf{c})$, $g_1(\mathbf{x}) - g_2(\mathbf{x}) = 0$.
Furthermore, since g_1 and g_2 are both continuous in $D_i^*(\mathbf{c})$, we can conclude that

$$g_1(\mathbf{x})|_{x_i=c_i} = \lim_{x_i \to c_i} g_1(\mathbf{x}) = \lim_{x_i \to c_i} g_2(\mathbf{x}) = g_2(\mathbf{x})|_{x_i=c_i}$$

Thus, we have proven that $g_1(\mathbf{x}) = g_2(\mathbf{x})$ for all $\mathbf{x} \in D$.

Lemma 3.3 (Combination Lemma) *Say function $f_1 : D \to \mathbb{R}$ can be split with respect to $x_i \in \mathbb{R}$ and $\mathbf{c} \in D$ into forcing function h_1 and squared frequency function g_1. Say function $f_2 : D \to \mathbb{R}$ can be split with respect to x_i and $\mathbf{c}$ into forcing function h_2 and squared frequency function g_2.*

- *For arbitrary $\alpha, \beta \in \mathbb{R}$, the sum $(\alpha f_1 + \beta f_2) : D \to \mathbb{R}$ can be split with respect to x_i and $\mathbf{c}$ into forcing function $(\alpha h_1 + \beta h_2)$ and squared frequency function $(\alpha g_1 + \beta g_2)$.*
- *The product $(f_1 f_2) : D \to \mathbb{R}$ can be split with respect to x_i into forcing function $(h_1 h_2)$ and squared frequency function $(h_1 g_2 + h_2 g_1 - (x_i - c_i) g_1 g_2)$.*
- *The quotient $(f_1/f_2) : D \to \mathbb{R}$ can be split with respect to x_i and $\mathbf{c}$ into forcing function (h_1/h_2) and squared frequency function $((h_2 g_1 - h_1 g_2)/(h_2 f_2))$, provided both $f_2(\mathbf{x})$ and $h_2(\mathbf{x}_i^*)$ are nonzero for all $\mathbf{x} \in D$.*

Proof Say function $f_1 : D \to \mathbb{R}$ can be split with respect to $x_i \in \mathbb{R}$ and $\mathbf{c} \in D$ into forcing function h_1 and squared frequency function g_1. Then for all $\mathbf{x} \in D$,

$$f_1(\mathbf{x}) = h_1(\mathbf{x}_i^*) - (x_i - c_i)g_1(\mathbf{x})$$

Say function $f_1 : D \to \mathbb{R}$ can be split with respect to x_i and $\mathbf{c}$ into forcing function h_2 and squared frequency function g_2. Then for all $\mathbf{x} \in D$,

$$f_2(\mathbf{x}) = h_2(\mathbf{x}_i^*) - (x_i - c_i)g_2(\mathbf{x})$$

First of all, notice that $D_i^*(\mathbf{c}) = \{\mathbf{x} \in D : x_i = c_i\}$ is automatically not empty since both f_1 and f_2 can be split on D.

Take $\alpha, \beta \in \mathbb{R}$ arbitrarily.

$$\begin{aligned}\alpha f_1(\mathbf{x}) + \beta f_2(\mathbf{x}) &= \alpha\left(h_1(\mathbf{x}_i^*) - (x_i - c_i)g_1(\mathbf{x})\right) + \beta\left(h_2(\mathbf{x}_i^*) - (x_i - c_i)g_2(\mathbf{x})\right)\\ &= \left(\alpha h_1(\mathbf{x}_i^*) + \beta h_2(\mathbf{x}_i^*)\right) - (x_i - c_i)\left(\alpha g_1(\mathbf{x}) - \beta g_2(\mathbf{x})\right)\end{aligned}$$

Notice that

- the linear combination $\alpha f_1 + \beta f_2$ is continuous over D in x_i since f_1 and f_2 are continuous over D in x_i;
- the linear combination $\alpha h_1 + \beta h_2$ is constant and finite over D in x_i since h_1 and h_2 are constant and finite over D in x_i;
- the linear combination $\alpha g_1 + \beta g_2$ is continuous over D in x_i since g_1 and g_2 are continuous over D in x_i.

Thus we constructed the splitting of $(\alpha f_1 + \beta f_2)$ with respect to x_i and $\mathbf{c}$.

We can also split the product of f_1 and f_2.

$$\begin{aligned}f_1(\mathbf{x})f_2(\mathbf{x}) &= \left(h_1(\mathbf{x}_i^*) - (x_i - c_i)g_1(\mathbf{x})\right)\left(h_2(\mathbf{x}_i^*) - (x_i - c_i)g_2(\mathbf{x})\right)\\ &= \left(h_1(\mathbf{x}_i^*)h_2(\mathbf{x}_i^*)\right) - (x_i - c_i)\left(h_1(\mathbf{x}_i^*)g_2(\mathbf{x}) + h_2(\mathbf{x}_i^*)g_2(\mathbf{x}) - (x_i - c_i)g_1(\mathbf{x})g_2(\mathbf{x})\right)\end{aligned}$$

Notice that

- the product $f_1 f_2$ is continuous over D in x_i since f_1 and f_2 are continuous over D in x_i;
- the product $h_1 h_2$ is constant and finite over D in x_i since h_1 and h_2 are constant and finite over D in x_i;
- the function $h_1(\mathbf{x}_i^*)g_2(\mathbf{x}) + h_2(\mathbf{x}_i^*)g_2(\mathbf{x}) - (x_i - c_i)g_1(\mathbf{x})g_2(\mathbf{x})$ is continuous over D in x_i since g_1 and g_2 are continuous and h_1 and h_2 are constant and finite over D in x_i.

Thus we constructed the splitting of $f_1 f_2$ with respect to x_i and $\mathbf{c}$.

We can also split the quotient of f_1 and f_2, provided $h_2(\mathbf{x}_i^*) \neq 0$ and $f_2(\mathbf{x}) \neq 0$ for all $\mathbf{x} \in D$.

$$\frac{f_1(\mathbf{x})}{f_2(\mathbf{x})} = \frac{h_1(\mathbf{x}_i^*) - (x_i - c_i)g_1(\mathbf{x})}{h_2(\mathbf{x}_i^*) - (x_i - c_i)g_2(\mathbf{x})}$$
$$= \left(\frac{h_1(\mathbf{x}_i^*)}{h_2(\mathbf{x}_i^*)}\right) - (x_i - c_i)\left(\frac{h_2(\mathbf{x}_i^*)g_1(\mathbf{x}) - h_1(\mathbf{x}_i^*)g_2(\mathbf{x})}{h_2(\mathbf{x}_i^*)f_2(\mathbf{x})}\right)$$

Notice that

- the quotient f_1/f_2 is continuous over D in x_i since f_1 and f_2 are continuous over D in x_i and $f_2(\mathbf{x}) \neq 0$ for all $\mathbf{x} \in D$;
- the quotient h_1/h_2 is constant and finite over D in x_i since h_1 and h_2 are constant and finite over D in x_i and $h_2(\mathbf{x}_i^*) \neq 0$ for all $\mathbf{x} \in D$;
- the function $(h_2 g_1 - h_1 g_2)/(h_2 f_2)$ is continuous over D in x_i since g_1 and g_2 are continuous over D in x_i, h_1 and h_2 are constant and finite over D in x_i, and $h_2(\mathbf{x}_i^*) \neq 0$ and $f_2(\mathbf{x}) \neq 0$ for all $\mathbf{x} \in D$.

Thus we have constructed the splitting of f_1/f_2 with respect to x_i and $\mathbf{c}$.

The following theorem is perhaps the most useful theorem concerning splittable functions.

Theorem 3.1 (Existence of Splittings for Differentiable Functions) *Say function $f : D \to \mathbb{R}$ is differentiable over D with respect to $x_i \in \mathbb{R}$. Take $\mathbf{c} \in D$. If the partial derivative $\partial f/\partial x_i$ is continuous with respect to x_i in c_i, then f can be split into proper forcing function h and proper squared frequency function g, defined as*

$$h(\mathbf{x}_i^*) = f(\mathbf{x})|_{x_i=c_i}$$
$$g(\mathbf{x}) = \begin{cases} \dfrac{f(\mathbf{x})|_{x_i=c_i} - f(\mathbf{x})}{x_i - c_i} & x_i \neq c_i \\ -\left.\dfrac{\partial f(\mathbf{x})}{\partial x_i}\right|_{x_i=c_i} & x_i = c_i \end{cases}$$

Proof Say function $f : D \to \mathbb{R}$ is differentiable over D with respect to x_i. Lets define functions h and g as above.

Since f is differentiable and thus continuous over D with respect to x_i, we know immediately from Lemma 3.1 that h is constant and finite in terms of x_i, given $x_1, \ldots x_{i-1}, x_{i+1}, \ldots x_n$.

Investigating the properties of g takes a bit more work. Lets take $x_i \neq c_i$, then g is continuous over D in x_i because f is differentiable and thus also continuous over D in x_i.
Lets take $x_i = c_i$, then we can conclude the following, using L'Hopital's Theorem and the prerequisite that the derivative $\partial f/\partial x_i$ must be continuous in c_i.

$$\begin{aligned}
\lim_{x_i \to c_i} g(\mathbf{x}) &= \lim_{x_i \to c_i} \left(\frac{f(\mathbf{x})|_{x_i=c_i} - f(\mathbf{x})}{x_i - c_i} \right) \\
&= -\lim_{x_i \to c_i} \frac{\partial f(\mathbf{x})}{\partial x_i} \\
&= -\frac{\partial f(\mathbf{x})}{\partial x_i}\bigg|_{x_i=c_i} \\
&= g(\mathbf{x})|_{x_i=c_i}
\end{aligned}$$

Thus, we have proven that g is continuous in D with respect to x_i.

Finally, we must prove that the equation

$$f(\mathbf{x}) = h(\mathbf{x}_i^*) - (x_i - c_i)g(\mathbf{x})$$

is valid in the first place. Take $\mathbf{x} \in D$ arbitrarily. We then have to consider two mutually exclusive cases.
Say $x_i \neq c_i$. Then

$$\begin{aligned}
h(\mathbf{x}_i^*) - (x_i - c_i)g(\mathbf{x}) &= f(\mathbf{x})|_{x_i=c_i} - (x_i - c_i) \left(\frac{f(\mathbf{x})|_{x_i=c_i} - f(\mathbf{x})}{x_i - c_i} \right) \\
&= f(\mathbf{x})|_{x_i=c_i} - \left(f(\mathbf{x})|_{x_i=c_i} - f(\mathbf{x}) \right) \\
&= f(\mathbf{x})
\end{aligned}$$

Say instead $x_i = c_i$. Then we know that $((x_i - c_i)g(\mathbf{x}))\,|_{x_i=c_i} = 0$ since $g(\mathbf{x})|_{x_i=c_i}$ is continuous and therefore finite. Thus

$$\begin{aligned}
h(\mathbf{x}_i^*) - (x_i - c_i)g(\mathbf{x}) &= f(\mathbf{x})|_{x_i=c_i} - 0 \\
&= f(\mathbf{x})
\end{aligned}$$

Thus, for any $\mathbf{x} \in D$, $h(\mathbf{x}_i^*) - (x_i - c_i)g(\mathbf{x}) = f(\mathbf{x})$. Thus, h is the forcing function and g is the squared frequency function of f.

Of course, a splitting of f can not be achieved without defining $\mathbf{c} \in D$ first. The constant $\mathbf{c}$ can of course be arbitrary, but we will primarily focus on one particular scenario. When a function f is split with respect to $\mathbf{c} = \mathbf{0}$, then we define this to be the proper splitting of f, with h defined to be the proper forcing function and g defined to be the proper squared frequency function.The reason for this is made clear with an example.

Lets say we have a multipolynomial second order ODE $\ddot{x}_i = f(\mathbf{x})$, where $f{:}D \to \mathbb{R}$. Previous literature (as far as the authors are aware) has strictly focused on gathering evidence for the Competitive Modes Conjecture from dynamical systems whose set of differential equations consist of these sorts of ODEs. It can be easily proven[1] that the proper splitting of f always exists, and that the resulting proper forcing function and proper squared frequency function are defined identically to the forcing functions and squared frequency functions defined in previous literature [1–6]. As a result, the theory of proper splittings is a direct expansion of Definition 3.1.

3.3 Example: The Wimol-Banlue Attractor

To show the applicability of this new theory of proper splittings, we will apply it to a modification of the system mentioned in [7], which we will call the Wimol-Banlue System. The original Wimol Banlue Dynamical System is given by

$$\begin{cases} \dot{x} &= y - x \\ \dot{y} &= -z\tanh(x) \\ \dot{z} &= -\alpha + xy + |y| \end{cases} \tag{3.5}$$

where $\alpha \in \mathbb{R}$. The reason we chose to work with the Wimol-Banlue System is because it is the most accessible non-multipolynomial system which has been proven to exhibit a chaotic attractor. An unfortunate property of this system is that $\dot{z}$ is not differentiable with respect to y at $y = 0$. To counterattack this, we introduce function ϕ, dependent on parameter $\beta > 0$, defined as

$$\phi(y; \beta) = \sqrt{y^2 + \beta^2} \tag{3.6}$$

First, notice that ϕ is a well-defined, positive, differentiable function over all $\mathbf{R}$, with its derivative being

$$\phi'(y; \beta) = \frac{y}{\sqrt{y^2 + \beta^2}}$$

We want to compare $\phi(y; \beta)$ to $|y|$; to that end, we construct the difference function $\varphi(y; \beta) = \phi(y; \beta) - |y|$. It is easy to prove that φ is a positive, continuous function for $y \in \mathbb{R}$. Furthermore φ is differentiable for $y \neq 0$, with its derivative being

$$\varphi'(y; \beta) = \begin{cases} \dfrac{\sqrt{y^2} - \sqrt{y^2 + \beta^2}}{\sqrt{y^2 + \beta^2}} & y > 0 \\ \dfrac{-\sqrt{y^2} + \sqrt{y^2 + \beta^2}}{\sqrt{y^2 + \beta^2}} & y < 0 \end{cases}$$

[1] The calculations needed to prove this are straightforward but cumbersome. For the sake of space, we chose to omit them.

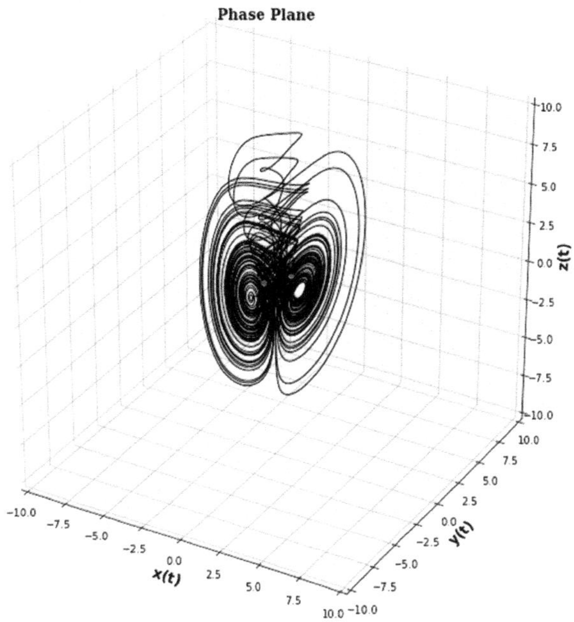

Fig. 3.1 The trajectory of our modified Wimol-Banlue Attractor as defined in Eq. (3.7) with initial condition $\mathbf{x}_0 = [1.32, -0.63, 1.91]^T$. The trajectory was approximating using 70,000 iterations of an adaptive RK4 method, using a time step of 0.01. Notice the presence of an attractor

Because of this, $\varphi'(y; \beta) < 0$ for $y > 0$ and $\varphi'(y; \beta) > 0$ for $y < 0$; we can then make the following inequality

$$|\varphi(y; \beta)| \leq |\varphi(0; \beta)| = \beta$$

Thus ϕ converges uniformly to $|y|$ as β goes to 0. Therefore, ϕ is a sufficiently accurate, differentiable approximation of $|y|$ and we can modify the Wimol-Banlue System slightly into

$$\begin{cases} \dot{x} &= y - x \\ \dot{y} &= -z \tanh(x) \\ \dot{z} &= -\alpha + xy + \sqrt{y^2 + \beta^2} \end{cases} \tag{3.7}$$

Let us first prove that this modified system still has a chaotic attractor. For the continuation of this example, lets say $\alpha = 2$ and $\beta = 0.001$. With arbitrary initial vector $\mathbf{x}_0 = [1.32, -0.63, 1.91]^T$, the resulting trajectory is presented in Fig. 3.1. As one can see, an attractor is still present in this system.

Through this trajectory, the Lyapunov Exponent is approximately equal to 0.228483. As further evidence of the attractor's chaotic nature, we provide the plot of the convergence of the Lyapunov Exponent in Fig. 3.2.

We consider this sufficient evidence to safely proven the presence of a chaotic attractor in our system.

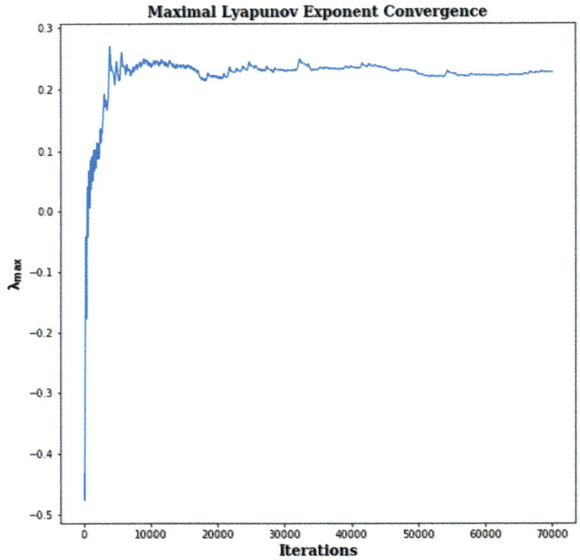

Fig. 3.2 The convergence of the maximal Lyapunov Exponent of our modified Wimol-Banlue Attractor, using a trajectory with initial condition $\mathbf{x}_0 = [1.32, -0.63, 1.91]^T$. The trajectory was approximating using 70,000 iterations of an adaptive RK4 method, using a time step of 0.01

To see if the modified system in Eq. (3.7) can be properly split, the system must first be differentiated in terms of time, which is done as follows.

$$\begin{aligned}
\ddot{x} &= -\dot{x} + \dot{y} \\
&= -(y - x) + (-z \tanh(x)) \\
&= x - y - z \tanh(x) \\
\ddot{y} &= -z \operatorname{sech}^2(x)\dot{x} - \tanh(x)\dot{z} \\
&= -z \operatorname{sech}^2(x)(y - x) - \tanh(x)\left(-\alpha + xy + \phi(y; \beta)\right) \\
&= (x - y)z \operatorname{sech}^2(x) + \left(\alpha - xy - \sqrt{y^2 + \beta^2}\right) \tanh(x) \\
\ddot{z} &= y\dot{x} + \left(x + \frac{y}{\sqrt{y^2 + \beta^2}}\right) \dot{y} \\
&= y(y - x) + \left(x + \frac{y}{\sqrt{y^2 + \beta^2}}\right) (-z \tanh(x)) \\
&= y^2 - xy - \left(x + \frac{y}{\sqrt{y^2 + \beta^2}}\right) z \tanh(x)
\end{aligned}$$

We can differentiate $\ddot{x}$ with respect to x, $\ddot{y}$ with respect to y, and $\ddot{z}$ with respect to z as follows.

$$\frac{\partial \ddot{x}}{\partial x} = 1 - z\,\text{sech}^2(x)$$
$$\frac{\partial \ddot{y}}{\partial y} = -z\,\text{sech}^2(x) - \left(x + \frac{y}{\sqrt{y^2+\beta^2}}\right)\tanh(x)$$
$$\frac{\partial \ddot{z}}{\partial z} = -\left(x + \frac{y}{\sqrt{y^2+\beta^2}}\right)\tanh(x)$$

Since sech and tanh are continuous and bounded over all $\mathbb{R}$, $\partial\ddot{x}/\partial x$, $\partial\ddot{y}/\partial y$, and $\partial\ddot{z}/\partial z$ exist and are continuous over all $\mathbb{R}^3$. Thus, we can use Theorem 3.1 to define the following proper forcing functions and proper squared frequency functions.

$$\ddot{x}(x,y,z) = h_x(y,z) - x g_x(x,y,z) \tag{3.8}$$
$$\ddot{y}(x,y,z) = h_y(x,z) - y g_y(x,y,z) \tag{3.9}$$
$$\ddot{z}(x,y,z) = h_z(x,y) - z g_y(x,y,z) \tag{3.10}$$
$$h_x(y,z) = -y \tag{3.11}$$
$$g_x(x,y,z) = \begin{cases} \dfrac{z\tanh(x)}{x} - 1 & x \neq 0 \\ z - 1 & x = 0 \end{cases} \tag{3.12}$$
$$h_y(x,z) = xz\,\text{sech}^2(x) + (\alpha - \beta)\tanh(x) \tag{3.13}$$
$$g_y(x,y,z) = \begin{cases} z\,\text{sech}^2(x) + x\tanh(x) + \dfrac{\left(\sqrt{y^2+\beta^2} - \beta\right)\tanh(x)}{y} & y \neq 0 \\ z\,\text{sech}^2(x) + x\tanh(x) & y = 0 \end{cases} \tag{3.14}$$
$$h_z(x,y) = y^2 - xy \tag{3.15}$$
$$g_z(x,y,z) = \left(x + \frac{y}{\sqrt{y^2+\beta^2}}\right)\tanh(x) \tag{3.16}$$

The forcing functions and the squared frequency functions over our trajectory plotted in Figs. 3.1 are shown in Figs. 3.3 and 3.4, respectively. Notice that the squared frequency functions are most definitely competitive. All in all, our theory of properly splittable functions concludes that the Competitive Modes Conjecture (Conjecture 3.1) is valid for our modified Wimol-Banlue Attractor, which is what we expected. This is significant since, as far as the authors know, this sort of Competitive Modes analysis has never been applied to these sorts of systems before.

3.4 Further Research: Improper Splittings

Notice the requisite in Definition 3.3 stating that $D_i^*(\mathbf{0}) = \{\mathbf{x} \in D : x_i = 0\} \neq$ for a proper splitting. In other words, for a function f to have a proper splitting in terms of x_i, it must be defined on $x_i = 0$. Obviously this is not the case for all functions, such as the logarithm and reciprocal functions.

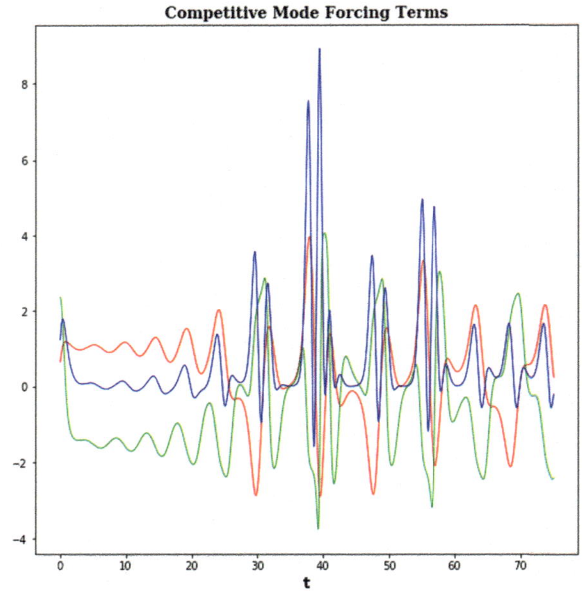

Fig. 3.3 The functions h_x (in red), h_y (in green), and h_z (in blue) of our modified Wimol-Banlue Attractor as defined in Eq. (3.7), using a trajectory with initial condition $\mathbf{x}_0 = [1.32, -0.63, 1.91]^T$. The trajectory was approximating using 7500 iterations of an adaptive RK4 method, using a time step of 0.01

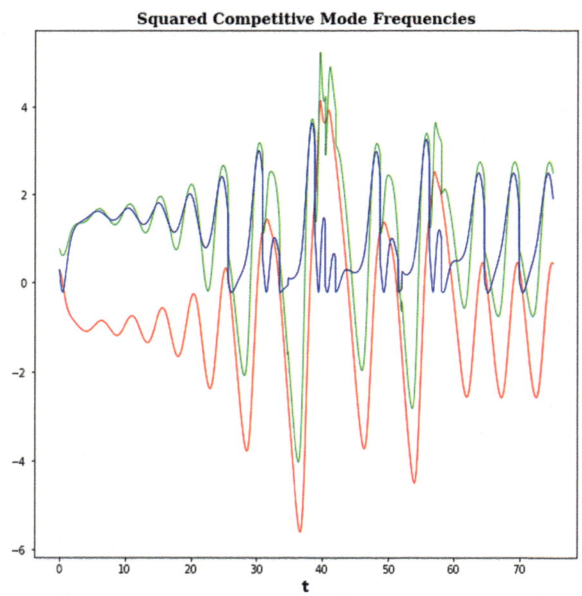

Fig. 3.4 The functions g_x (in red), g_y (in green), and g_z (in blue) based on the trajectory of our modified Wimol-Banlue Attractor as defined in Eq. (3.7), using a trajectory with initial condition $\mathbf{x}_0 = [1.32, -0.63, 1.91]^T$. The trajectory was approximating using 7500 iterations of an adaptive RK4 method, using a time step of 0.01

A work-around to this problem is the introduction of of an improper splitting, which is simply the splitting of a function with respect to $\mathbf{c} \in D \backslash D_i^*(0)$. How this will affect the resulting improper forcing function and improper squared frequency function is yet unclear and requires much more in-depth research to fully understand.

References

1. Davidson, M., Essex, C., Yao, W., Yu, P.: Competitive modes and their application. Int. J. Bifurcat. Chaos. **16**, 497–522 (2006). https://doi.org/10.1142/s0218127406014976
2. Chen, G., Yao, W., Yu, P.: Analysis on topological properties of the Lorenz and the Chen attractors using GCM. Intl. J. Bifurcat. Chaos. **17**, 2791–2796 (2007). https://doi.org/10.1142/s0218127407018762
3. Yu, P.: Chapter 1: Bifurcation, limit cycle and Chaos of nonlinear dynamical systems. In: Edited Series on Advances in Nonlinear Science and Complexity, vol. 1, pp. 1–125. Elsevier B.V. (2006). https://doi.org/10.1016/s157469090601001X
4. Essex, C., Yao, W., Yu, P.: Estimation of chaotic parameter regimes via generalized competitive modes approach. Commun. Nonlinear Sci. **7**, 197–205 (2002). https://doi.org/10.1016/s1007570402000217
5. Choudhury, S.R., Van Gorder, R.A.: Competitive modes as reliable predictors of chaos versus hyperchaos and as geometric mappings accurately delimiting attractors. Nonlinear Dynam. **69**, 2255–2267 (2012). https://doi.org/10.1007/s1107101204240
6. Choudhury, S.R., Van Gorder, R.A.: Classification of chaotic regimes in the T system by use of competitive modes. Int. J. Bifurcat. Chaos. **20**, 3785–3793 (2010). https://doi.org/10.1142/s0218127410028033
7. San-Um, W., Srisuchinwong, B.: A high-chaoticity high-complexity modified diffusionless Lorenz system. In: Proceedings of 2011 International Conference on Computer Applications and Network Security, pp. 561–565 (2011)

Chapter 4
The Construction of Stabilizing Regulators Sets for Nonlinear Control Systems with the Help of Padé Approximations

Yulia Danik and Mikhail Dmitriev

Abstract Here an overview of some stabilizing control synthesis algorithms based on the Padé approximation technique for nonlinear control regularly perturbed state dependent coefficients (SDC) systems are considered. The conditions for the existence of approximate symbolic constructions describing the sets of stabilizing regulators based on the asymptotic expansions of the solutions of the matrix Riccati equations arising in the optimal control theory are formulated for some classes of nonlinear control systems. The numerical calculations illustrating the obtained theoretical results are presented.

4.1 Introduction

Nowadays the asymptotic analysis can be characterized not only as a classical science but also as an interdisciplinary subject that is intensively developing by integrating ideas from various fields to fulfill its main task—providing researchers with qualitative analysis technologies for complex nonlinear processes. Its other task is the construction of approximate solutions, but it is less popular nowadays due to the fact that it is supposedly "cheaper" to perform numerical calculations with specific data than to construct an asymptotics which is nontrivial and requires some special skills. Asymptotic analysis makes it possible to obtain a set of solutions, approximate formulas for qualitative analysis and real time control. As a result, along with the qualitative information about the solution an asymptotics allows to obtain an approximate symbolic representation of the set of solutions that are functions of the parameter by which the asymptotic expansions are constructed. Such approximate

Y. Danik (✉) · M. Dmitriev
Federal Research Center "Computer Science and Control" of Russian Academy of Sciences (FRC CSC RAS), pr. 60-letiya Oktyabrya 9, 117312 Moscow, Russia
e-mail: danik@isa.ru

M. Dmitriev
e-mail: mdmitriev@mail.ru

A. K. Abramian et al. (eds.), *Nonlinear Dynamics of Discrete and Continuous Systems*,
Advanced Structured Materials 139,
https://doi.org/10.1007/978-3-030-53006-8_4

symbolic representations can replace the interpolation approximations in the intervals of parameters variation, where the asymptotic estimates are sufficiently small. This property of asymptotic expansions opens up new application possibilities for asymptotic analysis and its tools, in particular, the Padé approximation (PA) techniques [1–3], in the area of construction of approximate symbolic representations of the solutions in the problems of mechanics, mathematical physics, etc.

One of the most important tasks in the nonlinear dynamic systems control problems is the construction of synthesizing controls that ensure the stability of the system dynamics in the presence of external disturbances and, in particular, the stabilization of the system from any state position.

Here an overview of algorithms that use the Padé approximation technique for stabilizing feedback control laws construction in several classes of nonlinear control problems is presented. Moreover, it is assumed that the systems contain small or large parameters in the equations of motion. The sets of feedback control laws are constructed on the basis of the formal application of the Kalman algorithm for linear-quadratic optimal control problems [4, 5] with the help of the state dependent Riccati equations (SDRE) technique [6]. This technique has been developed for the approximate solution of nonlinear optimal control problems where the equations of motion are formally linear by state and control. The solution of the matrix algebraic state dependent Riccati equation is used for the control synthesis.

In [7] a nonlinear control correction technique has been proposed based on the selection of a quadratic quality criterion, where the weighting matrices coefficients are also state dependent.

In literature there are various examples of Padé approximations application for the solution of optimal control problems [8, 9]. In this work the stabilizing regulators are constructed using the SDRE approach and the gain matrices are the matrix Padé approximations constructed on the basis of asymptotic expansions by large and small parameters in the right-hand sides of the equations of motion.

4.2 Control Problems, SDRE Approach and Matrix Padé Approximations

The application of Padé approximations in the SDRE approach has been considered in a number of control problems, where the right-hand sides of the dynamics equations contain a parameter that takes either only small (here the one-point Padé approximations are used) [10–14], or both small and large values (two-point Padé approximations are applied) [10, 11]. Here only several problem statements for nonlinear continuous control systems with a parameter are considered and the results of the stabilizing regulator sets construction are described, where the gain coefficients matrices are one-point or two-point Padé approximations of the matrix Riccati equations solutions.

Firstly, let us consider a regular perturbed weakly nonlinear continuous system [11]

$$\dot{x} = A(x, \varepsilon)x + B(x, \varepsilon)u = (A_0 + \varepsilon A_1(x))x + (B_0 + \varepsilon B_1(x))u, \ x(0) = x^0, \tag{4.1}$$

$x(t) \in X \subset R^n$, $u(t) \in R^r$, $t \in (0, \infty)$, $X \subset R^n$ is a bounded state space subset, A_0, $A_1(x) \in R^{n\times n}$, B_0, $B_1(x) \in R^{n\times r}$, A_0, B_0 are constant matrices, $\varepsilon \in (0, \infty)$.

According to the SDRE approach the feedback control for (4.1) has the form of the Kalman regulator

$$u = -R^{-1}B^T(x, \varepsilon)P(x, \varepsilon)x, \tag{4.2}$$

where $P(x, \varepsilon)$ is the solution of the matrix algebraic Riccati equation

$$-A^T(x, \varepsilon)P(x, \varepsilon) - P(x, \varepsilon)A(x, \varepsilon) + P(x, \varepsilon)B(x, \varepsilon)R_0^{-1}B^T(x, \varepsilon)P(x, \varepsilon) - Q(x, \varepsilon) = 0, \tag{4.3}$$

for some $Q(x, \varepsilon) \in R^{n\times n}$, $Q(x, \varepsilon) > 0 \ \forall x \in X, \varepsilon \in (0, \infty)$, $R_0 \in R^{r\times r}$, $R_0 > 0$.

Let us introduce the condition

I. *The triple of matrices* $(A(x, \varepsilon), B(x, \varepsilon), Q^{1/2}(x, \varepsilon))$ *is pointwise controllable and observable for all* $x \in X, \varepsilon \in (0, \infty)$.

From condition *I* it follows [4, 5] that the Riccati equation (4.3) has a positive definite solution $P(x, \varepsilon)$.

The regulator (4.2) uses the structure of the Kalman regulator in the optimal control problem, but it is not optimal for nonlinear system (4.1) with criterion

$$\int_0^\infty \left(x^T Q(x, \varepsilon)x + u^T R_0 u\right) dt \to \inf_u, \tag{4.4}$$

as $P(x, \varepsilon)$ in (4.2) does not recover the exact control gain matrix, but is only its approximation. Here, as in [7], it is assumed that $Q(x, \varepsilon) = Q_0 + \varepsilon Q_1(x) + \varepsilon^2 Q_2(x) > 0$, $Q_0 > 0$ and Q_0, $Q_1(x)$, $Q_2(x)$ are selected in a special way for each control system.

The asymptotic expansion of the solution of the matrix algebraic state dependent Riccati equation (4.3) for small values of ε is defined as $\tilde{P}_2(x, \varepsilon) = \tilde{P}_0 + \varepsilon \tilde{P}_1(x) + \varepsilon^2 \tilde{P}_2(x)$ and for large parameter values as $\hat{P}_2(x, \varepsilon) = \hat{P}_0(x) + \frac{1}{\varepsilon}\hat{P}_1(x) + \frac{1}{\varepsilon^2}\hat{P}_2(x)$. In both cases the second order asymptotic expansions are constructed.

Firstly, the formal second-order asymptotic approximation $\tilde{P}_2(x, \varepsilon)$ for small ε is constructed. By substituting the expression for $\tilde{P}_2(x, \varepsilon)$ into (4.3) and equating the terms with the same powers of ε we obtain the equations

$$
\begin{aligned}
&-\tilde{P}_0 A_0 - A_0^T \tilde{P}_0 + \tilde{P}_0 B_0 R_0^{-1} B_0^T \tilde{P}_0 - Q_0 = 0,\\
&\tilde{P}_1(x)\left(A_0 - B_0 R_0^{-1} B_0^T \tilde{P}_0\right) + \left(A_0 - B_0 R_0^{-1} B_0^T \tilde{P}_0\right)^T \tilde{P}_1(x) +\\
&+\tilde{P}_0\left(A_1(x) - B_1(x) R_0^{-1} B_0^T \tilde{P}_0\right) + \left(A_1(x) - B_1(x) R_0^{-1} B_0^T \tilde{P}_0\right)^T \tilde{P}_0 + Q_1(x) = 0,\\
&\tilde{P}_2(x)\left(A_0 - B_0 R_0^{-1} B_0^T \tilde{P}_0\right) + \left(A_0 - B_0 R_0^{-1} B_0^T \tilde{P}_0\right)^T \tilde{P}_2(x) +\\
&+\tilde{P}_1(x)\left(A_1(x) - B_1(x) R_0^{-1} B_0^T \tilde{P}_0\right) + \left(A_1(x) - B_1(x) R_0^{-1} B_0^T \tilde{P}_0\right)^T \tilde{P}_1(x) -\\
&-\left(\tilde{P}_1(x) B_0 + \tilde{P}_0 B_1(x)\right) R_0^{-1} \left(\tilde{P}_1(x) B_0 + \tilde{P}_0 B_1(x)\right)^T + Q_2(x) = 0.
\end{aligned}
\tag{4.5}
$$

Here the first equation is the Riccati equation and the other two are the Lyapunov matrix equations. For $\varepsilon \gg 1$ we make a replacement $\varepsilon = 1/\mu$ and seek the solution of (4.3) in the form $\hat{P}_2(x, \mu) = \hat{P}_0(x) + \mu \hat{P}_1(x) + \mu^2 \hat{P}_2(x)$, after equating the terms with the same powers μ we obtain the following system for the terms of representation $\hat{P}_2(x, \mu)$

$$
\begin{aligned}
&\hat{P}_0(x) B_1(x) R_0^{-1} B_1^T(x) \hat{P}_0(x) - Q_2(x) = 0,\\
&-\hat{P}_1(x) B_1(x) R_0^{-1} B_1(x)^T \hat{P}_0(x) - \hat{P}_0(x) B_1(x) R_0^{-1} B_1(x)^T \hat{P}_1(x) +\\
&+\left(A_1^T(x) - \hat{P}_0(x) B_0 R_0^{-1} B_1^T(x)\right) \hat{P}_0(x) + \hat{P}_0(x)\left(A_1(x) - B_1(x) R_0^{-1} B_0^T \hat{P}_0(x)\right) +\\
&+Q_1(x) = 0,\\
&-\hat{P}_2(x) B_1(x) R_0^{-1} B_1^T(x) \hat{P}_0(x) - \hat{P}_0(x) B_1(x) R_0^{-1} B_1^T(x) \hat{P}_2(x) +\\
&+\left(A_0^T - \hat{P}_1(x) B_1(x) R_0^{-1} B_0^T\right) \hat{P}_0(x) + \hat{P}_0(x)\left(A_0 - B_0 R_0^{-1} B_1^T(x) \hat{P}_1(x)\right) +\\
&+\left(A_1^T(x) - \hat{P}_0(x) B_1(x) R_0^{-1} B_0^T\right) \hat{P}_1(x) + \hat{P}_1(x)\left(A_1(x) - B_0 R_0^{-1} B_1^T(x) \hat{P}_0(x)\right) -\\
&-\hat{P}_1(x) B_1(x) R_0^{-1} B_1^T(x) \hat{P}_1(x) - \hat{P}_0(x) B_0 R_0^{-1} B_0^T \hat{P}_0(x) + Q_0 = 0,
\end{aligned}
\tag{4.6}
$$

where the first one is also the Riccati type equation and the other two are the Lyapunov type matrix equations.

The next two statements take place

Theorem 4.1 *If for any $x \in X$ matrices A_0, $A_1(x)$, B_0, $B_1(x)$ and matrices $Q_0 > 0$, $Q_1(x) > 0$, $Q_2(x) > 0$ satisfy the conditions:*

1. *Rank* $\left[B_0, A_0 B_0, \ldots, A_0^{n-1} B_0\right] = n$, *rank* $\left[Q_0^{1/2}, A_0 Q_0^{1/2}, \ldots, A_0^{n-1} Q_0^{1/2}\right] = n$;

2. $\tilde{P}_0\left(A_1(x) - B_1(x) R_0^{-1} B_0^T \tilde{P}_0\right) + \left(A_1(x) - B_1(x) R_0^{-1} B_0^T \tilde{P}_0\right)^T \tilde{P}_0 + Q_1(x) > 0$;

3. $\tilde{P}_1(x)\left(A_1(x) - B_1(x) R_0^{-1} B_0^T \tilde{P}_0\right) + \left(A_1(x) - B_1(x) R_0^{-1} B_0^T \tilde{P}_0\right)^T \tilde{P}_1(x) -$

$$
-\left(\tilde{P}_1(x) B_0 + \tilde{P}_0 B_1(x)\right) R_0^{-1} \left(\tilde{P}_1(x) B_0 + \tilde{P}_0 B_1(x)\right)^T + Q_2(x) > 0,
$$

then for all $x \in X$, $\varepsilon > 0$ *the Riccati equation in* (4.5) *has a positive definite solution* $\tilde{P}_0$ *and Lyapunov equations in* (4.5) *have unique positive definite solutions* $\tilde{P}_1(x)$, $\tilde{P}_2(x)$.

Proof The equation for $\tilde{P}_0$ in (4.5) is a Riccati equation and by the first condition of Theorem 4.1 it has a positive definite solution $\tilde{P}_0$ and it follows that $Re\lambda(A_0 - B_0R_0^{-1}B_0^T\tilde{P}_0) < 0$ [5]. Lyapunov equations for $\tilde{P}_1(x)$ and $\tilde{P}_2(x)$ have positive definite solutions for all x in X, $x(t+1) = f(t, x(t))$ by conditions 2 and 3 [15]. Moreover, as $\tilde{P}_0 > 0$, $\tilde{P}_1(x) > 0$, $\tilde{P}_2(x) > 0$ for all $x \in X$, it follows that $\tilde{P}_2(x, \varepsilon) = \tilde{P}_0 + \varepsilon\tilde{P}_1(x) + \varepsilon^2\tilde{P}_2(x) > 0$, that concludes the proof.

Theorem 4.2 *If for any* $x \in X$ *the matrices* $A_0, A_1(x), B_0, B_1(x)$ *and* $Q_0 > 0$, $Q_1(x) > 0$, $Q_2(x) > 0$ *satisfy the conditions:*

$$1.\ Rank\ B_1(x) = n,\ rank\ Q_2^{1/2}(x) = n;\ 2.\ Re\,\lambda\,\{-\hat{P}_0(x)B_1(x)R_0^{-1}B_1(x)^T\} < 0;$$

$$3.\ \left(A_1^T(x) - \hat{P}_0(x)B_0R_0^{-1}B_1^T(x)\right)\hat{P}_0(x) +$$

$$\hat{P}_0(x)\left(A_1(x) - B_1(x)R_0^{-1}B_0^T\hat{P}_0(x)\right) + Q_1(x) > 0;$$

$$4.\ \left(A_0^T - \hat{P}_1(x)B_1(x)R_0^{-1}B_0^T\right)\hat{P}_0(x) + \hat{P}_0(x)\left(A_0 - B_0R_0^{-1}B_1^T(x)\hat{P}_1(x)\right) +$$

$$+\left(A_1^T(x) - \hat{P}_0(x)B_1(x)R_0^{-1}B_0^T\right)\hat{P}_1(x) + \hat{P}_1(x)\left(A_1(x) - B_0R_0^{-1}B_1^T(x)\hat{P}_0(x)\right) -$$

$$-\hat{P}_1(x)B_1(x)R_0^{-1}B_1^T(x)\hat{P}_1(x) - \hat{P}_0(x)B_0R_0^{-1}B_0^T\hat{P}_0(x) + Q_0 > 0,$$

then for all $x \in X$, $\varepsilon > 0$ *the Riccati equation in* (4.6) *has a positive definite solution* $\hat{P}_0(x)$ *and Lyapunov equations in* (4.6) *have unique positive definite solutions* $\hat{P}_1(x)$, $\hat{P}_2(x)$.

Proof The matrix equation for $\hat{P}_0(x)$ in (4.6) is a special case of an algebraic Riccati equation. The condition of its solvability and positive definiteness of its solution is the condition 1 of the theorem. Solutions $\hat{P}_1(x)$ and $\hat{P}_2(x)$ of the Lyapunov equations in (4.6) are positive definite for all x when conditions 2, 3 and 4 of Theorem 4.2 are satisfied. Moreover from the fact that $\hat{P}_0(x) > 0$, $\hat{P}_1(x) > 0$, $\hat{P}_2(x) > 0$ for all $x \in X$ it follows that $\hat{P}_2\,(x, \mu) = \hat{P}_0(x) + \mu\hat{P}_1(x) + \mu^2\hat{P}_2(x) > 0$, $\mu = 1/\varepsilon$, that concludes the proof.

Remark 4.1 Generally, for Padé approximation construction it is not necessary to simultaneously fulfill all the conditions of both theorems. It is necessary only to establish the existence of matrices $\tilde{P}_0$, $\tilde{P}_1(x)$, $\tilde{P}_2(x)$ and $\hat{P}_0(x)$, $\hat{P}_1(x)$, $\hat{P}_2(x)$.

4.2.1 Two-Point Padé Approximation of the Riccati Equation Solution

Here the two-point right Padé approximation (PA) of [2/2] order, or the Padé-bridge [2/2], is constructed based on the two asymptotics by small and large values of the parameter. The PA is called right because the matrix inverse is situated at the right of the expression. By matching the two asymptotic expansions $\tilde{P}_2(x, \varepsilon)$ and $\hat{P}_2(x, \mu)$ it is possible to construct a Padé approximation which is close to $\tilde{P}_2(x, \varepsilon)$ for small values of ε and is close to $\hat{P}_2(x, \mu)$ for large values of ε and may be closer to the exact solution for "middle" values of the parameter than any of these two asymptotics. We seek the Padé approximation in the form [1, 16]

$$PA_{bridge}^{[2/2]}(x, \varepsilon) = \left(\mathrm{M}_0(x) + \varepsilon\,\mathrm{M}_1(x) + \varepsilon^2\mathrm{M}_2(x)\right)\left(E + \varepsilon\,\mathrm{N}_1(x) + \varepsilon^2\mathrm{N}_2(x)\right)^{-1}, \tag{4.7}$$

where E—the $n \times n$ identity matrix.

The [2/2] order right Padé approximation is an example of a diagonal Padé approximation (the degree of the matrix polynomials in the "numerator" and the "denominator" are equal) [16].

The unknown matrices in (4.7) are found from the following system of equations

$$\begin{aligned}
&\left(\mathrm{M}_0(x) + \varepsilon\,\mathrm{M}_1(x) + \varepsilon^2\mathrm{M}_2(x)\right)\left(E + \varepsilon\,\mathrm{N}_1(x) + \varepsilon^2\mathrm{N}_2(x)\right)^{-1} = \\
&= \tilde{P}_0 + \varepsilon\tilde{P}_1(x) + \varepsilon^2\tilde{P}_2(x), \\
&\left(\mathrm{M}_0(x) + \varepsilon\,\mathrm{M}_1(x) + \varepsilon^2\mathrm{M}_2(x)\right)\left(E + \varepsilon\,\mathrm{N}_1(x) + \varepsilon^2\mathrm{N}_2(x)\right)^{-1} = \\
&= \hat{P}_0(x) + \tfrac{1}{\varepsilon}\hat{P}_1(x) + \tfrac{1}{\varepsilon^2}\hat{P}_2(x).
\end{aligned}$$

Multiplying both equalities on the right by $\left(E + \varepsilon N_1(x) + \varepsilon^2 N_2(x)\right)$ and equating the terms with the same powers of ε we get the following ten equations for the determination of five unknown matrices in (4.7) for the degrees of ε from -2 to 4

$$\begin{aligned}
&\varepsilon^{-2} : 0 = \hat{P}_2(x), \quad \varepsilon^{-1} : 0 = \hat{P}_1(x) + \hat{P}_2(x)\mathrm{N}_1(x), \\
&\varepsilon^{0} : \mathrm{M}_0(x) = \tilde{P}_0;\ \mathrm{M}_0(x) = \hat{P}_0 + \hat{P}_1(x)\mathrm{N}_1(x) + \hat{P}_2(x)\mathrm{N}_2(x), \\
&\varepsilon^{1} : \mathrm{M}_1(x) - \tilde{P}_0\mathrm{N}_1(x) - \tilde{P}_1(x) = 0,\ \mathrm{M}_1(x) = \hat{P}_0\mathrm{N}_1(x) + \hat{P}_1(x)\mathrm{N}_2(x), \\
&\varepsilon^{2} : \mathrm{M}_2(x) = \tilde{P}_1(x)\mathrm{N}_1(x) + \tilde{P}_2(x) + \tilde{P}_0\mathrm{N}_2(x),\ \mathrm{M}_2(x) = \hat{P}_0\mathrm{N}_2(x), \\
&\varepsilon^{3} : 0 = \tilde{P}_2(x)\mathrm{N}_1(x) + \tilde{P}_1(x)\mathrm{N}_2(x),\ \ \varepsilon^{4} : 0 = \tilde{P}_2(x)\mathrm{N}_2(x).
\end{aligned}$$

In the proposed control algorithm instead of this overdefined system of equations for definiteness we use an heuristic construction that contains only five equations for five unknown matrices $\mathrm{M}_0(x)$, $\mathrm{M}_1(x)$, $\mathrm{M}_2(x)$, $\mathrm{N}_1(x)$, $\mathrm{N}_2(x)$

$$
\begin{aligned}
&\mathrm{M}_0(x) = \tilde{P}_0,\\
&\mathrm{M}_0(x) = \hat{P}_0(x) + \hat{P}_1(x)\mathrm{N}_1(x) + \hat{P}_2(x)\mathrm{N}_2(x),\\
&\mathrm{M}_1(x) - \tilde{P}_0\mathrm{N}_1(x) - \tilde{P}_1(x) = 0,\\
&\mathrm{M}_1(x) = \hat{P}_0(x)\mathrm{N}_1(x) + \hat{P}_1(x)\mathrm{N}_2(x),\\
&\mathrm{M}_2 = \hat{P}_0\mathrm{N}_2
\end{aligned}
$$

or

$$
\begin{pmatrix}
E & 0 & 0 & 0 & 0\\
0 & E & 0 & -\tilde{P}_0 & 0\\
E & 0 & 0 & -\hat{P}_1(x) & -\hat{P}_2(x)\\
0 & -E & 0 & \hat{P}_0(x) & \hat{P}_1(x)\\
0 & 0 & E & 0 & -\hat{P}_0(x)
\end{pmatrix}
\begin{pmatrix}
\mathrm{M}_0(x)\\ \mathrm{M}_1(x)\\ \mathrm{M}_2(x)\\ \mathrm{N}_1(x)\\ \mathrm{N}_2(x)
\end{pmatrix}
=
\begin{pmatrix}
\tilde{P}_0\\ \tilde{P}_1(x)\\ \hat{P}_0(x)\\ 0\\ 0
\end{pmatrix}. \tag{4.8}
$$

Because of the possible non-symmetry of $PA^{[2/2]}_{bridge}(x, \varepsilon)$ we introduce a symmetric construction for the approximation of the Riccati equation solution—a Padé-bridge of [2/2] order

$$
K^{[2/2]}_{bridge}(x, \varepsilon) = \frac{\left(PA^{[2/2]}_{bridge}(x, \varepsilon) + \left(PA^{[2/2]}_{bridge}(x, \varepsilon)\right)^T\right)}{2}. \tag{4.9}
$$

The condition for the existence of (4.9) is based on the asymptotic approximations $\tilde{P}_2(x, \varepsilon),\ \hat{P}_2(x, \mu)$

II *Matrices* $\tilde{P}_0,\ \tilde{P}_1(x),\ \hat{P}_0(x),\ \hat{P}_1(x),\ \hat{P}_2(x)$ *exist, system* (4.8) *is uniquely solvable, matrix* $\mathrm{N}(x, \varepsilon) = E + \varepsilon\,\mathrm{N}_1(x) + \varepsilon^2\mathrm{N}_2(x)$ *is nonsingular and* $K^{[2/2]}_{bridge}(x, \varepsilon)$ *is a positive definite matrix for all* $x \in X,\ \varepsilon > 0$.

Remark 4.2 The examples show that condition *II* can be satisfied by the proper selection of matrices $Q_0,\ Q_1(x),\ Q_2(x)$, in particular, it is possible to obtain the terms of the asymptotic expansions and the Padé approximation with the required properties from the solution of a nonlinear programming problem for the selection of $Q_i,\ i = 0, 1, 2$ matrix coefficients.

With the help of the Padé-bridge (4.9) we may introduce the control gain coefficients matrix which is positive definite for all $x \in X,\ \varepsilon > 0$.

Thus, for the problem (4.1), (4.4) we may introduce a parametric set of regulators

$$
u(x, \varepsilon) = -R_0^{-1}B^T(x, \varepsilon)K^{[2/2]}_{bridge}(x, \varepsilon)\,x, \tag{4.10}
$$

where $K^{[2/2]}_{bridge}(x, \varepsilon)$ is positive definite matrix for all $x \in X,\ \varepsilon > 0$ due to *II*.

As $\mathrm{N}(x, \varepsilon)$ is nondegenerate, the Padé-bridge (4.9) for the Riccati equation solution is an interpolation matrix construction that approximates $P(x, \varepsilon)$ for all positive values of the parameter. The Padé-bridge is close to the corresponding asymptotic approximations in the neighborhood of small and large values of the parameter respectively.

So, for the correct application of the Padé approximations for the Riccati equation solutions it is necessary to obtain the corresponding asymptotic estimates. After the determination of the conditions for the existence of the terms of representations $\tilde{P}_2(x, \varepsilon)$ and $\hat{P}_2(x, \mu)$, we establish the asymptotic estimates of the obtained asymptotic approximations to the exact solution of the Riccati equation (4.3). The proof of such estimates for algebraic matrix Riccati equations can be carried out using the Newton–Kantorovich theorem [17]. Under the conditions for the existence of Riccati equation solution (condition I), all the terms of the asymptotic approximations $\tilde{P}_2(x, \varepsilon)$, $\hat{P}_2\left(x, \frac{1}{\varepsilon}\right)$ and the Padé constructions (condition II) for all $x \in X$, $\varepsilon > 0$ we can obtain the statements, that there is a sufficiently small $\varepsilon^* > 0$ such that the following estimates for asymptotic approximations $\tilde{P}_2(x, \varepsilon)$, $\hat{P}_2\left(x, \frac{1}{\varepsilon}\right)$ are true

$$\begin{aligned} \left\| P(x, \varepsilon) - \tilde{P}_2(x, \varepsilon) \right\| &= O(\varepsilon^3), \ 0 < \varepsilon \le \varepsilon^*, \ x \in X, \\ \left\| P(x, \varepsilon) - \hat{P}_2\left(x, \tfrac{1}{\varepsilon}\right) \right\| &= O\left(\tfrac{1}{\varepsilon^3}\right), \ \varepsilon \ge \tfrac{1}{\varepsilon^*}, \ x \in X. \end{aligned} \tag{4.11}$$

Here the main steps of the proof of the estimates (4.11) for small values of the parameter is presented. The equation for the discrepancy $Z = P(x, \varepsilon) - \tilde{P}_2(x, \varepsilon)$ of the solution of the Riccati equation for small values of the parameter is defined as

$$\begin{aligned} &-\left(\tilde{P}_0 + \varepsilon\tilde{P}_1(x) + \varepsilon^2\tilde{P}_2(x) + Z\right) A(x, \varepsilon) - A^T(x, \varepsilon)\left(\tilde{P}_0 + \varepsilon\tilde{P}_1(x) + \varepsilon^2\tilde{P}_2(x) + Z\right) + \\ &+ \left(\tilde{P}_0 + \varepsilon\tilde{P}_1(x) + \varepsilon^2\tilde{P}_2(x) + Z\right) \tilde{R}(x, \varepsilon)\left(\tilde{P}_0 + \varepsilon\tilde{P}_1(x) + \varepsilon^2\tilde{P}_2(x) + Z\right) - Q(x, \varepsilon) = 0, \end{aligned}$$

where $\tilde{R}(x, \varepsilon) = B(x, \varepsilon)R_0^{-1}B^T(x, \varepsilon)$.

The existence and uniqueness of the solution of the last equation in some neighborhood of the origin can be proved using the Newton–Kantorovich theorem [17]. In this case we linearize the equation and introduce a vector $z \in R^{n^2}$, $z = \left(Z_{11}\ Z_{12} \cdots Z_{21}\ Z_{22} \cdots Z_{n1}\ Z_{n2} \cdots Z_{nn}\right)$, where Z_{ij}, $i, j = 1, \ldots, n$ are the components of the matrix Z presented in a vector form, row by row. The following nonlinear operator equation for z is obtained

$$F(z, \varepsilon) = 0,$$

where $F(z, \varepsilon)$ is the Frechet differentiable operator for $0 < \varepsilon \le \varepsilon_0$ acting from some set $\mathfrak{M} = S(z_0, \sigma(\varepsilon)) = \{z \in R^{n^2} : \|z - z_0\| < \sigma(\varepsilon)\}$ of the Euclidean vector space E_1 of dimension n^2 into Euclidean space E_1.

This nonlinear operator equation is replaced with

$$z = J_0 z + J_1(z, \varepsilon),$$

where $J_0 z = z - \Gamma_0(0)F(z,0)$ is the linear part of the operator and $J_1(z,\varepsilon)$ is the nonlinear part (perturbation) which equals $J_1(z,\varepsilon) = -\Gamma_0(\varepsilon)F(z,\varepsilon) + \Gamma_0(0)$ $F(z,0)$ and is a function of ε, $\Gamma_0(\varepsilon) = [F'(z_0,\varepsilon)]^{-1}$ and $z_0 = 0$.

Using the Newton–Kantorovich method it is proved that operator $J_0 z$ is contracting on $\mathfrak{M}$ and $J_1(z,\varepsilon)$ is bounded, which allows us to establish the existence and uniqueness of the solution z in $\mathfrak{M}$ and find the estimate $\left\| P(x,\varepsilon) - \tilde{P}_2(x,\varepsilon) \right\|$. Here the following norm is applied $\|A\|_2 = \rho(AA^*)$, where $\rho(B) = \max\limits_{1\le i\le n} |\lambda_i|$ – the spectral radius of matrix B, λ_i–the eigenvalues of matrix B. In the proof the operator $J_1(z,\varepsilon)$ is expanded into the parameter power series. The following conditions are imposed on the operator $J_0 z = z - \Gamma_0(0)F(z,0)$: (1) operator $F(z,0)$ is Frechet differentiable acting from $\mathfrak{M} = S(z_0, \sigma(\varepsilon))$; (2) the derivative of $F(z,0)$ satisfies the Lipschitz condition on $\mathfrak{M}$ with constant $L < 1$; (3) $\|\Gamma_0(0)\| \le b_0$; (4) $\|\Gamma_0(0)F(z_0 = 0,0)\| \le h_0$, $h_0 = b_0 L \eta_0 < \frac{1}{2}$, $r_0 = \frac{1-\sqrt{1-2h_0}}{h_0}\eta_0 \le \sigma(\varepsilon)$.

As result, it is proved that $F(z,\varepsilon)$ is a contracting operator that maps each element of a set of radius $\sigma(\varepsilon)$ into the element of a set of radius $r_1(\varepsilon)$, $0 < \varepsilon \le \varepsilon_1$, $\varepsilon_1 \le \varepsilon_0$ which can be made sufficiently small by selecting a sufficiently small $\varepsilon_2 \le \varepsilon_1$, $0 < \varepsilon \le \varepsilon_2$. Now it can be stated that the unique discrepancy equation solution $z^* = vec(Z^*) = vec(P(x,\varepsilon) - \tilde{P}_2(x,\varepsilon))$ exists in the set $S(z_0 = 0, r_1(\varepsilon))$ with radius $r_1(\varepsilon)$, $0 < \varepsilon \le \varepsilon_2$, $\|z^*\| < r_1(\varepsilon)$. Here $vec()$ denotes the operation of matrix transformation into a vector row by row. Moreover, since the matrix norm for Z^* is less than the norm of the vector z^*, we have $\|Z^*\| \le r_1(\varepsilon)$.

Given that the equation for the residual term of the Riccati equation can be written in the form $\Pi(Z,x,\varepsilon) = \Phi(Z,x,\varepsilon) + \varepsilon^3 \Upsilon(x,\varepsilon) = 0$, where $\Phi(Z,x,\varepsilon) = -Z\overset{\frown}{A}$ $-\overset{\frown}{A}^T Z + Z\overset{\frown}{S}Z$, and $\Upsilon(x,\varepsilon)$ does not depend on Z and contains the terms with ε^3 and higher, we get $\left\| Z^* - \bar{Z} \right\| = \|\Upsilon(x,\varepsilon)\| = O(\varepsilon^3)$, $\|Z^*\| = O(\varepsilon^3)$ uniformly by $x \in X$ and $\bar{Z} = 0$ is the solution of equation $\Phi(Z,x,\varepsilon) = 0, 0 < \varepsilon \le \varepsilon_0$. From here, the required estimate is obtained. The scheme of the proof for large values of the parameter is similar.

A possible way to overcome the degeneracy of matrix $\mathrm{N}(x,\varepsilon)$ in the Padé construction is to select or construct matrices $\mathrm{M}(x,\varepsilon) = \mathrm{M}_0(x) + \varepsilon \mathrm{M}_1(x) + \varepsilon^2 \mathrm{M}_2(x)$ and $\mathrm{N}(x,\varepsilon) = E + \varepsilon\,\mathrm{N}_1(x) + \varepsilon^2\,\mathrm{N}_2(x)$ with the required properties. For example, matrix $\mathrm{N}_2(x)$ is defined as a constant positive definite matrix, and the search for the remaining Padé matrices is carried out by the least squares method with constraints on the positive definiteness of the matrices $\mathrm{M}(x,\varepsilon)$ and $\mathrm{N}(x,\varepsilon)$, i.e. the next problem is solved

$$(M_0 - \hat{P}_0 - \hat{P}_1(x)N_1 - \hat{P}_2(x)N_2)^2 + (M_1 - \tilde{P}_0 N_1 - \tilde{P}_1(x))^2 + \\ +(M_1 - \hat{P}_0 N_1 - \hat{P}_1(x)N_2)^2 \to \min_{\mathrm{M}_1(x),\mathrm{N}_1(x)},$$

where $M_0 = \tilde{P}_0$; $M_2 = \hat{P}_0 N_2$ with the following constraints $M_0 + \varepsilon M_1(x)$ $+ \varepsilon^2 M_2(x) > 0$, $E + \varepsilon N_1(x) + \varepsilon^2 N_2(x) > 0$. As a result, we may obtain a positive definite Padé-bridge for the Riccati equation (4.3) solution, which determines the coefficients of the regulator (4.10).

For the cases when the Padé-bridge is not positive definite or does not exist at some points (x, ε) due to the degeneracy of matrix $\mathrm{N}(x, \varepsilon)$, a cubic matrix spline can be used. For example $n = 2$, let us define a matrix cubic spline $S(\varepsilon) = \begin{bmatrix} s_1(\varepsilon) & s_2(\varepsilon) \\ s_2(\varepsilon) & s_3(\varepsilon) \end{bmatrix}$ that approximates a Padé-bridge [2/2] $K^{[2/2]}_{bridge}(\varepsilon) = \begin{bmatrix} p_1(\varepsilon) & p_2(\varepsilon) \\ p_2(\varepsilon) & p_3(\varepsilon) \end{bmatrix}$ on the interval $[\varepsilon_0, \varepsilon_1]$, where ε_0, ε_1 are two points, $\varepsilon_0 < \varepsilon_1$ that define an interval containing the zero eigenvalue of matrix $\mathrm{N}(x, \varepsilon)$. Here $s_j(\varepsilon)$, $j = 1, 2, 3$ are functions $s_j(\varepsilon) \in C^2[\varepsilon_0, \varepsilon_1]$ that are scalar polynomials of degree 3, i.e. $s_j(\varepsilon) = a_j + b_j\varepsilon + c_j\varepsilon^2 + d_j\varepsilon^3$, $d_j \neq 0$, $s_j(\varepsilon_i) = p_j(\varepsilon_i)$, $i = 0, 1$, $j = 1, 2, 3$, and $K^{[2/2]}_{bridge}(\varepsilon_i) = S(\varepsilon_i)$, $i = 0, 1$.

Example 4.1 Padé-bridge construction The coefficients of the system (4.1)–(4.4) have the form

$$A_0 = \begin{pmatrix} 1 & 0.5 \\ 0.7 & 0.2 \end{pmatrix}, \quad A_1(x) = \begin{pmatrix} 0.1 + 0.1\sin(x_1) & 0.1 \\ 0 & 0.1 + 0.1\sin(x_2) \end{pmatrix},$$

$$B_0 = \begin{pmatrix} 0.2 & 0.4 \\ 0.5 & 0.1 \end{pmatrix}, \quad B_1(x) = \begin{pmatrix} 0.1 & 0.1\sin(x_1) \\ 0.1\sin(x_2) & 0.2 \end{pmatrix},$$

$$Q_0 = \begin{pmatrix} 5 & 1 \\ 1 & 5 \end{pmatrix}, \quad R_0 = \begin{pmatrix} 1 & 0 \\ 0 & 1 \end{pmatrix}, \quad Q_1(x) = \begin{pmatrix} 45 + x_1^2 & 35 + 0.1x_1^2x_2^2 \\ 35 + 0.1x_1^2x_2^2 & 47 + x_2^2 \end{pmatrix},$$

$$Q_2(x) = \begin{pmatrix} 5 + x_1^2 & 1 + 0.1x_1^2x_2^2 \\ 1 + 0.1x_1^2x_2^2 & 5 + x_2^2 \end{pmatrix}, \quad x_0 = \begin{pmatrix} -1 & 1 \end{pmatrix}^T, \quad t \in [0, 10].$$

For such Q_k, $k = 0, 1, 2$ matrices $\tilde{P}_0, \tilde{P}_1, \hat{P}_0, \hat{P}_1, \hat{P}_2$, M_0 are symmetric, $\mathrm{M}_0 = \begin{pmatrix} 11.40 & 0.88 \\ 0.88 & 4.99 \end{pmatrix} > 0$ and matrices $\mathrm{M}_1(x)$, $\mathrm{M}_2(x)$, $\mathrm{N}_1(x)$, $\mathrm{N}_2(x)$ are not symmetric. The quality criterion values for the compared regulators are presented in Table 4.1.

Example 4.2 Padé approximation components with the specified properties Let us consider system (4.1), (4.4) where

$$A = \begin{pmatrix} -2 & -0.5 \\ -1 & -0.7 \end{pmatrix}, \quad B = \begin{pmatrix} 2 + 0.1x(1) & 0.4 \\ 0.5 & 1.4 + 0.7x(2) \end{pmatrix},$$

$$Q_0 = \begin{pmatrix} 5 & 0.5 \\ 0.5 & 5 \end{pmatrix},$$

$$Q_1 = \begin{pmatrix} 1 & 0 \\ 0 & 1 \end{pmatrix}, \quad Q_2 = \begin{pmatrix} 1 & 0 \\ 0 & 1 \end{pmatrix}, \quad R_0 = \begin{pmatrix} 1 & 0 \\ 0 & 1 \end{pmatrix}, \quad x_0 = \begin{pmatrix} -1 & 1 \end{pmatrix}^T.$$

Matrix N_2 is defined as $N_2 = \begin{pmatrix} 4 & 2 \\ 2 & 4 \end{pmatrix}$, and other components of the Padé approximation are found using the least squares method (cf. Table 4.2).

Example 4.3 Padé construction with a spline The spline application is demonstrated on the following example for a stationary system (4.1) with criterion (4.4), where $A(\varepsilon) = \begin{pmatrix} 1 + \varepsilon/0.1 & 0.5 + \varepsilon/0.2 \\ 1.5 + \varepsilon/0.1 & 2 + \varepsilon/0.1 \end{pmatrix}$, $B(\varepsilon) = \begin{pmatrix} 0.2 + \varepsilon/0.2 & 0.4 + \varepsilon/0.1 \\ 0.5 + \varepsilon/0.1 & 0.1 + \varepsilon/0.2 \end{pmatrix}$,

Table 4.1 Criterion values for different ε

ε	SDRE regulator (SDRE)	Padé-bridge [2/2] (Padé)	Asymptotics by small parameter ε	Asymptotics by large parameter ε
0.01	14.93	14.93	14.93	–
0.1	17.85	17.86	17.86	–
0.3	25.10	25.47	25.76	–
0.6	40.16	43.02	77.21	–
0.8	55.20	61.85	–	–
1	77.47	94.64	–	–
3	107.46	112.86	–	–
6	44.96	44.77	–	45.38
10	35.09	35.05	–	34.84
15	31.31	31.30	–	31.17

Table 4.2 Criterion values for different ε

ε	SDRE regulator (SDRE)	Padé-bridge [2/2] (Padé)
0.01	9.48	9.48
0.1	11.64	11.70
0.3	9.15	11.27
0.6	6.70	11.91
0.8	6.65	11.32
1	5.28	6.18
3	3.27	3.84
6	2.57	2.87
10	2.25	2.41

$$Q_0 = \begin{pmatrix} 5 & 1 \\ 1 & 5 \end{pmatrix}, \quad Q_1 = \begin{pmatrix} 45 & 35 \\ 35 & 47 \end{pmatrix}, \quad Q_2 = \begin{pmatrix} 2 & 1 \\ 1 & 2 \end{pmatrix}, \quad R = \begin{pmatrix} 1 & 0 \\ 0 & 1 \end{pmatrix}, \quad x_0 = \begin{pmatrix} -1 & 1 \end{pmatrix}^T.$$

Here $\tilde{P}_0, \tilde{P}_1, \hat{P}_0, \hat{P}_1 > 0$, $M_0 = \begin{pmatrix} 11.777 & 1.288 \\ 1.288 & 15.477 \end{pmatrix} > 0$, and matrices $\hat{P}_2$, $M_1 = \begin{pmatrix} -9.612 & -11.983 \\ -23.084 & -2.409 \end{pmatrix}$, $M_2 = \begin{pmatrix} 0.303 & -1.115 \\ -1.837 & 0.362 \end{pmatrix}$, $N_2 = \begin{pmatrix} -0.026 & -0.138 \\ -0.240 & 0.009 \end{pmatrix}$, $N_1 = \begin{pmatrix} -1.407 & -0.415 \\ -0.965 & -0.598 \end{pmatrix}$, $t \in [0, 2]$. The comparison of the two-point Padé regulator [2/2] with the SDRE regulator and regulators based on the asymptotic expansions by small and large values of the parameter by the criterion values is presented in Table 4.3.

For $\varepsilon = 0.6$ the $K^{[2/2]}_{bridge}(x, 0.6) = \frac{\left(PA_{[2/2]}(x,0.6)+PA^T_{[2/2]}(x,0.6)\right)}{2}$ has both positive and negative eigenvalues (is not positive definite as required by condition *II*). In the neighborhood of this parameter value the controller based on the Padé approximation

Table 4.3 Criterion values for the considered regulators for different values of the parameter

ε	SDRE regulator (SDRE)	Padé-bridge [2/2] (Padé)
0.01	23.87	23.87
0.1	26.51	26.52
0.3	32.37	32.6
0.6	42.05	8.343×10^3
0.8	49.58	62.09
1	58.39	63.98
3	360.4	296.4
6	71.89	76.78
10	41.67	41.84
15	32.70	32.72

of the Riccati equation solution does not stabilize the system (has a large overshoot). It can be seen that the value of the criterion for the Padé controller increases dramatically for $\varepsilon = 0.6$. After the spline construction the singularity in the neighborhood of this point disappears, we obtain the following result for $\varepsilon = 0.6$, $I(u) = 29.45$ instead of the previous value 8.343×10^3. And now the corresponding system trajectories converge to the equilibrium position.

4.2.2 Large Gain Systems and Weakly Controllable Systems

Let us consider another class of optimal control problems for continuous nonlinear systems with a parameter and a quadratic quality criterion [10, 11]

$$\dot{x} = A(x)x + \varepsilon B(x)u, \;\; x(0) = x^0, \tag{4.12}$$

$$\int_0^\infty \left(x^T Q(x,\varepsilon)x + u^T R_0 u\right) dt \to \inf_u, \tag{4.13}$$

where $x(t) \in X \subset R^n$, $u(t) \in R^r$, $t \in (0, \infty)$, $X \subset R^n$ is a bounded state space subset, $A(x) \in R^{n \times n}$, $B(x) \in R^{n \times r}$, $rank\, B(x) = r$, $\forall x \in X$, $Q(x,\varepsilon) > 0$, $R_0 > 0$, $\varepsilon \in (0\,, \infty)$ is a parameter that can take both large or small values, i.e. in the first case, we have a large gain system (4.12), and in the second case, the so-called weakly controllable system, all matrices are assumed to be sufficiently smooth.
Here, we will also use the SDRE approach scheme to construct the feedback control law

$$u = -\varepsilon R_0^{-1} B^T(x) P(x,\varepsilon)x, \tag{4.14}$$

where $P(x, \varepsilon)$ is a solution of the matrix algebraic Riccati equation for all $x \in X$, $\varepsilon \in (0, \infty)$

$$-A^T(x)P(x,\varepsilon) - P(x,\varepsilon)A(x) + \varepsilon^2 P(x,\varepsilon)B(x)R_0^{-1}B^T(x)P(x,\varepsilon) - Q(x,\varepsilon) = 0. \tag{4.15}$$

The following condition is introduced

III. *The triple of matrices* $(A(x), B(x), Q^{1/2}(x,\varepsilon))$ *is pointwise controllable and observable for any* $x \in X, \varepsilon \in (0, \infty)$.

Under condition *III* Riccati equation (4.15) has a positive definite solution $P(x, \varepsilon)$. At first, the second-order formal asymptotic approximation of $P(x, \varepsilon)$ for small values of ε is constructed, where Q is also selected in the form $Q(x, \varepsilon) = Q_0(x) + \varepsilon Q_1(x) + \varepsilon^2 Q_2(x)$. The following relations are obtained

$$\begin{aligned}&-A^T(x)\tilde{P}_0(x) - \tilde{P}_0(x)A(x) - Q_0(x) = 0, \ -A^T(x)\tilde{P}_1(x) - \tilde{P}_1(x)A(x) - Q_1(x) = 0,\\ &-A^T(x)\tilde{P}_2(x) - \tilde{P}_2(x)A(x) + \tilde{P}_0(x)B(x)R_0^{-1}B^T(x)\tilde{P}_0(x) - Q_2(x) = 0,\end{aligned} \tag{4.16}$$

which are the Lyapunov matrix equations. Next, we proceed to the construction of a formal second-order asymptotic approximation of the solution of Eq. (4.15) in case of large parameter values. We get

$$\begin{aligned}&\hat{P}_0(x)S(x)\hat{P}_0(x) - Q_2(x) = 0, \ \hat{P}_0(x)B(x)R_0^{-1}B^T(x)\hat{P}_1(x)+\\ &+\hat{P}_1(x)B(x)R_0^{-1}B^T(x)\hat{P}_0(x) - Q_1(x) = 0, \ -A^T(x)\hat{P}_0(x) - \hat{P}_0(x)A(x)+\\ &+\hat{P}_0(x)B(x)R_0^{-1}B^T(x)\hat{P}_2(x) + \hat{P}_2(x)B(x)R_0^{-1}B^T(x)\hat{P}_0(x)+\\ &+\hat{P}_1(x)B(x)R_0^{-1}B^T(x)\hat{P}_1(x) - Q_0(x) = 0.\end{aligned} \tag{4.17}$$

The following is true

Theorem 4.3 *If matrices* $A(x), B(x)$ *and* $R_0 > 0, Q_0(x) > 0, Q_1(x) > 0, Q_2(x) > 0$ *for all* $x \in X$ *satisfy the conditions*

$$1.\ Re\lambda\,(A(x)) < 0; \ 2.\ Q_2(x) - \tilde{P}_0(x)B(x)R_0^{-1}B^T(x)\tilde{P}_0(x) > 0;$$

$$3.\ rank\ B(x) = n, \ rank\ Q_2^{1/2}(x) = n; \ 4.\ Re\,\lambda(B(x)R_0^{-1}B^T(x)\hat{P}_0(x)) < 0;$$

$$5.\ Q_0(x) + A^T(x)\hat{P}_0(x) + \hat{P}_0(x)A(x) - \hat{P}_1(x)B(x)R_0^{-1}B^T(x)\hat{P}_1(x) > 0,$$

then for all $x \in X, \ \varepsilon > 0$ *the Riccati equation* (4.17) *has a positive definite solution* $\hat{P}_0(x)$ *and the Lyapunov equations in* (4.16) *and* (4.17) *have unique positive definite solutions* $\tilde{P}_0(x), \ \tilde{P}_1(x), \tilde{P}_2(x), \ \hat{P}_1(x)\hat{P}_2(x)$.

Proof Lyapunov equations for $\tilde{P}_0(x)$, $\tilde{P}_1(x)$ in (4.16) have positive definite solutions for all $x \in X$ (condition 1 of the theorem and $Q_0(x) > 0$, $Q_1(x) > 0$, $\forall x \in X$). The Lyapunov equation for $\tilde{P}_2(x) > 0$ in (4.16) has a solution $\tilde{P}_2(x) > 0$ under conditions 1 and 2. The matrix equation for $\hat{P}_0(x)$ in (4.17) is a special case of the algebraic matrix Riccati equation. Condition 3 is the condition for its solvability and positive definiteness of its solution. The other two relations in (4.17) are the Lyapunov-type equations. For the solvability of these equations for $\hat{P}_1(x)$ and $\hat{P}_2(x)$, the fulfillment of conditions 4 and $Q_1(x) > 0$, $\forall x \in X$, and conditions 4, 5, respectively, is required. That concludes the proof.

It clearly follows that the asymptotic approximations $\tilde{P}_2(x, \varepsilon)$, $\hat{P}_2(x, \mu)$ from (4.16) and (4.17) are positive definite matrices.

Then, on the basis of the obtained asymptotic expansions an approximate solution of (4.15) for the whole interval of parameter variation is constructed using a two-point Padé-bridge (4.9). Now we can propose a regulator for nonlinear systems (4.12), (4.13) in the form

$$u(x, \varepsilon) = -\varepsilon R_0^{-1} B^T(x) K_{bridge}^{[2/2]}(x, \varepsilon)\, x, \tag{4.18}$$

for all $x \in X$, $\varepsilon > 0$. Thus, we obtain a possible stabilizing regulator for system (4.12), (4.13).

We consider problem (4.12), (4.13) in the stationary case when all matrices are constant. We introduce the Lyapunov function $V(x, \varepsilon) = x^T K(\varepsilon)\, x$. The total time derivative along the closed-loop system (4.12), (4.18) trajectory has the form

$$\begin{aligned}
&\tfrac{dV}{dt} = \dot{x}^T K(\varepsilon)\, x + x^T K(\varepsilon)\, \dot{x} = (Ax + \varepsilon Bu)^T K(\varepsilon)\, x + x^T K(\varepsilon)\, (Ax + \varepsilon Bu) = \\
&= \left(Ax - \varepsilon^2 B R_0^{-1} B^T K(\varepsilon)\, x\right)^T P_{PA} x + x^T P_{PA} \left(Ax - \varepsilon^2 B R_0^{-1} B^T K(\varepsilon)\, x\right) = \\
&= x^T A^T K(\varepsilon)\, x + x^T K(\varepsilon)\, Ax - \varepsilon^2 x^T K(\varepsilon)\, B R_0^{-1} B^T K(\varepsilon)\, x - \\
&-\varepsilon^2 x^T K(\varepsilon)\, B R_0^{-1} B^T K(\varepsilon)\, x = x^T [A^T K(\varepsilon) + K(\varepsilon)\, A] x - \\
&-\varepsilon^2 x^T \left(K(\varepsilon)\, B R_0^{-1} B^T K(\varepsilon) + K(\varepsilon)\, B R_0^{-1} B^T K(\varepsilon)\right) x.
\end{aligned}$$

According to the Lyapunov lemma [15], matrices $D_1 = -A^T K(\varepsilon) - K(\varepsilon)\, A > 0$, $D_2 = 2K(\varepsilon)\, B R_0^{-1} B^T K(\varepsilon) > 0$, then $\frac{dV(x,\varepsilon)}{dt} = -x^T D_1 x - \varepsilon^2 x^T D_2 x < 0$, $\forall \varepsilon > 0$, $x \neq 0$.

Thus, we have

Theorem 4.4 *If all matrices in* (4.12), (4.13) *are constant, then under condition II regulator* (4.18) *stabilizes system* (4.12) *for all* $\varepsilon \in (0, \infty)$.

Thus, in the stationary case, provided that condition *II* is satisfied, the stabilizing controller (4.18) for (4.12) has the property of robustness with respect to ε, since the asymptotic stability of the closed system along this controller is preserved for any intervals of positive parameter ε variation. The constructed set of regulators is an approximate symbolic description of the parametric set of stabilizing controls. The use of such representation of stabilizing regulators is computationally efficient

Table 4.4 Criterion values for the considered regulators for different values of the parameter ε

	ε	0.01	0.3	1	6	15
$I(u)$	SDRE regulator (SDRE)	9.109	7.385	3.697	1.870	1.728
	Padé-bridge(Padé)	9.109	7.652	3.769	1.870	1.728
	Asymptotics by large parameter ε	1.253×10^4	17.702	3.947	1.870	1.728
	Asymptotics by small parameter ε	9.109	8.180	7.425	13.433	24.700

due to the fact that it is not necessary to solve the Riccati equation in real time for specific values of the parameter.

Example 4.4 Here a stationary control system (4.12), (4.13) with a vector control and a quadratic quality criterion is considered, where $A = \begin{pmatrix} -2 & -0.5 \\ -1 & -0.7 \end{pmatrix}$, $B = \begin{pmatrix} 2 & 0.4 \\ 0.5 & 1.4 \end{pmatrix}$, $Q_0 = \begin{pmatrix} 5 & 0.5 \\ 0.5 & 5 \end{pmatrix}$, $Q_1 = \begin{pmatrix} 1 & 0 \\ 0 & 1 \end{pmatrix}$, $Q_2 = \begin{pmatrix} 1 & 0 \\ 0 & 1 \end{pmatrix}$, $R_0 = \begin{pmatrix} 1 & 0 \\ 0 & 1 \end{pmatrix}$, $x_0 = \begin{pmatrix} -1 & 1 \end{pmatrix}^T$. A stabilizing regulator with a positive parameter $\varepsilon \in (0, \infty)$ based on the matrix Padé-bridge $K_{bridge}^{[2/2]}(x, \varepsilon)$ (4.9) is constructed for this system. We get symmetric and positive definite matrices $\tilde{P}_0$, $\tilde{P}_1$, $\hat{P}_0$, $\hat{P}_1$, $\hat{P}_2$, M_0, the Padé-bridge matrices are $\mathrm{M}_0 = \begin{pmatrix} 2.315 & -2.130 \\ -2.130 & 5.093 \end{pmatrix}$, $\mathrm{M}_1 = \begin{pmatrix} 0.602 & -0.635 \\ -0.509 & 1.130 \end{pmatrix}$, $\mathrm{M}_2 = \begin{pmatrix} 1.137 & -1.184 \\ -1.032 & 2.262 \end{pmatrix}$, $\mathrm{N}_1 = \begin{pmatrix} 0.070 & -0.086 \\ 0.024 & -0.022 \end{pmatrix}$, $\mathrm{N}_2 = \begin{pmatrix} 1.786 & -1.317 \\ -0.936 & 2.656 \end{pmatrix}$. A series of numerical experiments was carried out for different values of ε for the Padé regulator (4.18) and the SDRE controller (4.14). The comparison of these two controls, as well as the algorithms based on the asymptotics by large and small parameters of the Riccati equation solution by quality criterion (4.13) values is presented in Table 4.4 and Fig. 4.1. From Table 4.4 it can be seen that the Padé regulator almost coincides with the exact solution for all $\varepsilon \in (0, \infty)$ (see Fig. 4.1).

4.2.3 Weakly Coupled Systems

Let us consider a problem for a weakly coupled continuous control system [12–14]

$$\begin{gathered} \dot{x} = A(x, \varepsilon)x + B(x, \varepsilon)u, \ x(0) = x_0, \\ \int_0^\infty \left(x^T Q(x, \varepsilon)x + u^T R_0 u\right) dt \to \inf_u, \end{gathered}$$

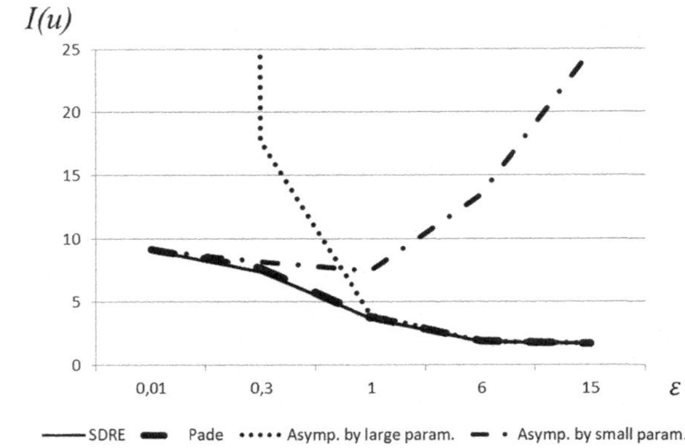

Fig. 4.1 Comparison of regulators by criterion values for different ε

where $A(x,\varepsilon)=\begin{bmatrix} A_1(x_1) & \varepsilon A_2(x_2) \\ \varepsilon A_3(x_1) & A_4(x_2)\end{bmatrix}$, $B(x,\varepsilon)=\begin{bmatrix} B_1(x_1) & \varepsilon B_2(x_2) \\ \varepsilon B_3(x_1) & B_4(x_2)\end{bmatrix}$, $Q(x,\varepsilon)=\begin{bmatrix} Q_1(x_1) & \varepsilon Q_2(x_1,x_2) \\ \varepsilon Q_2^T(x_1,x_2) & Q_3(x_2)\end{bmatrix}\geq 0,\ R_0=\begin{bmatrix} R_1 & 0 \\ 0 & R_2\end{bmatrix}>0,$

$x=\begin{pmatrix} x_1 \\ x_2\end{pmatrix}\in X,\ x_1\in R^n,\ x_2\in R^m,\ u=\begin{pmatrix} u_1 \\ u_2\end{pmatrix}\in R^r,\ u_1\in R^{r_1},\ u_2\in R^{r_2},$ $r_1+r_2=r,\ t\in[0,\infty),\ 0<\varepsilon\leq\varepsilon_0\ll 1.$

According to the SDRE approach the suboptimal regulator for this problem is found in the form $u=-R^{-1}B^T(x,\varepsilon)P(x,\varepsilon)x$, where $P(x,\varepsilon)$ is the solution of the corresponding Riccati equation

$-P(x,\varepsilon)A(x,\varepsilon)-A^T(x,\varepsilon)P(x,\varepsilon)+P(x,\varepsilon)B(x,\varepsilon)R^{-1}(x,\varepsilon)B^T(x,\varepsilon)P(x,\varepsilon)$ $-Q(x,\varepsilon)=0$, where $P(x,\varepsilon)$ is presented as [18]

$$P(x,\varepsilon)=\begin{bmatrix} P_1(x_1,x_2,\varepsilon) & \varepsilon P_2(x_1,x_2,\varepsilon) \\ \varepsilon P_2^T(x_1,x_2,\varepsilon) & P_3(x_1,x_2,\varepsilon)\end{bmatrix}=$$

$$=\begin{bmatrix} P_{10}+\varepsilon P_{11}+\varepsilon^2 P_{12} & \varepsilon P_{20}+\varepsilon^2 P_{21}+\varepsilon^3 P_{22} \\ \varepsilon P_{20}^T+\varepsilon^2 P_{21}^T+\varepsilon^3 P_{22}^T & P_{30}+\varepsilon P_{31}+\varepsilon^2 P_{32}\end{bmatrix}=$$

$$=\begin{bmatrix} P_{10} & 0 \\ 0 & P_{30}\end{bmatrix}+\varepsilon\begin{bmatrix} P_{11} & P_{20} \\ P_{20}^T & P_{31}\end{bmatrix}+\varepsilon^2\begin{bmatrix} P_{12} & P_{21} \\ P_{21}^T & P_{32}\end{bmatrix}.$$

Equations for K_{10} and K_{30} are the Lyapunov type equations and the equation for K_{20} is a two-term Sylvester type equation, where $K_{20}H-FK_{20}=Y,\ H\in R^{m\times m},\ F\in R^{n\times n},\ \in R^{n\times m},\ \ H=B_4(x_2)R_2^{-1}B_4^T(x_2)K_{30}-A_4(x_2),\ F=K_{10}B_1(x_1)R_1^{-1}B_1^T(x_1)-A_1^T(x_1),\ K_{20}\in R^{n\times m}$—a rectangular matrix. The detailed calculations are presented in [13]. Let us define the spectra of matrices H and F by $\sigma(H)=\{\mu_1,\ldots,\mu_n\}$, $\sigma(F)=\{\lambda_1,\ldots,\lambda_m\}$, respectively. According to [19], if $\sigma(H)\subset\{\lambda:\ Re\lambda<0\}$, $\sigma(F)\subset\{\lambda:\ Re\lambda>0\}$, then there exist a unique solution of this equation $K_{20}=-\int_0^\infty e^{-tF}Ye^{tH}dt$. For $\varepsilon,\varepsilon^2$ we get the Lyapunov equations for $K_{11},\ K_{31}$ and $K_{12},\ K_{32}$, and the Sylvester equations for $K_{21},\ K_{22}$.

Here, only a one-point Padé approximation of [1/2] order is constructed on the basis of the asymptotics by small values of the parameter, which allows one to expand the interval of the parameter variation where the corresponding Padé regulator is stabilizing.

Remark 4.3 Some of mentioned results have been used for discrete weakly coupled systems [14].

4.3 Conclusions

The numerical experiments show that two-point Padé regulators based on two asymptotic approximations are more efficient than the controllers based on individual asymptotic expansions. The sufficient conditions used in theorems restrict the classes of admissible systems where the proposed technique is applicable. The weakening of these conditions, however, is possible with some modification of the approach by selection of specific classes of system matrices and application of other asymptotic constructions.

Acknowledgements Research supported by the Russian Foundation for Basic Research (project No. 17-29-07053).

References

1. Baker, G.A., Graves-Morris, P.: Padé Approximants: Encyclopedia of Mathematics and It's Applications. Cambridge University Press (1996)
2. Andrianov, I.V., Awrejcewicz, J.: New trends in asymptotic approaches: summation and interpolation methods. Appl. Mech. Rev. **54**, 69–91 (2001)
3. Andrianov, I.V., Awrejcewicz, J.: Methods of Asymptotic Analysis and Synthesis in the Nonlinear Dynamics and Mechanics of a Deformable Solid. ICS, Izhevsk (2013) (in Russian)
4. Afanasyev, V.N., Kolmanovsky, V.B., Nosov, V.R.: Mathematical Theory of Control System Design. Springer, Netherlands (1996)
5. Kvakernaak, H., Sivan, R.: Linear Optimal Control Systems. Wiley-Interscience, New York (1972)
6. Cimen, T.: Survey of state-dependent Riccati equation in nonlinear optimal feedback control synthesis. AIAA J. Guid. Control Dyn. **35**, 1025–1047 (2012)
7. Dmitriev, M.G., Makarov, D.A.: Smooth nonlinear controller in a weakly nonlinear control system with state depended coefficients. Trudy ISA RAN. **64**, 53–58 (2014) (in Russian)
8. Pal, J.: Suboptimal control using Padé approximation techniques. IEEE Trans. Automatic Control **AC-25**, 1007–1008 (1980)
9. Belyaeva, N., Dmitriev, M., Komarova, E.: Padé-approximation as a "bridge" between two parametric boundary asymptotics. IFAC Proc. Vols. **34**, 635–639 (2001)
10. Danik, Y.E., Dmitriev, M.G., Komarova, E.V., Makarov, D.A.: Application of Pade-approximations to the solution of nonlinear control problems. In: Proceedings of International Conference on Dynamical Systems: Theory and Applications (DSTA 2017), pp. 155–164. Poland (2017)

11. Danik, Yu.E., Dmitriev, M.G.: Construction of parametric regulators for nonlinear control systems based on the Pade approximations of the matrix Riccati equation solution. IFAC-PapersOnLine **51**, 815–820 (2018)
12. Danik, Y.E., Dmitriev, M.G., Makarov, D.A.: Stabilizing regulators for nonlinear continuous systems of large dimension using the asymptotics of the matrix algebraic Riccati equations solutions. In: Proceedings of 2018 11th International Conference "Management of Large-Scale System Development", MLSD 2018, pp. 1–4. V.A. Trapeznikov Institute of Control Sciences Moscow, Russian Federation (2018)
13. Danik, Y.E. The construction of a parametric set of regulators for one class of weakly coupled nonlinear systems. In: Materials of the International Conference Voronezh Spring Mathematical School "Pontryagin Readings XXX", pp. 114–117. Voronezh State University Publishing House, Voronezh (2019) (in Russian)
14. Danik, Y.E., Dmitriev, M.G.: Stabilizing regulator for nonlinear discrete weakly coupled systems based on the Pade approximation. In: Proceedings of 2019 12th International Conference "Management of Large-Scale System Development", MLSD 2019. V.A. Trapeznikov Institute of Control Sciences Moscow, Russian Federation (2019)
15. Balandin, D.V., Kogan, M.M.: Controllers Synthesis on the Basis of Linear Matrix Inequalities. Litres (2017) (in Russian)
16. Baker, G.A.: The Padé approximant method and some related generalizations. Mathemat. Sci. Eng. **71**, 1–39 (1970)
17. Krasnosel'skii, M.A., Vainikko, G.M., Zabreiko, P.P., Rutitcki, J.B., Stecenko, V.J.: Approximated Solutions of Operator Equations. Walters—Noordhoff, Groningen (1972)
18. Gajic, Z., Petkovski, D., Shen, X.: Singularly Perturbed and Weakly Coupled Linear Control Systems: A Recursive Approach. Springer (1990)
19. Demidenko, G.V.: Matrix Equations. Publishing House Novosib. uni, Novosibirsk (2009) (in Russian)

Chapter 5
Galerkin's Method was not Developed by Ritz, Contrary to the Timoshenko's Statement

I. Elishakoff, J. Kaplunov, and E. Kaplunov

Abstract In the context of its title, this paper we discuss two letters sent to S. P. Timoshenko (1878–1972), as well as a letter of response sent by Timoshenko to Grigolyuk. B. G. Galerkin (1871–1945) is the author of the first letter to S. P. Timoshenko. Second letter to him is by E. I. Grigolyuk. The letters are concerned with the method known as the Galerkin method (in the West), or the Bubnov-Galerkin method or the Bubnov method (in Russia). The letters are fully reproduced here in English translation. Their originals in Russian language are stored at the Timoshenko Archive at Stanford University. The copies of the originals are also obtainable from the authors. Galerkin's letter appears to be the only document until now where B. G. Galerkin relates to this method, apart from his 1915 paper. The author of the second letter is E. I. Grigolyuk (1923–2005). Grigolyuk suggests to Timoshenko to co-author a paper on the priority associated with the Galerkin method, claiming that it belongs solely to I. G. Bubnov (1872–1919). Although a joint paper by Timoshenko and Grigolyuk was never written, Timoshenko expressed an interest in such an endeavor. These correspondents, namely, B. G. Galerkin, S. P. Timoshenko, and E. I. Grigolyuk have made important contributions to theoretical and applied mechanics of the last century, and their interaction appears to be of interest to the mechanics community. This paper is devoted to their correspondence concerning the priority of authorship, that was questioned by both S. P. Timoshenko and E. I. Grigolyuk, albeit in a different manner.

I. Elishakoff (✉)
Department of Ocean and Mechanical Engineering, Florida Atlantic University, 777 Glades Road, Boca Raton, FL 33431-0991, USA
e-mail: elishako@fau.edu

J. Kaplunov
Department of Computer Science and Mathematics, Keele University, Keele ST5 5BG, UK
e-mail: j.kaplunov@keele.ac.uk

Institute for Problems in Mechanical Engineering, RAS, St. Petersburg, Russia

E. Kaplunov
Whizz-Kidz Charity, 4th Floor, Portland House, Bressenden Place, London SW1E 5BH, UK
e-mail: ekaplunov@hotmail.co.uk

A. K. Abramian et al. (eds.), *Nonlinear Dynamics of Discrete and Continuous Systems*, Advanced Structured Materials 139,
https://doi.org/10.1007/978-3-030-53006-8_5

5.1 Introduction: Origins of the Method

Grigolyuk [22, pp. 3–4] tells the story of the origins of the method (translated from Russian): "In 1910, S. P. Timoshenko submitted his monograph, "On Stability of Elastic Systems" (published in Proceedings of the Kiev Polytechnic Institute) for the Zhuravsky Prize named after the famous Russian engineer Dmitrii Ivanovich Zhuravsky (1820–1891). In his monograph, Timoshenko presented solutions of a number of stability problems for different structures under various boundary conditions using Ritz's method. He did not deal with the convergence study of the method but compared the two-term approximations with a single term result, using a variant of the energy method, and showed improvement of the results.

The Zhuravsky prize was established in 1902, and was equal to the yearly professorial salary (in monetary value), it was not awarded to just anyone. This work was sent to six Professors for review: I. G. Bubnov, N. A. Beloliubsky, G. I. Belzetsky, V. L. Kirpichev, G. V. Kolosoff, G. N. Soloviev. In May 1911 the prize Committee…decided the prize to be awarded to S. P. Timoshenko, who was one of the five candidates. The reviews of Timoshenko's work were submitted in the written form. Four of the six reviews were published, two years later, in volume 81, in 1913, of the Proceedings of the Railway Engineering University under the general title "Reviews of Professors Kirpichev, Belzetsky, Bubnov and Kolosoff on the work of Professor Timoshenko, the winner of D. I. Zhuravsky prize." (Fig. 5.1).

Bubnov wrote [6] about the Ritz method: "the essence of this method consists in that expressing the deflections of the body's points in a series $w = a_1\varphi_1 + a_2\varphi_2 + a_3\varphi_3 + \ldots$ where φ are normal functions of coordinates x and y, or of only a single coordinate x (for beams), author forms an expression for V-change of energy of elastic bodies of the system and T-the work performed by external forces; making these expressions equal, author finds expressions for the critical load P, choosing in them the relations between coefficients a_1, a_2, a_3 … so that P takes a minimal value…".

Later in the review, Bubnov disagrees with Timoshenko who stated that the "method used by him makes it easier obtaining results in comparison with other methods which he (Timoshenko) or his predecessors used earlier. This statement appears to me not totally correct." Bubnov also writes on a particular mistake made

Fig. 5.1 Ivan Grigoryevich Bubnov

by Timoshenko by using a trial function that did not reflect correctly the plate's behavior.

Bubnov further made the following comment: "Quite simple solutions can be obtained also by usual method, i.e. by not resorting to system's energy, if only the convergence of the series for w is sufficiently great. By simple substitution into the differential equation of equilibrium; then by multiplying the obtained expression by *dxdy* and integrating over the entire volume, we obtain equation that connects the coefficient a_k along with all others, if the functions φ are chosen so that

$$\iint \varphi_n \varphi_k \mathrm{dxdy} = 0 \quad \text{when } n \neq k$$

This equality is valid in almost all problems of the considered work, since the author usually takes the expression for w in the form of trigonometric series, where this condition is satisfied…Writing from the obtained relationship as many equations as we want to keep, and equating the determinant consisting of factors in front of coefficients, we will reduce the problem of determining the critical load to the determination of the smallest root of a rational function, the highest degree of which equals the number of kept terms.".

According to Grigolyuk [22, p. 21], "I. G. Bubnov contrasts the orthogonalization method to the energy method. He formulates an alternative method, as it were. It is not surprising, that in this context I. G. Bubnov does not connect his method with variational problem, and thus formulates his method as a method of pure orthogonalization. The straightforward solution of the differential equation by the method of orthogonalization is the principal achievement, it is a difficult and decisive step, since before that mechanicians and mathematicians were under a great influence of Lord Rayleigh and W. Ritz.".

Galerkin's paper appeared in 1915. According to Grigolyuk, Galerkin was the "first one who applied the Bubnov method.". However, Galerkin's paper mentions that it is "devoted to the development and application of a different method of the approximate solution of some problems of equilibrium, proceeding directly from the equation of elastic curve or surface…".

One has to pay attention to the words "development and application.". Thus, Galerkin claims the method to belong to him. Who does this method belong to? To Bubnov? Or to Galerkin?

Grigolyuk arrives at the conclusion [23, p. 43]: "In the paper by Galerkin, [the name of] Bubnov is not mentioned as the predecessor and the author of the orthogonalization method. One recalls the folk-Latin word "plagiatus" "Grigolyuk translates "plagiatus" as "robbery" in this case. Grigolyuk continues: "Galerkin passed away in 1945—the question of I. G. Bubnov as the founder of the orthogonalization method was never raised by him.".

Below we highlight that in the letter to Timoshenko [17–20], Galerkin discusses the method, and claims it fully as belonging to him (Fig. 5.2).

Filin in [14, pp. 21–23] reports on key details of the scientific conference in honour of the 100th Anniversary of B. G. Galerkin organized by the Leningrad Scientific

Fig. 5.2 Boris Grigoryevich Galerkin

Technological Society for Civil Engineering in 1971. This conference was separate from the memorial event organised by the Institute for Problems in Mechanics at the USSR Academy of Sciences. The former took place in the Oak Hall at the House of Scientists on the embankment of Neva river (former palace of Prince Vladimir Alexandrovich). The meeting included three lectures by A. I. Lurie, S. G. Mikhlin and L. N. Mkrtychan.

The authorship of the method since recently known as the Galerkin method was of utmost interest for all the conference participants. S. G. Mikhlin started from the assumption that Galerkin may have not been aware of the I. G. Bubnov review of the book called "Stability of Elastic Systems" by S. P. Timoshenko, published in the local proceedings of the Railway Engineering University. In the aforementioned review, a method called the Bubnov-Galerkin method" had been suggested as an alternative to the energy approach adapted by S. P. Timoshenko. However, I. G. Bubnov only announced the method in the review but did not follow up with its implementation. In contrast, B. G. Galerkin not only clearly formulated the idea behind the method, but also applied it to a variety of specific problems.

5.2 V. V. Novozhilov's Views on the Method

In his article dedicated to Bubnov, Novozhilov writes [35, p. 380]: "each scientist has, as a rule, his beloved mathematical methods. Thus, S. P. Timoshenko and Yu. A. Shimansky were getting the majority of their results by using the energy method, i.e. they were reducing the boundary-value problems of structural mechanics to variational treatments. I. G. Bubnov preferred solving boundary-value problems using differential equations. The discovery of the approximate method that carries his name is associated with his predilection to differential equations.".

Thus, in the above passage, V. V. Novozhilov seemingly concurs with Grigolyuk in stating that the method ought to be referred to as "Bubnov method".

In another article, in the same book, dedicated to B. G. Galerkin, he writes about [36, p. 381]: "Development by him of the famous and widely used method of the approximate solution of mathematical physics.". From this passage it sounds as though Novozhilov is attributing the method under discussion to Galerkin! A natural question arises as to whether Novozhilov trying to have it both ways.

Let us continue quoting him: "One has to touch upon the history of this method, in order to avoid any misinterpretation, since the priority side of the question is associated with some details that demand precision. The method under discussion was given in a rough form by our outstanding naval engineer and mechanician Ivan Grigorievich Bubnov in his review on Timoshenko's work that was submitted in 1911 for the Zhuravsky prize.

Timoshenko considered many stability problems for columns and plates, using the Ritz energy method. But Bubnov never used energy methods in his works and thus could not agree with the statement that Timoshenko's problems cannot be solved differently. In his review, he showed how one can solve these problems, utilizing solely differential equations and boundary conditions. The review [6] was published in 1913, in the Proceedings of the Railway Engineering University, and one year later, the second volume of his book "Structural Mechanics of Ships" appeared, in which the new method was used to solve two problems of plate stability. But while suggesting this method, Bubnov did not reveal its essence in the problems of mechanics and thus made the possibility of its use considerably difficult. In particular, if one has only the above works of Bubnov, then it remains unclear if one can solve two- or three-dimensional problems via the method, and if yes, then how? And even while solving one-dimensional problems (if one has the information available in the works of Bubnov), one can arrive at incorrect results.

Full clarity on how to use in the problems of mechanics what was suggested by Bubnov, was brought by B. G. Galerkin, connecting this method with the method of virtual displacements. After Galerkin's work, published in the very beginning of 1915, it became clear how to also extend the method to inhomogeneous problems, and how to avoid mistakes in choosing the functions utilized in projecting the differential equation.

The paper by Galerkin soon became known, and from the mid-twenties it found a wide dissemination. As far as Bubnov's review of Timoshenko's work is concerned, it was exposed in a way, in a rough blueprint, and for a long time remained forgotten. It was remembered about 25 years ago, and if before that the method was associated only with Galerkin, after that some researchers connected this method only with Bubnov. It is clear from the above that one of the actions is unjustified. Undoubtedly, the idea was suggested by I. G. Bubnov, and the first examples of its application for solving concrete problems belong to him. But it is no less undoubted that the key for conscious employment of Bubnov's idea was given by B. G. Galerkin, and with exhaustive clarity. It is sufficient to read his paper of 1915, in order to utilize, without errors, the method to arbitrary boundary-value problem of mechanics. Hence it is justified to call this wonderful method the Bubnov-Galerkin method, as it is currently done by the majority of authors.".

This comment was surely a rebuke to Grigolyuk who was obsessed with idea that Galerkin resorted to plagiarism. In fact, Grigolyuk's first appearance on this topic was on December 31, 1970 when he gave a seminar at the Moscow Aviation Institute. His seminar was titled "Towards Development of Bubnov's Orthogonalization Method.".

5.3 Boris Grigorievich Galerkin and Heinrich Hencky

It appears that the life of Boris Grigorievich Galerkin was closely intertwined with that of the German engineer Heinrich Hencky whose paper on the Galerkin method was the first harbinger, informing the West of this method. As stated by Altenbach and Bruhns, Heinrich Hencky (1885–1951), was "an engineer with contributions to plasticity, continuum mechanics, and plate theory, among others…He received a doctorate in 1913 with a thesis on the numerical calculation of stresses in thin plates. After completing his doctorate, he sought for a new position in railway engineering (in those days one of the most attractive branches in civil engineering) and in 1914 moved to Kharkov (Kharkiv) in the Ukraine, an emerging industrial and commercial center. His career was abruptly terminated when World War I broke out, he was interned in the Ural region and sent back to Germany after the war ended" [2]. Tanner and Tanner inform [47]: "During this time he met his Russian wife, Aleksandra Yuditskaya; they got married in January 1918.".

To digress, let us cite here again [47] on Hencky's doctoral dissertation: "In 1913 he received his doctorate of engineering from Technical University of Darmstadt. The title of his thesis was "Über den Spannungszustand in rechteckigen ebenen Platten bei gleichmäßig verteilter und bei konzentrierter Belastung" (On the stress state in rectangular flat plates under uniformly distributed and concentrated loading). Hencky's thesis [24] used a numerical method to study the stresses in flat plates; the thesis has been cited at least 74 times since 1974–2003 when the paper [47] was published. The authors state that Hencky's thesis "was a substantial contribution to the elastic plate theory, which became one of his favourite subjects. He published his findings in the paper [25].

Grigolyuk provides an additional detail [22, p. 30]: "Heinrich Hencky, former prisoner of war during the WWI, with knowledge of the Russian language, had a good acquaintance with Galerkin; he was the first, see [26], to cite the paper [16] …". The acquaintance with B. G. Galerkin, while he was in Russia as a POW, was prompted, by all probability, by the keen interest of both in plate theory.

The paper [26] appeared in the volume dedicated to Professor August Föppl (1854–1924), famous German mechanician and author, in the book devoted to his 70th birthday. Incidentally, Föppl also died the same year. It should be mentioned that S. P. Timoshenko attended Föppl's lectures during his summer visit to Munich in early 1900s, and was also invited to contribute to Föppl's volume. Another connection with Timoshenko was through Ludwig Prandtl (1875–1953), whose lectures S. P. Timoshenko also attended during his visits to Germany, receiving the topic of his doctoral dissertation from Prandtl, who became the son-in-law of Föppl.

Hencky later moved to Delft University of Technology following not securing a permanent position in Dresden. According to [47], "he joined the department headed by Professor C. B. Biezeno, situated in the old Mechanical engineering building which now is an apartment block.". There, Hencky brought the Galerkin method to the attention of Biezeno.

In 1924, the First Congress in Theoretical and Applied Mechanics took place in Delft, where Biezeno contributed the paper on the Galerkin method [3]. Thus, the misfortune of being the POW in Russia lead to Hencky getting to know both Galerkin and his method. Hencky's failure to obtain a permanent position in Germany brought him to the Netherlands and supported the propagation of the method in the West, with further papers being contributed by [7, 8] as well as many others. Thus, the Galerkin method owes its celebrated status to Hencky's lack of luck. By the way, this unlucky chain of events continued throughout Hencky's life—on the 26th of July Hencky left Delft and took a position at the Massachusetts Institute of Technology in June 1930, serving there as an Associate Professor. According to [2] "Unfortunately, this position again was not permanent and ended in 1932, when the MIT was reorganized...".

One could wonder if Hencky even wrote to S. P. Timoshenko, who at that time was at Stanford University, to solicit help in obtaining another position in the United States. He might have not applied to S. P. Timoshenko, if he knew of Timoshenko's negative attitude to the Galerkin method.

The authors of [2] inform that "As of 1935 an offer arrived from Boris G. Galerkin, whom he knew from his Delft period, he accepted a position as professor of Engineering mechanics at the Kharkov Polytechnic Institute and later at the Institute of Mechanics of Lomonosov University in Moscow with Aleksei A. Ilyushin. Unfortunately, information about Hencky in the Soviet Union is very rare. This is certainly due to the secrecy of those days. However, it is belied that with the help of Galerkin, Hencky was hired to improve the Soviet Union lightweight (airplane) construction. His deformation theory could contribute a lot to this matter, and it is known that Ilyushin was very much impressed by this theory. It is not clear what really happened in those pre-war days in Moscow, but in 1938, he and his family had to leave the Soviet Union within 24 h (the latter fact, leaving the former Soviet Union, and alive, perhaps were, these authors should interject, the happiest moment in his life, IE, JK, EK)." There, in Germany, he is "suspiciously observed by the security service (SD) of the Nazi regime, but under the protection of his supervisor".

According to [47, p. 98], Galerkin and Hencky "had become acquainted at the International Conference on Mechanics held in Delft in 1924 and Galerkin had shown an interest in Hencky's work." According to Wikipedia entry on Galerkin [60], "In 1924 he made his last trip abroad—he participated in the Congress on applied mechanics in the Netherlands," or the first IUTAM Congress. By the way, the fiftieth anniversary of the first congress took place in the city of Delft again, now in 1976, with one of us (IE) being its participant.

About Hencky's misfortunes Tanner and Tanner write [47]: "One must admit that his life was not easy—the early Russian internment and later problems with the Soviets, his teaching problems in Dresden, his difficulties with Biezeno in Delft, and ultimately his loss of the MIT position, need to be born in mind.". His end was tragic too: Hencky died in a climbing accident in the Alps in Tyrol, Austria, in 1951.

5.4 Interrelation Between Ritz and Galerkin Method

Hencky wrote apparently the first paper outside Russia, on the Galerkin method [27]. Leipholz devoted many papers to the application of the Galerkin method to various problems in mechanics, e.g. see [29]. In this paper he writes [29, p. 315]: "Galerkin's method is shown to be independent of Ritz method and applicable to non-conservative, non-self-adjoint problems.". He also emphasizes that [29, p. 317] "...for conservative, self-adjoint problems there is no essential difference between Galerkin and Ritz method [4]. For this reason, Galerkin's method has been considered for a long time to be only a special case of Ritz's method, a point of view which is not very favourable to the independent development of Galerkin's method. This situation changed when it was required to solve new engineering problems, mainly in the field of aeroelasticity, which are non-conservative and non-self-adjoint. For such problems, Ritz's method in its classical form cannot be used, and a different method has to be applied. Since it is obvious that the basic condition...of Galerkin's method does not refer to any variational principle, Galerkin's method seemed to be adequate for non-conservative, non-self-adjoint problems. This Galerkin's method was proved to be in fact more general than Ritz's method, or, in other words, Ritz's method was seen to be a restricted, special case of Galerkin's method. Indeed, investigations by Repman [43], Petrov [37], Keldysh [28], and Mikhlin [34] and later on of Pflueger [38] and Leipholz [30] indicated clearly the applicability of Galerkin's method to quite general problems...". The interested reader can consult also with the paper by Afendikova [1] on the role of Galerkin's method in work of academician M. B. Keldysh (1911–1978), who made numerous outstanding contributions to science and technology, including being "one of the key figures behind Soviet space program.", see Wikipedia [61].

Leipholz summarized that [29, p. 318] "Galerkin's method, being now a very general method, had to have a foundation of its own, and theorems on the convergence of the approximate solution. This requirement naturally led to a strong development of the theory involved with this method." In conclusion of his study, he reiterated [29, p. 328]: "it was shown that Galerkin's method is independent of Ritz's method and is applicable to non-conservative, non-self-adjoint problems.".

One of the strengths of Galerkin's paper was the fact that he considered some problems where he summed up the contributions of all the terms. An analogous approach was taken by Elishakoff and Lee [11], as well as Elishakoff and Zingales [12, 13]. Interested readers can also consult with the book by Svirsky [46], along with a review by Vorovich [56] dedicated to 100th anniversary of B. G. Galerkin's birth, as well as a more recent paper by Repin [42] devoted to a centenary of the method's discovery.

5.5 Relationship Between S. P. Timoshenko and B. G. Galerkin

Filin writes [14, p. 40]: "I remember that A. I. Lurie (who held S. P. Timoshenko in high regard) said, that B. G. Galerkin did not have an amicable relationship with neither S. P. Timoshenko nor with P. F. Papkovich. B. G. Galerkin apparently felt that Timoshenko thought of him as a theoretician as opposed to a more practically oriented scientist. Galerkin could not agree with this. In fact, he could provide strong evidence to the contrary (for example, his involvement and supervision of the engineering design of the large thermal power station building in Leningrad using a metallic fachwerk (truss)). Namely, S. P. Timoshenko expressed his skeptical view of Galerkin's specialization during the conversation at the Leningrad Railway Engineering University.".

It cannot be said with full certainty whether S. P. Timoshenko was objective. Karl S. Pister, Professor at the University of California at Berkley, writes in his paper [39]: "On the 8th October, 1926 at the meeting of the structural group of the American Civil Engineering Society held in Philadelphia, PA, Professor H. Westergaard (1988–1950), who was a professor of theoretical and applied mechanics at the University of Illinois at Urbana Champaign during that time, delivered the paper titled "On hundred and fifty years of advances in structural analysis".". Westergaard briefly mentioned both Galerkin and Timoshenko in his survey. Specifically, he [emphasized that [59, p. 240] "Dr. Timoshenko's presence in this country is a reminder, at the same time of the international character of the science of structural mechanics.". He discussed, specifically, contributions made by the French Tradition [59, pp. 232–233], the English tradition [59, p. 233] as well as contributions made by scientists of Germany, United States, Italy, Denmark, and others. He specifically mentioned that "there is a Russian tradition which may be traced back to the influence of Euler.".

The discussion of the paper involved several interesting comments. S. P. Timoshenko, Professor of the Michigan University at the time, wrote that "It is said that the ancient project of the Babylon tower resulted in the language barriers that have slowed down the progress in science. These barriers were but a minor hindrance for English speakers with regards to many languages, especially French and German, but, unfortunately, it is not easy to overcome the Russian barrier. The author of these remarks wants to fix this issue and reference the Russian tradition.".

Next, Pister notes that [39] "Timoshenko then lists the names and contributions of the famous figures in the history of mechanics: D. Bernoulli, Euler (no one of them were Russian, both were Swiss), Zhuravsky (who developed tangential stresses in beams), Golovin (curved roads), Krylov and Bubnov (the theory for naval structures)." He then emphasizes that [39] "It is worth noting that Timoshenko does not mention Galerkin's name.".

It would seem that Timoshenko is biased in this case, as he was not fully reflecting on the role of the achievements of the Russian tradition in applied mechanics. The pertinent question arises, on whether or not Timoshenko considered Galerkin as belonging to the Russian tradition in the first place.

5.6 Timoshenko's Sensitivity to the Priority Question: His Letter to the Editor of Philosophical Magazine

In 1921, on 30 August, Timoshenko sent a letter to the editor of the journal *Philosophical Magazine.* He writes: "Gentlemen, –In connection with two papers on the Buckling of Deep Beams published in your periodical (see Dr. J. Prescott, xxxvi, p. 297 and xxxix, p. 194), I beg to communicate the following:–

The question of the buckling of beams can be regards as solved a long time already. The first paper on this subject was published in your periodical by A. G. M. Michell, in 1899, vol. xlviii. The same problem was investigated with more details by Prof. L. Prandtl, 1899 (Munich Dissert.). In both these papers, as also in the paper by Dr. J. Prescott, the bending of the flanges of girders by sideways buckling is neglected, and in consequence of that results cannot be applied to the calculation of I girders.

The influence of the flexion of flanges of the girder was studied by me, and the results were published in Russia, 1905 (Bulletins of the Polytechnical Institute, Petersburg).

The translation of this paper in German can be found in *Zeitschr. f. Math. u. Phys.*, Bd. lviii (1910). Other manner of solution is given in my memoir published in French (*Annales des ponte et chaussees,* Fasc. iii, -v). There are given the numerical tables, which enable us to calculate very easily I girders under different conditions of loading and fastening of ends.". As it can be seen, Timoshenko brings to the attention of readers some work that was performed by him, and written in Russian, and because of inaccessibility of the journal, unappreciated by the community. To avoid the narrow distribution to his results he submitted these to German and French journals. As we saw in the previous section, he neglected to bring the work of B. G. Galerkin to the attention of the readers.

5.7 Galerkin's Letter to S. P. Timoshenko

Leningrad, 26 February 1932

Dear Stephan Prokofyevich!

I received your letter from 16th of January. I would be very grateful if you could send me your book on the Theory of Elasticity when it is released. I am reporting to you my most important developments on plates.

Rectangular plates supported along the edges (Proceedings of St. Petersburg Polytechnic Institute, 1915, v. XXIV, pp. 279–282). You referred to this paper in part II of your Theory of Elasticity, published in Russian. The solution for a simply supported plate under a uniformly distributed load is also given here; besides, this solution, thanks to the selected function f(x) for the load in the form of a fourth-order polynomial, satisfying the equation $N\nabla^2\nabla^2 f = p$, reduces to rather rapidly converging series. The cases of plates on elastic supports are investigated and the solution for a plate simply supported at 4 points is presented as a particular example. In the same place the solution for a plate with fixed edges on inelastic supports is given. Then the solution for an infinite (in width and length) plate under uniform load, supported in points, is given. In this work the Table is also presented.

Besides, I have now developed a solution for "Pilzdecke" (a plate supported along rectangular regions), the solution is given in terms of rapidly converging series, virtually for an arbitrary load. A solution is also obtained for the case when a plate is infinite in two directions, and when it is bounded in one direction (simply supported by walls or beams, or clamped, along two edges). Now Tables are being generated.

In the paper "Bending of rectangular plates and walls" (Proceedings of St. Petersburg Polytechnic Institute, 1916, v. XXVI, pp. 124–254 and 1918, v. XXVII, pp. 187–319) plates under a fairly general load and at various support clamping types are considered (the general solution for a continuous plate, as well as the Tables are given for bending characteristics (transverse displacement, moments, shear forces, and reaction forces). Among particular cases, a plate simply supported along three sides (the fourth one is free, i.e. without support), a plate clamped along two sides, and a plate clamped along one side, are considered. Tables are given for a uniformly distributed load, as well as a load in the form of a triangle (hydrostatic pressure); in addition, examples for other loads are presented, such as a line load or a point load. In the paper "Deformation and stresses in rectangular plates under point loads" (a collection of articles called "Engineering structures and structural mechanics", Leningrad, 1924, pp. 3–23), I give the solution for the case when a load is distributed along a rectangular located arbitrarily along the plate. This solution gives an opportunity to obtain an arbitrary load and, also a point load. There are detailed Tables given to show when a part of the plate is subjected to a uniform load. I published roughly the same paper, but without the Tables, in "Messenger of Mathematics", Cambridge, 1925, June, pp. 26–39 ("Equilibrium of thin rectangular elastic plates under the action of continuous and concentrated loads"). In the paper "Contribution à la théorie des plaques continues" (Le Génie civil, 1928, t. 92, pp. 181–184) the solution for a continuous plate (with one span along one of the directions, and an arbitrary number of spans along other directions); the problem is reduced to the solution of a set of equations with three unknowns (the solution in this respect looks like that for a continuous beam).

I gave the solution for triangular plates (Proceedings of the Russian Academy of Sciences, 1919, pp. 111–118, C.R. de l" Acad. des Sc. de Paris 1925, t. 181, p. 369, Journal of Leningrad Physical and Mathematical Society, 1926, vol. 1). Tables for various types of loading are given (Proceedings of the Petrograd Polytechnic Institute, 1919, vol. XXVIII, pp. 1–57). The solution for a point load is given in the collection of articles of Leningrad Institute of Railway Engineering, 1927, issue 94).

I gave the solution for a plate bounded by two arcs of concentric circles and two radii (Proceedings of the Russian Academy of Sciences, 1919, pp. 415–426) for the case when the plate is simply supported along the radii, with arbitrary supports along the arcs. The Tables are composed for the case of a full sector: (a) the arc is clamped—Proceedings of the Leningrad Polytechnic Institute, 1925, vol. XXIX, pp. 271–334, (b) the arc is simply supported—same journal, 1927, vol. XXX, pp. 461–485 (c) the arc is free (suspended freely)—same journal, 1928, vol. XXXI, pp. 229–246. Then I give the solution for the case of a point load ("Plaques minces blastiques, limitées par deux ares de cercles concentriques et deux rayons sous l''action des forces concentrées" in C.R., 1924, t. 178, p. 919). The stability issue of such a plate under the compressive stresses is considered by myself in the paper 'sur la stabilité d''une plaque uniformément comprimée parallélement à la surface, limitée pas deux ares...." in C.R. 1924, t. 178, p. 1392.

In the paper "Adaptation of curvilinear isothermal coordinates to integrate the equations of equilibrium of elastic plates". In Mess. of Math., 1922, Nov., pp. 99–109) I apply isothermal coordinates for investigating displacements and stresses in plates; circular, elliptic, and semi-elliptic plates are presented as examples. In the paper "Berechnung der frei gelagerten elliptischen Platte auf Biegung" (Z. t. angev. Math. Mech., 1923, S. 113–117) I use isothermal coordinates for investigating a simply supported elliptic plate under uniform load. Here the tables for displacement, bending moments, and shear forces are presented.

The last contributions on thick plates are known to you. In the Reports of the USSR Academy of Sciences, 1931, I give the solution for simply supporting thick rectangular and triangular plates under an arbitrary load. The solution is presented in double series. It has a significant disadvantage—the series are slowly convergent.

In the C. R. paper, 1931, I give the solution for a rectangular plate in single series, this solution may be extended to arbitrary support clamping. For some loads it is not difficult to compile Tables as well, since the series converge. It would appear that I have presented everything which may be of a certain interest to you. I have only forgotten about the paper "Rods and Plates. Series in some problems of elastic equilibrium of rods and plates.". Engineering Herald, 1915, pp. 897–908. A particular interest in this paper is presented by an approximate method of integration; I consider, and apparently some others as well, that it may substitute the Ritz method. I believe that this is the case. My method is simpler, even considerably simpler, than the Ritz method, and leads to the same results. There are several publications about this method (Biezeno wrote about this in two papers and also in the lecture at the Delft Congress (Proceedings of the 1st International Congress for Applied Mechanics, 1924), Hencky (Z. f. angew. Math. Mech., 1927, S. 80–81). Presently, this method is included in the Handbuch der Physik, B. VI, S. 345. However, the proof, with the help of the variation calculus, belongs to Hencky; I approached this issue in a simpler way.

Should you require something from what I have listed above, write to me; I do not have everything, but anything I have I will readily send to you, I inform you of my new address: Zhukovskii Street 4, apartment 7.

Yours,

B. Galerkin.

5.8 Discussion of Galerkin's Letter to Timoshenko

As was written by Grigolyuk [22]: "Galerkin passed away in 1945—the question on I. G. Bubnov as the founder of the orthogonalization method was never raised by him.". Fortunately, in the above letter [17–20], Galerkin comments on his method, this letter being, up to now, the only document where he makes such comments. Galerkin emphasizes: "My method is simpler, even considerably simpler, than the Ritz method, and leads to the same results.". He uses the term "my method" not "Bubnov's method". Galerkin informs Timoshenko what the latter probably already knew, that Biezeno devoted two papers to Galerkin's method, as well as Hencky [27], Galerkin gives credit to Hencky in providing the proof, thus he is giving credit when it is due. Galerkin writes these not without some pride, that his method, although developed in Russia, became known in the West.

Galerkin emphasizes that his method "may substitute the Ritz method". He repeats, as it were: "I believe that this is the case. My method is simpler, even considerably simpler, than the Ritz method, and leads to the same results.".

Galerkin suggests that Timoshenko may contact him in case Timoshenko has any questions. We do not know if Timoshenko ever responded to Galerkin. Unfortunately, Galerkin Archive, if it exists, is not available to us.

The pertinent question arises: how does S. P. Timoshenko, if at all, present the method of Galerkin?

Timoshenko does "not provide" his "opinion about this method" in his "works". This is indeed correct. In his book on history of strength of materials [52], in the sections entitled, respectively as "Progress in Strength of Materials during the Twentieth Century" [52, pp. 354–388] and "Theory of Elasticity during the Period 1900–1950" [52, pp. 389–421], he exposed his beloved method of Ritz he refrains from mentioning the Galerkin method, although he does quote some works of Bubnov, as well as of Galerkin, in other contexts.

Timoshenko's book on vibration problems in engineering [49], again reviews the Rayleigh and Ritz methods but is silent about Galerkin's method. Second and third editions of his book [51]—written after the correspondence he had with Galerkin, see [17–20]—again omits even the mentioning of the Galerkin method. Then, the fourth edition of the book [53], now co-authored with Donovan Harold Young (1904–1980) does include the method. Specifically, on p. 390, after exposition of Ritz's method, the authors note: "the calculations can be simplified by using the second form of the Ritz method*, in which, instead of calculating strain energy and kinetic energy of a vibrating system, we use directly the differential equation of vibration.". Authors use the two-term Galerkin method and conclude, on p. 391: "from this example, it may be appreciated that the second form of the Ritz method represents a considerable simplification, since it does not require the calculation of the strain energy which was used in our preceding examples.". As is seen above the word method was supplied by Timoshenko and Young by an asterisk, the associated footnote mentions: "The method is sometimes attributed to Galerkin, but was introduced by W. Ritz, see p. 228, *loc. cit.*". Now, p. 228 does not mention Ritz at all! But this is a mild error.

The grand error of both Timoshenko and Young is that Walther Ritz (1878–1909) had nothing to do with the Galerkin method, except, possibly providing an inspiration!

Who wrote this passage, Timoshenko or Young? If even it was written by Timoshenko, then Young could have read the papers by Ritz, which do not contain Galerkin method, or papers by [7, 8] if he could not read Galerkin's paper [16] in Russian, or Hencky's paper [27] in German! The same remark applies to Professor Weaver, who was a co-author of Timoshenko and Young. Later, he was the first author of the book by Weaver, Timoshenko and Young [58] with the same statement unchanged. Also, what is this 'second form of the Ritz method'?! This term is a mislabeling of Timoshenko and Young. If Timoshenko is the one who misnomered it, Young should have opposed to the term. He could have read papers by Ritz (1908,1909) and told Timoshenko that these papers do not contain the "second form of Ritz method.".

5.9 Letter from Grigolyuk to Timoshenko

10 October 1965

Dear Stepan Prokofievich!

I am sending you a list of your original works. What is your opinion about the current division of the papers by volumes? Do you feel that all of these papers should be included? Do you disagree with the inclusion of any of them? Are there any gaps? If a decision was made about the publication of the volumes, would you agree to write an introduction?

How would you feel about the proposed review of your contribution into the development of deformable systems? This review could be written by me and improved by you in the future.

I am awaiting your swift reply.

I have another question for you. You remember of course the review of Bubnov I. G. about your work which received the Zhuravsky prize in 1912 (Bubnov I. G. "Review of the work of Professor Timoshenko S. P. "On the stability of elastic systems." Proceedings of St. Petersburg Railway Engineering Institute. Volume 81, 1912, pp. 33–36. Reprint: Bubnov I. G. Selected publications of Sudpromgiz. Leningrad 1956, pp. 136–139). The solution for differential equations proposed by Bubnov I. G. constitutes a powerful tool in mathematical physics. Sadly, you did not provide your opinion about this method in your works. The work of Bubnov I. G. is not known in the West. This method is known under a different name there. Your opinion is very influential for both scientists and within wider engineering circles. Could you please take some public steps in this regard, such as by publishing a paper. I have all the necessary materials and perhaps it might be possible to publish a joint paper about this issue.

I wish you health and great vigor.

E. I. Grigolyuk

Corresponding member of Academy of Sciences of the USSR

Moscow, B-333, 34/4 Vavilova Street, Unit B, Flat 391.

5.10 Discussion of Grigolyuk's Letter to Timoshenko

Grigolyuk edited two volumes in the Russian language collecting Timoshenko's papers. His letter starts with a question pertaining to the distribution of Timoshenko's papers in these two volumes. He also wrote an extensive review of Timoshenko's scientific works. In this letter he asks Timoshenko about the feasibility of writing such a review. Then he asks about the Zhuravsky prize S. P. Timoshenko was awarded in 1911. Grigolyuk mentions Bubnov's review of Timoshenko's work submitted for a prize. Grigolyuk emphasizes that 'sadly", Timoshenko does "not provide" his "opinion about this method" in his "works". This is indeed correct: In his book on the history of strength of materials [52], Timoshenko does not refer to the Galerkin method when he discusses progress made during years 1900–1950.

5.11 Letter from Timoshenko to Grigolyuk

In response to the letter sent to him by Grigolyuk, Timoshenko responded: "Next, the question about preparation work on the papers about Bubnov's method is of great interest to me. I wrote my work under the Raleigh influence. By the time this piece of work was ready to be combined with the Ritz paper, it was already too late to change it (I was taking part in the competition for the Zhuravsky Prize at that time) and so I only mentioned it in the introduction. As for me, the methods by Bubnov and later by Galerkin correspond to one of the Ritz's approaches. Neither myself nor Bubnov attended the Scientific Congress in Mechanics in 1924 in Delft. Our works and that of Ritz were not mentioned and it seemed that the methods for approximate evaluation of the critical load fully belonged to the presenter.".

5.12 Personal Inputs of V. V. Novozhilov, V. V. Bolotin and A. L. Goldenveiser

One of the authors (IE) discussed the issue with V. V. Bolotin during his visit to the Florida Atlantic University by the invitation of Y. K. Lin who was the Director of the Centre for Applied Stochastic Research at the Florida Atlantic University at the time [5]. Bolotin's (rather strong, we ought to say) opinion (translated as close to the original source as possible) was that "the E. I. Grigolyuk's strong focus, seemingly an obsession, with this subject is not worthy the attention of a Corresponding member of the Academy of Sciences. At the same time, I visited V. V. Novozhilov (1910–1987) in Leningrad with the special purpose of clarifying the authorship of the method. Novozhilov stated that it was almost impossible that Galerkin did not know about Bubnov's review of Timoshenko's book as Bubnov was Galerkin's supervisor at the time.". Bolotin then provided his own insight on the issue at hand: "As for me, I would suggest that Bubnov as Galerkin's supervisor had asked him to write the review.". Such a possibility is feasible. terSee Spassky's contribution to the book about V. V. Novozhilov [45, p. 21], where he writes the following about Bubnov: "Ivan Grigorievich Bubnov had a strong personality and in his office, he adhered to a full autocracy." If indeed Bolotin's conclusion is valid, then the method should be attributed to Galerkin only.

One of us (IE) recalls the public defense of the doctoral habilitation dissertation of Boris Petrovitch Makarov, in the early 1970s. Professor Eduard Ivanovich Grigolyuk was supposed to serve as the so-called Official Opponent. The scientific secretary, Professor Yury Nikolaevich Novichkov announced that Grigolyuk will not be in attendance. However, his extremely short letter was read. It stated that the dissertation was on the highest scientific level, and that Makarov fully deserved the sought degree. Grigolyuk had one critical comment, however. Instead of the term used by Makarov, namely "Galerkin method", one should use the term "Bubnov method". Later, after a successful defense, Novichkov informed us that Bolotin had a chance to read Grigolyuk's letter prior to the habilitation defense. Bolotin stated that it was a very personal judgment, also see [14, p. 726] where it is stated that Grigolyuk related to others by "views on people projected by national (i.e. religious—IE, JK, EK) belonging". And again, in Grigolyuk's letter to Filin, see [14, p. 725], Grigolyuk wrote: "I don't agree with you in one terminological question: still, it is better [to use the term]—Bubnov's method.".

Likewise, one of us (JK) recalls that Aleksei Lvovich Goldenveiser (1911–2003) was rather ironic regarding Grigolyuk's long-term focus on the authorship of the method in question. Goldenveiser also believed that Grigolyuk might have something personal against Galerkin. It is interesting that in 1952 Grigolyuk lost the competition for the Galerkin prize by the Supreme Council of the All-Union Civil Engineering Society awarded for the best contribution to structural mechanics; in fact, the 1952 Galerkin prize winner was Goldenveiser.

5.13 Conclusion

The above discussed letters by B. G. Galerkin and E. I. Grigolyuk, both directed to S. P. Timoshenko, are stored at Stanford University, and were discovered by the first author at the S. P. Timoshenko Archive. We have no access to the letter which Timoshenko had seemingly sent on January 1932 to B. G. Galerkin, as it is mentioned in Galerkin's above letter. Neither we know if Timoshenko responded to Galerkin. It might be potentially useful to establish the Galerkin Archive based on the correspondences he received, as the Full Member of the Soviet Academy of Sciences of the former USSR. Naturally, such an endeavor must be extremely difficult task since Galerkin passed away in 1945, and his correspondences might well get lost in the passage of time. Likewise, it appears to be desirable to establish E. I. Grigolyuk's Archive collecting correspondences he had received over the years, as the Corresponding Member of the Soviet (and later) Russian Academy of Sciences, provided that consent of the families of these late scientists was obtained. Such correspondence could elucidate numerous interesting issues within the history of mechanics.

We conclude that the orthogonalization method could be referred to as the "Galerkin method' or as the "Bubnov-Galerkin method" but surely not as the "Bubnov method," contrary to the claim of E. I. Grigolyuk; neither is this method a "second form of the Ritz method" as wrongly dubbed by S. T. Timoshenko. The other relevant papers are Refs. 9, 10, 15, 21, 31–33, 40, 44, 48, 50, 54, 55 and 57.

ADDENDUM: During the proofreading of this paper a new study (Ref. 62) by Reddy and Srinivasa was published. The authors maintain, correctly, about the Ritz and the Galerkin methods, that "The two methods are distinctly different" (Ref. 62, p. 288), supporting our conclusions.

Acknowledgements One of the authors (IE) is grateful to the Computation and Mechanics group at the Mechanical Engineering Faculty of Stanford University for appointing him as a S. P. Timoshenko Visiting Scholar in June 2019. The support of the Timoshenko Archive, and especially the permission to publish the letters of B. G. Galerkin and E. I. Grogolyuk, was extremely valuable for working on the paper. He also thanks Dr. V. Kornecki for making available the book containing E. I. Grigolyuk's paper, and Prof. V. I. Mamai of Moscow State University's Institute of Mechanics, for making available the book by Grigolyuk. We thank Professor Igor Andrianov for his constructive criticisms and his tolerance, if not an agreement, with ideas expressed in this paper. The copies of originals of the letters of (a) Galerkin to Timoshenko, of (b) Grigolyuk to Timoshenko, and of (c) Timoshenko to Grigolyuk are obtainable from the authors. JK acknowledges spport by the Russian Science Foundation (grant number 20-11-20133).

References

1. Afendikova, N.G.: History of Galerkin's Method and Its Role in M.B. Keldysh's Work, Preprint No. 77, Institute of Applied Mathematics Named after M.B. Keldysh. Russian Academy of Sciences, Moscow (2014). http://library.keldysh.ru/preprint.asp?id=2014-77. Accessed 18 March 2020
2. Altenbach, H., Bruhns, O.T.: Hencky, Heinrich, Encyclopaedia in Continuum Mechanics. Springer, Berlin (2020)
3. Biezeno, C.B.: Graphical and numerical methods for solving stress problems. In: Biezeno, C.B., Burgers, J.M. (eds.) Proceedings of the First International Congress for Applied Mechanics, pp. 3–17. Technische Bockhandel en Drukkerij J. Waltman, Jr., Delft (1925)
4. Biezeno, C.B., Grammel, R.: Technische Dynamik, vol. 1, 2nd ed., pp. 145–147. Springer, Berlin (1953) (in German) (see also English edition: Engineering Dynamics, Blackie, 1954)
5. Bolotin, V.V.: Personal communication to I.E., Florida Atlantic University (2001)
6. Bubnov, I.G.: Reviews of Professor Kirpichev, Belzetskii, Bubnov and Kolosoff on works of Professor Timoshenko, Awarded the D. I. Zhuravskii Prize. Sbornik St. Peterburgskogo

Instituta Inzhenerov Putei Soobchenia (Collection of St. Petersburg Institute of Transportation Engineering) **81**, 1–40 (1913) (see also Bubnov, I.G.: Selected Works, pp. 136–139. Sudpromgiz Publishers, Leningrad (1956), in Russian)
7. Duncan, W.J.: Galerkin's method in mechanics and differential equations, Aeronautical Research Committee, Reports and Memoranda Technical Report N 1798 (1937)
8. Duncan, W.J.: The principles of the Galeskin's method, Aeronautical Research Committee, Reports and Memoranda, Technical Report N 1848 (1938)
9. Duffing, G.: Erzwungene Schwingungen bei veränderlicher Eigenfrequenz und ihre Technische Bedeutung (1919). https://hdl.handle.net/2027/wu.89080439300
10. Elishakoff, I., Arvan, A.P., Marzani, A.: Rigorous versus naïve Implementation of the Galerkin method for stepped beams. Acta Mech. **230**(11), 3861–3873 (2019)
11. Elishakoff, I., Lee, L.H.N.: Equivalence of the Galerkin and Fourier series methods for one class of problems. J. Sound Vib. **119**, 174–177 (1986)
12. Elishakoff, I., Zingales, M.: Coincidence of Bubnov-Galerkin and exact solution in an applied mechanics problem. J. Appl. Mech. **70**, 777–779 (2003)
13. Elishakoff, I., Zingales, M.: Convergence of Bubnov-Galerkin method exemplified. AIAA J. **42**(9), 1931–1933 (2004)
14. Filin, A.P.: Essays About Scientists in Mechanics. Publishing House "Strategiia", Moscow (2007) (in Russian)
15. Finlayson, B.A., Scriven, L.E.: The method of weighted residuals—a review. Appl. Mech. Rev. **19**, 735–748 (1966)
16. Galerkin, B.G.: Rods and plates, series in some questions of elastic equilibrium of rods and plates. Vestnik Inzhenerov i Technikov **19**, 897–908 (1915); Reprint: Galerkin, B.G.: Collected Publications, pp. 168–195. Publishing House of the USSR Academy of Sciences (1952) (in Russian). (English Translation, TTF-63-18924, National Technical Information Service, US Department of Commerce, Springfield, VA 22161, 1968; see also Rodden, W.P.: Theoretical and Computational Aeroelasticity, pp. 700–745. Crest Publishing (2011))
17. Galerkin, B.G.: Letter to S.P. Timoshenko, p. 1, Document SC0641_b3-f07-Letters_in_Russian.010.tif, Timoshenko Archive, Stanford University (1932)
18. Galerkin, B.G.: Letter to S.P. Timoshenko, p. 2–3, Document SC0641_b3-f07-Letters_in_Russian.011.tif, Timoshenko Archive, Stanford University (1932)
19. Galerkin, B.G.: Letter to S.P. Timoshenko, p. 4, Document SC0641_b3-f07-Letters_in_Russian.012.tif, Timoshenko Archive, Stanford University (1932)
20. Galerkin, B.G.: Letter to S.P. Timoshenko, p. 5, Document SC0641_b3-f07-Letters_in_Russian.013.tif, Timoshenko Archive, Stanford University (1932)
21. Grigolyuk, E.I.: Towards Development of Bubnov's orthogonalization method, Moscow Aviation Institute, Seminars, 31 December. Izvestiya Akademii Nauk SSSR, Mekhanika Tverdogo Tela, Issue **3**, 205 (1970) (in Russian)
22. Grigolyuk, E.I.: On Bubnov's method: towards sixty years of its creation. In: Galimov, K.Z. (ed.) Investigations in Theory of Plates and Shells, vol. 11, pp. 3–41. Kazan University Press, Kazan (1975) (in Russian)
23. Grigolyuk, E.I.: The Bubnov Method: Origins, Formulation, Development. Moscow State University, The Institute of Mechanics, Moscow (1996) (in Russian)
24. Hencky, H.: Über den Spannungszustand in rechteckigen ebenen Platten bei gleichmäßig verteilter und bei konzentrierter Belastung, Ph.D., R. Oldenbourg (1913) (in German)
25. Hencky, H.: Uber den Spannungszustand in kreisrunden Platten mit verschwindender Biegungssteifigkeit. Zeitschrift fur Mathematik und Physik **63**, 311–317 (1915) (in German)
26. Hencky, H.: Über ein einfaches Näherungsverfahren zur Bestimmung des Spannungszustandes in rechteckig begrenzten Scheiben, auf deren Umfang nur Normalspannungen wirken. In: Beiträge zur Technischen Mechanik und Technischen Physik, August Foppl zum 70en Geburstage, pp. 62–73. Springer, Berlin (1924) (in German)
27. Hencky, H.: Eine wichtige Vereinfachung der Methode von Ritz zur angenäehrten Behandlung von variations Problemen. Zeitschrift für angewandte Mathematik und Mechanik **7**, 80–81 (1927) (in German)

28. Keldysh, M.W.: On Galerkin's method applied to boundary-value problems. Izvestiya Akademii Nauk SSSR, Seriya Matematika **6**(6) (1942) (in Russian)
29. Leipholz, H.H.E.: Recent trends in Galerkin method. In: Computer-Aided Engineering, Proceedings of the Symposium held at the University of Waterloo, May 11–13, pp. 315–331 (1971)
30. Leipholz, H.H.E.: Grundzuege einer Stabilitaetstheorie fuer elastische systeme unter nichtkonservtiver Belastung. Ingenieur-Archiv **34**, 58–62 (1965) (in German)
31. Leipholz, H.H.E.: Use of Galerkin's method for vibration problems. Shock Vib. Probl. **8**, 3–18 (1976)
32. Leipholz, H.H.E.: The Galerkin formulation and the Hamilton-Ritz formulation: a critical review. Acta Mech. **47**(3–4), 283–290 (1983)
33. Meleshko, V.V.: Selected topics in the history of the two-dimensional biharmonic problem. Appl. Mech. Rev. **56**(1), 33–85 (2003)
34. Mikhlin, S.G.: Variational Methods of Mathematical Physics, Moscow (1957) (in Russian) (English Translation: Pergamon Press, Oxford, 1964)
35. Novozhilov, V.V.: Remembrances: I.G. Bubnov. In: Novozhilov, V.V. (ed.) Questions of Mechanics of Solids, pp. 377–381. "Sudostroenie" Publishing House, Leningrad (1989)
36. Novozhilov, V.V.: Remembrances: B.G. Galerkin. In: Novozhilov, V.V. (ed.) Questions of Mechanics of Solids, pp. 381–382. "Sudostroenie" Publishing House, Leningrad (1989)
37. Petrov, G.I.: Application of Galerkin's method to the stability problem of viscous fluids. J. Appl. Math. Mech. **4**(3) (1940) (in Russian)
38. Pflüger, A.: Stabilitätsprobleme der Elastostatik. Springer, Berlin (1950) (in German)
39. Pister, K.S.: Trends in computational structural mechanics. After dinner talk delivered by Karl S. Pister on 23 May 2001, at the Trends in Computational Mechanics Conference Organised by Wolfgang A. Wall and colleagues under the auspices of the German Association of Computational Mechanics, IACM, International Association of Computational Mechanics (2001). http://www.cimne.com/iacm/News/Expressions11.pdf. Accessed 16 April 2017
40. Prescott, J.: XXX. The buckling of deep beams. Lond. Edinb. Dublin Philos. Mag. J. Sci. **36**(214), 297–314 (1918)
41. Prescott, J., Carrington, H.: XIX. The buckling of deep beams (Second Paper). Lond. Edinb. Dublin Philos. Mag. J. Sci. **39**(230), 194–223 (1920)
42. Repin, S.: One hundred years of the Galerkin method. Comput. Methods Appl. Math. **17**(3), 351–357 (2017)
43. Repman, J.W.: On mathematical foundation of Galerkin method for the solution of stability problems of elastic systems. J. Appl. Math. Mech. (2) (1940) (in Russian)
44. Singer, J.: On the equivalence of the Galerkin and Rayleigh-Ritz methods. Aeronaut. J. **66**(621), 592–592 (1962)
45. Spassky, I.D.: Beginning of the creative roadmap. In: Zubov, V.I. (ed.) Academician V.V. Novozhilov–Scientist, Pedagogue, Citizen, pp. 19–36. Leningrad University Press, Leningrad (1990) (in Russian)
46. Svirsky, I.V.: Methods of Bubnov-Galerkin Type and of Successive Approximations. "Nauka" Publishing House, Moscow (1968) (in Russian)
47. Tanner, R.I., Tanner, E.: Heinrich Hencky: a rheological pioneer. Rheol. Acta **42**, 93–101 (2003)
48. Timoshenko, S.P.: CVII. On the buckling of deep beams: to the editors of the Philosophical Magazine. Lond. Edinb. Dublin Philos. Mag. J. Sci. **43**(257), 1023–1024 (1922)
49. Timoshenko, S.P.: Vibration Problems in Engineering, p. 231. Constable and Company, London (1928)
50. Timoshenko, S.P.: Discussion of paper by H.M. Westergaard. Trans. Am. Soc. Civil Eng. **94**(1), 241–242 (1930)
51. Timoshenko, S.P.: Vibration Problems in Engineering, 2nd edn. Van Nostrand Reinhold Company, New York (1937)
52. Timoshenko, S.P.: History of Strength of Materials with a Brief Account of Theory and Elasticity and Theory of Structures. McGraw-Hill, New York (1953)

53. Timoshenko, S.P., Young, D.H.: Vibration Problems in Engineering, 3rd edn. Van Nostrand Reinhold Company, New York (1955)
54. Timoshenko, S.P., Young, D.H., Weaver Jr., W.: Vibration Problems in Engineering, 4th edn. Wiley, New York (1974)
55. Vol'mir, A.S.: Development of I. G. Bubnov's ideas in the modern theory for plates and shells. Struct. Mech. Anal. Struct. (4), 67–69 (1972) (in Russian)
56. Vorovich, I.I.: The Bubnov-Galerkin method, its development and role in applied mathematics. In: Advances in Mechanics of Deformable Media, pp. 121–133. "Nauka" Publishing House, Moscow (1975) (in Russian)
57. Vorovich, I.I.: Nonlinear Theory of Shallow Shells, vol. 133. Springer Science & Business Media (1998)
58. Weaver, W., Jr., Timoshenko, S.P., Young, D.M.: Vibration Problems in Engineering. Wiley, New York (1990)
59. Westergaard, H.M.: One hundred fifty years advance in structural analysis. Trans. Am. Soc. Civ. Eng. **94**(1), 226–240 (1930)
60. Wikipedia: Boris Galerkin (2020). https://en.wikipedia.org/wiki/Boris_Galerkin. Accessed 9 March 2020
61. Wikipedia: Mstislav Vsevolodovich Keldysh (2020). https://en.wikipedia.org/wiki/Mstislav_Keldysh. Accessed 18 March 2020
62. Reddy, J.N., Srinivasa, A.R.: Misattributions and misnomers in mechanics: Why they matter in the search for insight and precision of thought. Vietnam J. Mech. **42**(3), 283–291 (2020)

Chapter 6
Global Bifurcation Analysis of Polynomial Dynamical Systems

Valery A. Gaiko

Abstract We carry out a global bifurcation analysis of planar polynomial dynamical systems. In particular, using a bifurcational geometric approach, we study the global dynamics and solve the problem on the maximum number and distribution of limit cycles in a polynomial Euler–Lagrange–Liénard type mechanical system. We consider also a rational endocrine system carrying out a global bifurcation analysis of a reduced planar quartic Topp system which models the dynamics of diabetes. Studying global bifurcations and applying the Wintner–Perko termination principle, we prove that such a system can have at most two limit cycles.

6.1 Introduction

We carry out a global bifurcation analysis of planar polynomial dynamical systems and, first of all, we would like to recall some basic facts on their singular points and limit cycles. In particular, the study of singular points of polynomial systems will use two index theorems by H. Poincaré; see [2]. The definition of the Poincaré index is the following [2].

Definition 1 Let S be a simple closed curve in the phase plane not passing through a singular point of the system

$$\dot{x} = P(x, y), \quad \dot{y} = Q(x, y), \tag{6.1}$$

where $P(x, y)$ and $Q(x, y)$ are continuous functions (for example, polynomials), and M be some point on S. If the point M goes around the curve S in the positive direction (counterclockwise) one time, then the vector coinciding with the direction of a tangent to the trajectory passing through the point M is rotated through the angle

V. A. Gaiko (✉)
United Institute of Informatics Problems, National Academy of Sciences of Belarus,
Surganov Str. 6, Minsk 220012, Belarus
e-mail: valery.gaiko@gmail.com

A. K. Abramian et al. (eds.), *Nonlinear Dynamics of Discrete and Continuous Systems*,
Advanced Structured Materials 139,
https://doi.org/10.1007/978-3-030-53006-8_6

$2\pi j$ ($j = 0, \pm 1, \pm 2, \ldots$). The integer j is called the *Poincaré index* of the closed curve S relative to the vector field of system (6.1) and has the expression

$$j = \frac{1}{2\pi} \oint_S \frac{P\,dQ - Q\,dP}{P^2 + Q^2}. \tag{6.2}$$

According to this definition, the index of a node or a focus, or a center is equal to $+1$ and the index of a saddle is -1. The following Poincaré index theorems are valid [2].

Theorem 1 *The indices of singular points in the plane and at infinity sum to* $+1$.

Theorem 2 *If all singular points are simple, then along an isocline without multiple points lying in a Poincaré hemisphere which is obtained by a stereographic projection of the phase plane, the singular points are distributed so that a saddle is followed by a node or a focus, or a center and vice versa. If two points are separated by the equator of the Poincaré sphere, then a saddle will be followed by a saddle again and a node or a focus, or a center will be followed by a node or a focus, or a center.*

Consider a polynomial system in the vector form

$$\dot{\boldsymbol{x}} = \boldsymbol{f}(\boldsymbol{x}, \boldsymbol{\mu}), \tag{6.3}$$

where $\boldsymbol{x} \in \mathbf{R}^2$; $\boldsymbol{\mu} \in \mathbf{R}^n$; $\boldsymbol{f} \in \mathbf{R}^2$ ($\boldsymbol{f}$ is a polynomial vector function).

Recall some basic facts concerning limit cycles of (6.3). Assume that system (6.3) has a limit cycle

$$L_0 : \boldsymbol{x} = \boldsymbol{\varphi}_0(t)$$

of minimal period T_0 at some parameter value $\boldsymbol{\mu} = \boldsymbol{\mu}_0 \in \mathbf{R}^n$.

Let l be the straight line normal to L_0 at the point $\boldsymbol{p}_0 = \boldsymbol{\varphi}_0(0)$ and s be the coordinate along l with s positive exterior to L_0. It then follows from the implicit function theorem that there is a $\delta > 0$ such that the Poincaré map $h(s, \boldsymbol{\mu})$ is defined and analytic for $|s| < \delta$ and $\|\boldsymbol{\mu} - \boldsymbol{\mu}_0\| < \delta$. The displacement function for system (6.3) along the normal line l to L_0 is defined as the function

$$d(s, \boldsymbol{\mu}) = h(s, \boldsymbol{\mu}) - s.$$

We denote derivatives of d with respect to s or components of μ by subscripts, and the m-th derivative of d with respect to s by $d_s^{(m)}$. In terms of the displacement function, a multiple limit cycle can be defined as follows [9].

Definition 2 A limit cycle L_0 of (6.3) is a *multiple limit cycle* iff

$$d(0, \boldsymbol{\mu}_0) = d_s(0, \boldsymbol{\mu}_0) = 0.$$

It is a *simple limit cycle* (or hyperbolic limit cycle) if it is not a multiple limit cycle; furthermore, L_0 is a limit cycle of multiplicity m iff

$$d(0, \boldsymbol{\mu}_0) = d_s(0, \boldsymbol{\mu}_0) = \cdots = d_s^{(m-1)}(0, \boldsymbol{\mu}_0) = 0,$$

$$d_s^{(m)}(0, \boldsymbol{\mu}_0) \neq 0.$$

Note that the multiplicity of L_0 is independent of the point $\boldsymbol{p}_0 \in L_0$ through which we take the normal line l.

Let us write down also the following formulae which have already become classical ones and determine the derivatives of the displacement function in terms of integrals of the vector field $\boldsymbol{f}$ along the periodic orbit $\boldsymbol{\varphi}_0(t)$ [9]:

$$d_s(0, \boldsymbol{\mu}_0) = \exp \int_0^{T_0} \nabla \cdot \boldsymbol{f}(\boldsymbol{\varphi}_0(t), \boldsymbol{\mu}_0)\, \mathrm{d}t - 1$$

and

$$d_{\mu_j}(0, \boldsymbol{\mu}_0) = \frac{-\omega_0}{\|\boldsymbol{f}(\boldsymbol{\varphi}_0(0), \boldsymbol{\mu}_0)\|} \times$$

$$\int_0^{T_0} \exp\left(-\int_0^t \nabla \cdot \boldsymbol{f}(\boldsymbol{\varphi}_0(\tau), \boldsymbol{\mu}_0)\, \mathrm{d}\tau\right) \times \boldsymbol{f} \wedge \boldsymbol{f}_{\mu_j}(\boldsymbol{\varphi}_0(t), \boldsymbol{\mu}_0)\, \mathrm{d}t$$

for $j = 1, \ldots, n$, where $\omega_0 = \pm 1$ according to whether L_0 is positively or negatively oriented, respectively, and where the wedge product of two vectors $\boldsymbol{x} = (x_1, x_2)$ and $\boldsymbol{y} = (y_1, y_2)$ in $\mathbf{R}^2$ is defined as

$$\boldsymbol{x} \wedge \boldsymbol{y} = x_1\, y_2 - x_2\, y_1.$$

Similar formulae for $d_{ss}(0, \boldsymbol{\mu}_0)$ and $d_{s\mu_j}(0, \boldsymbol{\mu}_0)$ can be derived in terms of integrals of the vector field $\boldsymbol{f}$ and its first and second partial derivatives along $\boldsymbol{\varphi}_0(t)$.

Now we can formulate the Wintner–Perko termination principle [31] for polynomial system (6.3).

Theorem 3 *Any one-parameter family of multiplicity-m limit cycles of relatively prime polynomial system (6.3) can be extended in a unique way to a maximal one-parameter family of multiplicity-m limit cycles of (6.3) which is either open or cyclic.*

If it is open, then it terminates either as the parameter or the limit cycles become unbounded; or, the family terminates either at a singular point of (6.3), which is typically a fine focus of multiplicity m, or on a (compound) separatrix cycle of (6.3) which is also typically of multiplicity m.

The proof of this principle for general polynomial system (6.3) with a vector parameter $\boldsymbol{\mu} \in \mathbf{R}^n$ parallels the proof of the planar termination principle for the system

$$\dot{x} = P(x, y, \lambda), \quad \dot{y} = Q(x, y, \lambda) \tag{6.4}$$

with a single parameter $\lambda \in \mathbf{R}$ (see [9, 31]), since there is no loss of generality in assuming that system (6.3) is parameterized by a single parameter λ; i.e., we can assume that there exists an analytic mapping $\boldsymbol{\mu}(\lambda)$ of $\mathbf{R}$ into $\mathbf{R}^n$ such that (6.3) can be written as (6.4) and then we can repeat everything that had been done for system (6.4) in [31]. In particular, λ is said to be a *field-rotation parameter* if it rotates the vectors of the field in one direction [2, 9, 31]. If λ is a field rotation parameter of (6.4), the following Perko's theorem on monotonic families of limit cycles is valid; see [31].

Theorem 4 *If L_0 is a nonsingular multiple limit cycle of (6.4) for $\lambda = \lambda_0$, then L_0 belongs to a one-parameter family of limit cycles of (6.4); furthermore:*

(1) if the multiplicity of L_0 is odd, then the family either expands or contracts monotonically as λ increases through λ_0;

(2) if the multiplicity of L_0 is even, then L_0 bifurcates into a stable and an unstable limit cycle as λ varies from λ_0 in one sense and L_0 disappears as λ varies from λ_0 in the opposite sense; i. e., there is a fold bifurcation at λ_0.

We use these theorems and develop our methods for studying limit cycle bifurcations of polynomial dynamical systems. In Sect. 6.2, applying canonical systems with field rotation parameters and using geometric properties of the spirals filling the interior and exterior domains of limit cycles, we solve the problem on the maximum number and distribution of limit cycles in an Euler–Lagrange–Liénard type mechanical system. In Sect. 6.3, we consider an endocrine system model carrying out a global qualitative analysis of a reduced planar quartic Topp system which models the dynamics of diabetes; in particular, studying global bifurcations and applying the Wintner–Perko termination principle, we prove that such a system can have at most two limit cycles. This is related to the solution of Hilbert's sixteenth problem on the maximum number and distribution of limit cycles in planar polynomial dynamical systems [9].

6.2 Polynomial Mechanical System

6.2.1 Euler–Lagrange–Liénard Type Model

We study an Euler–Lagrange–Liénard type equation

$$\ddot{x} + h(x)\,\dot{x}^2 + f(x)\,\dot{x} + g(x) = 0 \tag{6.5}$$

and the corresponding dynamical system

$$\dot{x} = y, \quad \dot{y} = -g(x) - f(x)\,y - h(x)\,\dot{x}^2. \tag{6.6}$$

Equation (6.5) is a composition of two equations. One of them is

$$\alpha(q)\,\ddot{q} + \beta(q)\,\dot{q}^2 + \gamma(q) = 0, \tag{6.7}$$

where $q \in R$; $\alpha(q)$, $\beta(q)$ and $\gamma(q)$ are scalar functions, which represents a generic form of dynamics for an n-degree of freedom Euler–Lagrange system

$$\frac{\mathrm{d}}{\mathrm{d}t}\left(\frac{\partial L}{\partial \dot{Q}}\right) - \frac{\partial L}{\partial Q} = B(Q)\,u, \tag{6.8}$$

where $L(Q, \dot{Q})$ is a Lagrangian, $Q \in R^n$ is a vector of generalized coordinates, $u \in R^{n-1}$ and $B(Q)$ is $n \times (n-1)$ matrix function of full rank for each Q. Equation (6.7) can be used, in particular, for solving the periodic motion problem in mechanical systems; see, e. g., [35] and the references therein.

The other one is the Liénard equation

$$\ddot{x} + f(x)\,\dot{x} + g(x) = 0 \tag{6.9}$$

with the corresponding dynamical systems in the form

$$\dot{x} = y, \quad \dot{y} = -g(x) - f(x)\,y, \tag{6.10}$$

particular cases of which we have considered in [10–17]; see also [7, 8, 24, 28, 29, 32, 36]. There are many examples in the natural sciences and technology in which this and related systems are applied [1, 2, 30, 34]. Such systems are often used to model either mechanical or electrical, or biomedical systems, and in the literature, many systems are transformed into Liénard type to aid in the investigations. They can be used, e. g., in certain mechanical systems, where $f(x)$ represents a coefficient of the damping force and $g(x)$ represents the restoring force or stiffness, when modeling wind rock phenomena and surge in jet engines [1, 30]. Such systems can be also used to model resistor-inductor-capacitor circuits with non-linear circuit elements. Recently, e. g., the Liénard system has been shown to describe the operation of an optoelectronics circuit that uses a resonant tunnelling diode to drive a laser diode to make an optoelectronic voltage controlled oscillator [34].

There are also a number of examples of technical systems which are modelled with quadratic damping: a term in the second-order dynamics model, which is quadratic with respect to the velocity state variable. These examples include bearings, floating off-shore structures, vibration isolation and ship roll damping models [6, 25]. In robotics, quadratic damping appears in feed-forward control and in nonlinear impedance devices, such as variable impedance actuators [3]. Variable impedance actuators are of particular interest for collaborative robotics [33].

We suppose that system (6.6), where $g(x)$, $h(x)$ and $f(x)$ are arbitrary polynomials, has an anti-saddle (a node or a focus, or a center) at the origin and write it in the form

$$\begin{aligned} \dot{x} &= y, \\ \dot{y} &= -x(1 + a_1 x + \ldots + a_{2l} x^{2l}) + y(\alpha_0 + \alpha_1 x + \ldots + \alpha_{2k} x^{2k}) \\ &\quad + y^2(c_0 + c_1 x + \ldots + c_{2n} x^{2n}). \end{aligned} \tag{6.11}$$

Note that for $g(x) \equiv x$ and $h(x) \equiv 0$, by the change of variables $X = x$ and $Y = y + F(x)$, where $F(x) = \int_0^x f(s)\,\mathrm{d}s$, (6.11) is reduced to an equivalent system

$$\dot{X} = Y - F(X), \quad \dot{Y} = -X \tag{6.12}$$

which can be written in the form

$$\dot{x} = y, \quad \dot{y} = -x + F(y) \tag{6.13}$$

or

$$\dot{x} = y, \quad \dot{y} = -x + \gamma_1 y + \gamma_2 y^2 + \gamma_3 y^3 + \cdots + \gamma_{2k} y^{2k} + \gamma_{2k+1} y^{2k+1}. \tag{6.14}$$

In [10–13], we have presented a solution of Smale's thirteenth problem [36] proving that the Liénard system (6.14) with a polynomial of degree $2k + 1$ can have at most k limit cycles and we can conclude now that our results [10–13] agree with the conjecture of [28] on the maximum number of limit cycles for the classical Liénard polynomial system (6.14). There were some attempts to construct counterexamples to this conjecture, e. g., in [7, 8]. But that "counterexamples" were completely wrong.

In [14–17], we have studied the general Liénard polynomial system ($h(x) \equiv 0$)

$$\dot{x} = y, \quad \dot{y} = -x(1 + a_1 x + \cdots + a_{2l} x^{2l}) + y(\alpha_0 + \alpha_1 x + \cdots + \alpha_{2k} x^{2k}). \tag{6.15}$$

In [14–16], under some assumptions on the parameters of (6.15), and in [17], in the general case, we have found the maximum number of limit cycles and their possible distribution for system (6.15).

6.2.2 *Limit Cycles of the Euler–Lagrange–Liénard System*

Consider system (6.11) supposing that $a_1^2 + \cdots + a_{2l}^2 \neq 0$. Its finite singularities are determined by the algebraic system

$$x\,(1 + a_1 x + \cdots + a_{2l} x^{2l}) = 0, \quad y = 0. \tag{6.16}$$

This system always has an anti-saddle at the origin and, in general, can have at most $2l + 1$ finite singularities which lie on the x-axis and are distributed so that a saddle (or saddle-node) is followed by a node or a focus, or a center and vice versa [2]. For studying the infinite singularities, the methods applied in [2] for Rayleigh's and van der Pol's equations and also Erugin's two-isocline method developed in [9] can be used; see [10–17].

Following [9], we will study limit cycle bifurcations of (6.11) by means of canonical systems containing field rotation parameters of (6.11) [2, 9].

Theorem 5 *The Euler–Lagrange–Liénard polynomial system (6.11) with limit cycles can be reduced to one of the canonical forms:*

$$\begin{aligned}\dot{x} &= y,\\ \dot{y} &= -x\,(1 + a_1 x + \cdots + a_{2l} x^{2l})\\ &+ y(\alpha_0 - \beta_1 - \cdots - \beta_{2k-1} + \beta_1 x + \alpha_2 x^2 + \cdots + \beta_{2k-1} x^{2k-1} + \alpha_{2k} x^{2k})\\ &+ y^2(c_0 + c_1\,x + \cdots + c_{2n}\,x^{2n})\end{aligned} \tag{6.17}$$

or

$$\begin{aligned}\dot{x} &= y \equiv P(x, y),\\ \dot{y} &= x(x-1)(1 + b_1 x + \cdots + b_{2l-1} x^{2l-1})\\ &+ y(\alpha_0 - \beta_1 - \cdots - \beta_{2k-1} + \beta_1 x + \alpha_2 x^2 + \cdots + \beta_{2k-1} x^{2k-1} + \alpha_{2k} x^{2k})\\ &+ y^2(c_0 + c_1\,x + \cdots + c_{2n}\,x^{2n}) \equiv Q(x, y),\end{aligned} \tag{6.18}$$

where $1 + a_1 x + \cdots + a_{2l} x^{2l} \neq 0$, $\alpha_0, \alpha_2, \ldots, \alpha_{2k}$ *are field rotation parameters and* $\beta_1, \beta_3, \ldots, \beta_{2k-1}$ *are semi-rotation parameters.*

Proof Let us compare system (6.11) with (6.17) and (6.18). It is easy to see that system (6.17) has the only finite singular point: an anti-saddle at the origin. System (6.18) has at list two singular points including an anti-saddle at the origin and a saddle which, without loss of generality, can be always putted into the point $(1, 0)$. Instead of the odd parameters $\alpha_1, \alpha_3, \ldots, \alpha_{2k-1}$ in system (6.11), also without loss of generality, we have introduced new parameters $\beta_1, \beta_3, \ldots, \beta_{2k-1}$ into (6.17) and (6.18).

We will study now system (6.18) (system (6.17) can be studied absolutely similarly). Let all of the parameters $\alpha_0, \alpha_2, \ldots, \alpha_{2k}$ and $\beta_1, \beta_3, \ldots, \beta_{2k-1}$ vanish in this system,

$$\begin{aligned}\dot{x} &= y,\\ \dot{y} &= x(x-1)(1 + b_1 x + \cdots + b_{2l-1} x^{2l-1})\\ &+ y^2(c_0 + c_1\,x + \cdots + c_{2n}\,x^{2n})\end{aligned} \tag{6.19}$$

and consider the corresponding equation

$$\begin{aligned}\frac{dy}{dx} &= \frac{x(x-1)(1 + b_1 x + \cdots + b_{2l-1} x^{2l-1}) + y^2(c_0 + c_1\,x + \cdots + c_{2n}\,x^{2n})}{y}\\ &\equiv F(x, y).\end{aligned} \tag{6.20}$$

Since $F(x, -y) = -F(x, y)$, the direction field of (6.20) (and the vector field of (6.19) as well) is symmetric with respect to the x-axis. It follows that for arbitrary values of the parameters $b_1, \ldots, b_{2l-1}$ system (6.19) has centers as anti-saddles and cannot have limit cycles surrounding these points. Therefore, we can fix the parameters $b_1, \ldots, b_{2l-1}$ in system (6.18), fixing the position of its finite singularities on the x-axis.

To prove that the even parameters $\alpha_0, \alpha_2, \ldots, \alpha_{2k}$ rotate the vector field of (16), let us calculate the following determinants:

$$\begin{aligned}\Delta_{\alpha_0} &= P\, Q'_{\alpha_0} - QP'_{\alpha_0} = y^2 \geq 0,\\ \Delta_{\alpha_2} &= P\, Q'_{\alpha_2} - QP'_{\alpha_2} = x^2 y^2 \geq 0,\\ &\ldots\ldots\ldots\ldots\ldots\ldots\ldots\ldots\ldots\\ \Delta_{\alpha_{2k}} &= P\, Q'_{\alpha_{2k}} - QP'_{\alpha_{2k}} = x^{2k} y^2 \geq 0.\end{aligned}$$

By definition of a field rotation parameter [2, 9], for increasing each of the parameters $\alpha_0, \alpha_2, \ldots, \alpha_{2k}$, under the fixed others, the vector field of system (6.18) is rotated in the positive direction (counterclockwise) in the whole phase plane; and, conversely, for decreasing each of these parameters, the vector field of (6.18) is rotated in the negative direction (clockwise).

Calculating the corresponding determinants for the parameters $\beta_1, \beta_3, \ldots, \beta_{2k-1}$, we can see that

$$\begin{aligned}\Delta_{\beta_1} &= P\, Q'_{\beta_1} - QP'_{\beta_1} = (x-1)\, y^2,\\ \Delta_{\beta_3} &= P\, Q'_{\beta_3} - QP'_{\beta_3} = (x^3-1)\, y^2,\\ &\ldots\ldots\ldots\ldots\ldots\ldots\ldots\ldots\ldots\\ \Delta_{\beta_{2k-1}} &= P\, Q'_{\beta_{2k-1}} - QP'_{\beta_{2k-1}} = (x^{2k-1}-1)\, y^2.\end{aligned}$$

It follows [2, 9] that, for increasing each of the parameters $\beta_1, \beta_3, \ldots, \beta_{2k-1}$, under the fixed others, the vector field of system (6.18) is rotated in the positive direction (counterclockwise) in the half-plane $x > 1$ and in the negative direction (clockwise) in the half-plane $x < 1$ and vice versa for decreasing each of these parameters. We will call these parameters as semi-rotation ones.

Thus, for studying limit cycle bifurcations of (6.11), it is sufficient to consider the canonical systems (6.17) and (6.18) containing the field rotation parameters α_0, $\alpha_2, \ldots, \alpha_{2k}$ and the semi-rotation parameters $\beta_1, \beta_3, \ldots, \beta_{2k-1}$. The theorem is proved.

By means of the canonical systems (6.17) and (6.18), we will prove the following theorem.

Theorem 6 *The Euler–Lagrange–Liénard polynomial system (6.11) can have at most $k+l+1$ limit cycles, $k+1$ surrounding the origin and l surrounding one by one the other singularities of (6.11).*

Proof According to Theorem 5, for the study of limit cycle bifurcations of system (6.11), it is sufficient to consider the canonical systems (6.17) and (6.18) containing the field rotation parameters $\alpha_0, \alpha_2, \ldots, \alpha_{2k}$ and the semi-rotation parameters $\beta_1, \beta_3, \ldots, \beta_{2k-1}$. We will work with (6.18) again (system (6.17) can be considered in a similar way).

Vanishing all of the parameters $\alpha_0, \alpha_2, \ldots, \alpha_{2k}$ and $\beta_1, \beta_3, \ldots, \beta_{2k-1}$ in (6.18), we will have system (6.19) which is symmetric with respect to the x-axis and has centers as anti-saddles. Its center domains are bounded by either separatrix loops or digons of the saddles or saddle-nodes of (6.19) lying on the x-axis.

Let us input successively the semi-rotation parameters $\beta_1, \beta_3, \ldots, \beta_{2k-1}$ into system (6.19) beginning with the parameters at the highest degrees of x and alternating with their signs. So, begin with the parameter β_{2k-1} and let, for definiteness, $\beta_{2k-1} > 0$:

$$\begin{aligned} \dot{x} &= y, \\ \dot{y} &= x(x-1)(1+b_1x+\cdots+b_{2l-1}x^{2l-1}) \\ &+ y(-\beta_{2k-1}+\beta_{2k-1}x^{2k-1}) + y^2(c_0+c_1x+\cdots+c_{2n}x^{2n}). \end{aligned} \tag{6.21}$$

In this case, the vector field of (6.21) is rotated in the negative direction (clockwise) in the half-plane $x < 1$ turning the center at the origin into a rough stable focus. All of the other centers lying in the half-plane $x > 1$ become rough unstable foci, since the vector field of (6.21) is rotated in the positive direction (counterclockwise) in this half-plane [2, 9].

Fix β_{2k-1} and input the parameter $\beta_{2k-3} < 0$ into (21):

$$\begin{aligned} \dot{x} &= y, \\ \dot{y} &= x(x-1)(1+b_1x+\cdots+b_{2l-1}x^{2l-1}) \\ &+ y(-\beta_{2k-3}-\beta_{2k-1}+\beta_{2k-3}x^{2k-3}+\beta_{2k-1}x^{2k-1}) \\ &+y^2(c_0+c_1x+\cdots+c_{2n}x^{2n}). \end{aligned} \tag{6.22}$$

Then the vector field of (6.22) is rotated in the opposite directions in each of the half-planes $x < 1$ and $x > 1$. Under decreasing β_{2k-3}, when $\beta_{2k-3} = -\beta_{2k-1}$, the focus at the origin becomes nonrough (weak), changes the character of its stability and generates a stable limit cycle. All of the other foci in the half-plane $x > 1$ will also generate unstable limit cycles for some values of β_{2k-3} after changing the character of their stability. Under further decreasing β_{2k-3}, all of the limit cycles will expand disappearing on separatrix cycles of (6.22) [2, 9].

Denote the limit cycle surrounding the origin by Γ_0, the domain outside the cycle by D_{01}, the domain inside the cycle by D_{02} and consider logical possibilities of the appearance of other (semi-stable) limit cycles from a "trajectory concentration" surrounding this singular point. It is clear that, under decreasing the parameter β_{2k-3}, a semi-stable limit cycle cannot appear in the domain D_{02}, since the focus spirals filling this domain will untwist and the distance between their coils will increase because of the vector field rotation [10–17].

By contradiction, we can also prove that a semi-stable limit cycle cannot appear in the domain D_{01}. Suppose it appears in this domain for some values of the parameters $\beta^*_{2k-1} > 0$ and $\beta^*_{2k-3} < 0$. Return to system (6.19) and change the inputting order for the semi-rotation parameters. Input first the parameter $\beta_{2k-3} < 0$:

$$\begin{aligned} \dot{x} &= y, \\ \dot{y} &= x(x-1)(1+b_1x+\cdots+b_{2l-1}x^{2l-1}) \\ &+ y(-\beta_{2k-3}+\beta_{2k-3}x^{2k-3}) + y^2(c_0+c_1x+\cdots+c_{2n}x^{2n}). \end{aligned} \tag{6.23}$$

Fix it under $\beta_{2k-3} = \beta^*_{2k-3}$. The vector field of (6.23) is rotated counterclockwise and the origin turns into a rough unstable focus. Inputting the parameter $\beta_{2k-1} > 0$ into (6.23), we get again system (6.22) the vector field of which is rotated clockwise. Under this rotation, a stable limit cycle Γ_0 will appear from a separatrix cycle for some value of β_{2k-1}. This cycle will contract, the outside spirals winding onto the cycle will untwist and the distance between their coils will increase under increasing β_{2k-1} to the value β^*_{2k-1}. It follows that there are no values of $\beta^*_{2k-3} < 0$ and $\beta^*_{2k-1} > 0$ for which a semi-stable limit cycle could appear in the domain D_{01}.

This contradiction proves the uniqueness of a limit cycle surrounding the origin in system (6.22) for any values of the parameters β_{2k-3} and β_{2k-1} of different signs. Obviously, if these parameters have the same sign, system (6.22) has no limit cycles surrounding the origin at all. On the same reason, this system cannot have more than l limit cycles surrounding the other singularities (foci or nodes) of (6.22) one by one.

It is clear that inputting the other semi-rotation parameters $\beta_{2k-5}, \ldots, \beta_1$ into system (6.22) will not give us more limit cycles, since all of these parameters are rough with respect to the origin and the other anti-saddles lying in the half-plane $x > 1$. Therefore, the maximum number of limit cycles for the system

$$\begin{gathered}\dot{x} = y, \\ \dot{y} = x(x-1)(1 + b_1 x + \cdots + b_{2l-1}x^{2l-1}) \\ +y(-\beta_1 - \cdots - \beta_{2k-3} - \beta_{2k-1} + \beta_1 x + \cdots + \beta_{2k-3}x^{2k-3} + \beta_{2k-1}x^{2k-1}) \\ +y^2(c_0 + c_1\,x + \cdots + c_{2n}\,x^{2n})\end{gathered} \tag{6.24}$$

is equal to $l+1$ and they surround the anti-saddles (foci or nodes) of (24) one by one.

Suppose that $\beta_1 + \cdots + \beta_{2k-3} + \beta_{2k-1} > 0$ and input the last rough parameter $\alpha_0 > 0$ into system (6.24):

$$\begin{gathered}\dot{x} = y, \\ \dot{y} = x(x-1)(1 + b_1 x + \cdots + b_{2l-1}x^{2l-1}) \\ +\,y(\alpha_0 - \beta_1 - \cdots - \beta_{2k-1} + \beta_1 x + \cdots + \beta_{2k-1}x^{2k-1}) \\ +y^2(c_0 + c_1\,x + \cdots + c_{2n}\,x^{2n}).\end{gathered} \tag{6.25}$$

This parameter rotating the vector field of (6.25) counterclockwise in the whole phase plane also will not give us more limit cycles, but under increasing α_0, when $\alpha_0 = \beta_1 + \cdots + \beta_{2k-1}$, we can make the focus at the origin nonrough (weak), after the disappearance of the limit cycle Γ_0 in it. Fix this value of the parameter α_0 ($\alpha_0 = \alpha_0^*$):

$$\begin{gathered}\dot{x} = y, \\ \dot{y} = x(x-1)(1 + b_1 x + \cdots + b_{2l-1}x^{2l-1}) \\ +\,y(\beta_1 x + \cdots + \beta_{2k-1}x^{2k-1}) + y^2(c_0 + c_1\,x + \cdots + c_{2n}\,x^{2n}).\end{gathered} \tag{6.26}$$

Let us input now successively the other field rotation parameters $\alpha_2, \ldots, \alpha_{2k}$ into system (6.26) beginning again with the parameters at the highest degrees of x and alternating with their signs; see [10–17]. So, begin with the parameter α_{2k} and let $\alpha_{2k} < 0$:

$$\begin{aligned}\dot{x} &= y,\\ \dot{y} &= x(x-1)(1+b_1x+\cdots+b_{2l-1}x^{2l-1})\\ &\quad + y(\beta_1x+\cdots+\beta_{2k-1}x^{2k-1}+\alpha_{2k}x^{2k})\\ &\quad +y^2(c_0+c_1\,x+\cdots+c_{2n}\,x^{2n}).\end{aligned} \tag{6.27}$$

In this case, the vector field of (6.27) is rotated clockwise in the whole phase plane and the focus at the origin changes the character of its stability generating again a stable limit cycle. The limit cycles surrounding the other singularities of (6.27) can also still exist. Denote the limit cycle surrounding the origin by Γ_1, the domain outside the cycle by D_1 and the domain inside the cycle by D_2. The uniqueness of a limit cycle surrounding the origin (and limit cycles surrounding the other singularities) for system (6.27) can be proved by contradiction like we have done above for (6.22); see also [10–17].

Let system (6.27) have the unique limit cycle Γ_1 surrounding the origin and l limit cycles surrounding the other antisaddles of (6.27). Fix the parameter $\alpha_{2k} < 0$ and input the parameter $\alpha_{2k-2} > 0$ into (6.27):

$$\begin{aligned}\dot{x} &= y,\\ \dot{y} &= x(x-1)(1+b_1x+\cdots+b_{2l-1}x^{2l-1})\\ &\quad + y(\beta_1x+\cdots+\beta_{2k-1}x^{2k-1}+\alpha_{2k-2}x^{2k-2}+\alpha_{2k}x^{2k})\\ &\quad +y^2(c_0+c_1\,x+\cdots+c_{2n}\,x^{2n}).\end{aligned} \tag{6.28}$$

Then the vector field of (6.28) is rotated in the opposite direction (counterclockwise) and the focus at the origin immediately changes the character of its stability (since its degree of nonroughness decreases and the sign of the field rotation parameter at the lower degree of x changes) generating the second (unstable) limit cycle Γ_2. The limit cycles surrounding the other singularities of (28) can only disappear in the corresponding foci (because of their roughness) under increasing the parameter α_{2k-2}. Under further increasing α_{2k-2}, the limit cycle Γ_2 will join with Γ_1 forming a semi-stable limit cycle, Γ_{12}, which will disappear in a "trajectory concentration" surrounding the origin. Can another semi-stable limit cycle appear around the origin in addition to Γ_{12}? It is clear that such a limit cycle cannot appear either in the domain D_1 bounded on the inside by the cycle Γ_1 or in the domain D_3 bounded by the origin and Γ_2 because of the increasing distance between the spiral coils filling these domains under increasing the parameter [10–17].

To prove the impossibility of the appearance of a semi-stable limit cycle in the domain D_2 bounded by the cycles Γ_1 and Γ_2 (before their joining), suppose the contrary, i. e., that for some values of these parameters, $\alpha_{2k}^* < 0$ and $\alpha_{2k-2}^* > 0$, such a semi-stable cycle exists. Return to system (6.26) again and input first the parameter $\alpha_{2k-2} > 0$:

$$
\begin{gathered}
\dot{x} = y, \\
\dot{y} = x(x-1)(1 + b_1 x + \cdots + b_{2l-1} x^{2l-1}) \\
+ y(\beta_1 x + \cdots + \beta_{2k-1} x^{2k-1} + \alpha_{2k-2} x^{2k-2}) \\
+ y^2 (c_0 + c_1 x + \cdots + c_{2n} x^{2n}).
\end{gathered} \tag{6.29}
$$

This parameter rotates the vector field of (6.29) counterclockwise preserving the origin as a nonrough stable focus.

Fix this parameter under $\alpha_{2k-2} = \alpha^*_{2k-2}$ and input the parameter $\alpha_{2k} < 0$ into (6.29) getting again system (26). Since, by our assumption, this system has two limit cycles surrounding the origin for $\alpha_{2k} > \alpha^*_{2k}$, there exists some value of the parameter, α^{12}_{2k} ($\alpha^{12}_{2k} < \alpha^*_{2k} < 0$), for which a semi-stable limit cycle, Γ_{12}, appears in system (6.28) and then splits into a stable cycle Γ_1 and an unstable cycle Γ_2 under further decreasing α_{2k}. The formed domain D_2 bounded by the limit cycles Γ_1, Γ_2 and filled by the spirals will enlarge since, on the properties of a field rotation parameter, the interior unstable limit cycle Γ_2 will contract and the exterior stable limit cycle Γ_1 will expand under decreasing α_{2k}. The distance between the spirals of the domain D_2 will naturally increase, which will prevent the appearance of a semi-stable limit cycle in this domain for $\alpha_{2k} < \alpha^{12}_{2k}$ [10–17].

Thus, there are no such values of the parameters, $\alpha^*_{2k} < 0$ and $\alpha^*_{2k-2} > 0$, for which system (6.28) would have an additional semi-stable limit cycle surrounding the origin. Obviously, there are no other values of the parameters α_{2k} and α_{2k-2} for which system (6.28) would have more than two limit cycles surrounding this singular point. On the same reason, additional semi-stable limit cycles cannot appear around the other singularities (foci or nodes) of (6.28). Therefore, $l + 2$ is the maximum number of limit cycles in system (6.28).

Suppose that system (6.28) has two limit cycles, Γ_1 and Γ_2, surrounding the origin and l limit cycles surrounding the other antisaddles of (6.28) (this is always possible if $-\alpha_{2k} \gg \alpha_{2k-2} > 0$). Fix the parameters α_{2k}, α_{2k-2} and consider a more general system inputting the third parameter, $\alpha_{2k-4} < 0$, into (6.26):

$$
\begin{gathered}
\dot{x} = y, \\
\dot{y} = x(x-1)(1 + b_1 x + \cdots + b_{2l-1} x^{2l-1}) \\
+ y(\beta_1 x + \cdots + \beta_{2k-1} x^{2k-1} + \alpha_{2k-4} x^{2k-4} + \alpha_{2k-2} x^{2k-2} + \alpha_{2k} x^{2k}) \\
+ y^2 (c_0 + c_1 x + \cdots + c_{2n} x^{2n}).
\end{gathered} \tag{6.30}
$$

For decreasing α_{2k-4}, the vector field of (6.30) will be rotated clockwise and the focus at the origin will immediately change the character of its stability generating a third (stable) limit cycle, Γ_3. With further decreasing α_{2k-4}, Γ_3 will join with Γ_2 forming a semi-stable limit cycle, Γ_{23}, which will disappear in a "trajectory concentration" surrounding the origin; the cycle Γ_1 will expand disappearing on a separatrix cycle of (6.30).

Let system (6.30) have three limit cycles surrounding the origin: Γ_1, Γ_2, Γ_3. Could an additional semi-stable limit cycle appear with decreasing α_{2k-4} after splitting of which system (6.30) would have five limit cycles around the origin? It is clear that such a limit cycle cannot appear either in the domain D_2 bounded by the cycles Γ_1

and Γ_2 or in the domain D_4 bounded by the origin and Γ_3 because of the increasing distance between the spiral coils filling these domains after decreasing α_{2k-4}. Consider two other domains: D_1 bounded on the inside by the cycle Γ_1 and D_3 bounded by the cycles Γ_2 and Γ_3. As before, we will prove the impossibility of the appearance of a semi-stable limit cycle in these domains by contradiction.

Suppose that for some set of values of the parameters $\alpha^*_{2k} < 0$, $\alpha^*_{2k-2} > 0$ and $\alpha^*_{2k-4} < 0$ such a semi-stable cycle exists. Return to system (26) again inputting first the parameters $\alpha_{2k-2} > 0$ and $\alpha_{2k-4} < 0$:

$$\begin{aligned} \dot{x} &= y, \\ \dot{y} &= x(x-1)(1 + b_1 x + \cdots + b_{2l-1} x^{2l-1}) \\ &+ y(\beta_1 x + \cdots + \beta_{2k-1} x^{2k-1} + \alpha_{2k-4} x^{2k-4} + \alpha_{2k} x^{2k}) \\ &+ y^2(c_0 + c_1 x + \cdots + c_{2n} x^{2n}). \end{aligned} \tag{6.31}$$

Fix the parameter α_{2k-2} under the value α^*_{2k-2}. With decreasing α_{2k-4}, a separatrix cycle formed around the origin will generate a stable limit cycle Γ_1. Fix α_{2k-4} under the value α^*_{2k-4} and input the parameter $\alpha_{2k} > 0$ into (6.31) getting system (6.30).

Since, by our assumption, (6.30) has three limit cycles for $\alpha_{2k} > \alpha^*_{2k}$, there exists some value of the parameter α^{23}_{2k} ($\alpha^{23}_{2k} < \alpha^*_{2k} < 0$) for which a semi-stable limit cycle, Γ_{23}, appears in this system and then splits into an unstable cycle Γ_2 and a stable cycle Γ_3 with further decreasing α_{2k}. The formed domain D_3 bounded by the limit cycles Γ_2, Γ_3 and also the domain D_1 bounded on the inside by the limit cycle Γ_1 will enlarge and the spirals filling these domains will untwist excluding a possibility of the appearance of a semi-stable limit cycle there [10–17].

All other combinations of the parameters α_{2k}, α_{2k-2}, and α_{2k-4} are considered in a similar way. It follows that system (6.30) can have at most $l + 3$ limit cycles.

If we continue the procedure of successive inputting the field rotation parameters, $\alpha_{2k}, \ldots, \alpha_2$, into system (6.26),

$$\begin{aligned} \dot{x} &= y, \\ \dot{y} &= x(x-1)(1 + b_1 x + \cdots + b_{2l-1} x^{2l-1}) \\ &+ y(\beta_1 x + \cdots + \beta_{2k-1} x^{2k-1} + \alpha_2 x^2 + \cdots + \alpha_{2k} x^{2k}) \\ &+ y^2(c_0 + c_1 x + \cdots + c_{2n} x^{2n}), \end{aligned} \tag{6.32}$$

it is possible to obtain k limit cycles surrounding the origin and l surrounding one by one the other singularities (foci or nodes) $(-\alpha_{2k} \gg \alpha_{2k-2} \gg -\alpha_{2k-4} \gg \alpha_{2k-6} \gg \ldots)$.

Then, by means of the parameter $\alpha_0 \neq \beta_1 + \cdots + \beta_{2k-1}$ ($\alpha_0 > \alpha^*_0$, if $\alpha_2 < 0$, and $\alpha_0 < \alpha^*_0$, if $\alpha_2 > 0$), we will have the canonical system (6.18) with an additional limit cycle surrounding the origin and can conclude that this system (i. e., the Euler–Lagrange–Liénard polynomial system (6.11) as well) has at most $k + l + 1$ limit cycles, $k + 1$ surrounding the origin and l surrounding one by one the antisaddles (foci or nodes) of (6.18) (and (6.11) as well). The theorem is proved.

6.3 Rational Endocrine System

6.3.1 *Topp Model of Diabetes Dynamics*

In [37], a novel model of coupled β-cell mass, insulin, and glucose dynamics was presented, which is used to investigate the normal behavior of the glucose regulatory system and pathways into diabetes. The behavior of the model is consistent with the observed behavior of the glucose regulatory system in response to changes in blood glucose levels, insulin sensitivity, and β-cell insulin secretion rates.

In the post-absorptive state, glucose is released into the blood by the liver and kidneys, removed from the interstitial fluid by all the cells of the body, and distributed into many physiological compartments, e. g., arterial blood, venous blood, cerebral spinal fluid, interstitial fluid [37].

Since we are primarily concerned with the evolution of fasting blood glucose levels over a time-scale of days to years, glucose dynamics are modeled with a single-compartment mass balance equation

$$\dot{G} = a - (b + cI)G. \tag{6.33}$$

Insulin is secreted by pancreatic β-cells, cleared by the liver, kidneys, and insulin receptors, and distributed into several compartments, e. g., portal vein, peripheral blood, and interstitial fluid. The main concern is the long-time evolution of fasting insulin levels in peripheral blood. Since the dynamics of fasting insulin levels on this time-scale are slow, we use a single-compartment equation given by

$$\dot{I} = \frac{\beta G^2}{1 + G^2} - \alpha I. \tag{6.34}$$

Despite a complex distribution of pancreatic β cells throughout the pancreas, β-cell mass dynamics have been successfully quantified with a single-compartment model

$$\dot{\beta} = (-l + mG - nG^2)\beta. \tag{6.35}$$

Finally, the Topp model (a rational endocrine system) is

$$\begin{aligned} \dot{G} &= a - (b + cI)G, \\ \dot{I} &= \frac{\beta G^2}{1 + G^2} - \alpha I, \\ \dot{\beta} &= (-l + mG - nG^2)\beta \end{aligned} \tag{6.36}$$

with parameters as in [37].

On the short timescale, β is approximately constant and, relabelling the variables, the fast dynamics is a planar system

$$\begin{aligned} \dot{x} &= a - (b + c\,y)x, \\ \dot{y} &= \frac{\beta x^2}{1 + x^2} - \alpha\,y \end{aligned} \tag{6.37}$$

By rescaling time, this can be written in the form of a quartic dynamical system:

$$\begin{aligned} \dot{x} &= (1 + x^2)(a - (b + c\,y)x) \equiv P, \\ \dot{y} &= \beta x^2 - \alpha\,y(1 + x^2) \equiv Q. \end{aligned} \tag{6.38}$$

Together with (38), we will also consider an auxiliary system (see [2, 9, 31])

$$\dot{x} = P - \gamma Q, \qquad \dot{y} = Q + \gamma P, \tag{6.39}$$

applying to these systems new bifurcation methods and geometric approaches developed in [5, 9–21] and carrying out the qualitative analysis of (6.38).

6.3.2 *Global Bifurcation Analysis of the Topp System*

Consider system (6.38). Its finite singularities are determined by the algebraic system

$$\begin{aligned} &(1 + x^2)(a - (b + c\,y)x) = 0, \\ &\beta x^2 - \alpha\,y(1 + x^2) = 0 \end{aligned} \tag{6.40}$$

which can give us at most three singular points in the first quadrant: a saddle S and two antisaddles (non-saddles), A_1 and A_2, according to the second Poincaré index theorem (Theorem 2). Suppose that with respect to the x-axis they have the following sequence: A_1, S, A_2. System (6.38) can also have one singular point (an antisaddle) or two singular points (an antisaddle and a saddle-node) in the first quadrant.

To study singular points of (6.38) at infinity, consider the corresponding differential equation

$$\frac{dy}{dx} = \frac{\beta x^2 - \alpha\,y(1 + x^2)}{(1 + x^2)(a - (b + c\,y)x)}. \tag{6.41}$$

Dividing the numerator and denominator of the right-hand side of (6.41) by $x^4 (x \neq 0)$ and denoting y/x by u (as well as dy/dx), we will get the equation

$$u^2 = 0, \quad \text{where} \quad u = y/x, \tag{6.42}$$

for all infinite singularities of (6.41) except when $x = 0$ (the "ends" of the y-axis); see [2, 9]. For this special case we can divide the numerator and denominator of the right-hand side of (6.41) by y^4 $(y \neq 0)$ denoting x/y by v (as well as dx/dy) and consider the equation

$$v^2 = 0, \quad \text{where} \quad v = x/y. \tag{6.43}$$

According to the Poincaré index theorems (Theorems 1 and 2), the equations (6.42) and (6.43) give us two double singular points (saddle-nodes) at infinity for (6.41): on the "ends" of the x and y axes.

Using the obtained information on singular points and applying geometric approaches developed in [5, 9–21], we can study now the limit cycle bifurcations of system (6.38).

Applying the definition of a field rotation parameter [2, 9, 31], to system (6.38), let us calculate the corresponding determinants for the parameters a, b, c, α, and β, respectively:

$$\Delta_a = PQ'_a - QP'_a = -(1+x^2)(\beta x^2 - \alpha\, y(1+x^2)), \tag{6.44}$$

$$\Delta_b = PQ'_b - QP'_b = x(1+x^2)(\beta x^2 - \alpha\, y(1+x^2)), \tag{6.45}$$

$$\Delta_c = PQ'_c - QP'_c = xy(1+x^2)(\beta x^2 - \alpha\, y(1+x^2)), \tag{6.46}$$

$$\Delta_\alpha = PQ'_\alpha - QP'_\alpha = -y(1+x^2)^2(a - (b + c\, y)x), \tag{6.47}$$

$$\Delta_\beta = PQ'_\beta - QP'_\beta = x^2(1+x^2)(a - (b + c\, y)x). \tag{6.48}$$

It follows from (6.44)–(6.46) that in the first quadrant the signs of Δ_a, Δ_b, Δ_c depend on the sign of $\beta x^2 - \alpha\, y(1+x^2)$ and from (6.47) and (6.48) that the signs of Δ_α and Δ_β depend on the sign of $a - (b + c\, y)x$ on increasing (or decreasing) the parameters a, b, c, α, and β, respectively.

Therefore, to study limit cycle bifurcations of system (6.38) it makes sense together with (6.38) to consider also the auxiliary system (6.39) with field-rotation parameter γ :

$$\Delta_\gamma = P^2 + Q^2 \geq 0. \tag{6.49}$$

Using system (6.39) and applying Perko's results, we prove the following theorem [21].

Theorem 7 *The reduced Topp system (6.38) can have at most two limit cycles.*

Proof In [4, 5, 27, 38], where a similar quartic system was studied, it was proved that the cyclicity of singular points in such a system is equal to two and that the system can have at least two limit cycles; see also [18, 20, 23, 26] with similar results.

Consider systems (6.38)–(6.39) supposing that the cyclicity of singular points in these systems is equal to two and that the systems can have at least two limit cycles. Let us prove now that these systems have at most two limit cycles. The proof is carried out by contradiction applying Catastrophe Theory; see [9, 31].

We will study more general system (6.39) with three parameters: α, β, and γ (the parameters a, b, and c can be fixed, since they do not generate limit cycles). Suppose that (6.39) has three limit cycles surrounding the singular point A_1, in the first quadrant. Then we get into some domain of the parameters α, β, and γ being restricted by definite conditions on three other parameters, a, b, and c. This domain is bounded by two fold bifurcation surfaces forming a cusp bifurcation surface of multiplicity-three limit cycles in the space of the parameters α, β, and γ.

The corresponding maximal one-parameter family of multiplicity-three limit cycles cannot be cyclic, otherwise there will be at least one point corresponding to the limit cycle of multiplicity four (or even higher) in the parameter space.

Extending the bifurcation curve of multiplicity-four limit cycles through this point and parameterizing the corresponding maximal one-parameter family of multiplicity-four limit cycles by the field rotation parameter, γ, according to Theorem 4, we will obtain two monotonic curves of multiplicity-three and one, respectively, which, by the Wintner–Perko termination principle (Theorem 3), terminate either at the point A_1 or on a separatrix cycle surrounding this point. Since on our assumption the cyclicity of the singular point is equal to two, we have obtained a contradiction with the termination principle stating that the multiplicity of limit cycles cannot be higher than the multiplicity (cyclicity) of the singular point in which they terminate.

If the maximal one-parameter family of multiplicity-three limit cycles is not cyclic, using the same principle (Theorem 3), this again contradicts the cyclicity of A_1 not admitting the multiplicity of limit cycles to be higher than two. This contradiction completes the proof in the case of one singular point in the first quadrant.

Suppose that system (6.39) with three finite singularities, A_1, S, and A_2, has two small limit cycles around, for example, the point A_1 (the case when limit cycles surround the point A_2 is considered in a similar way). Then we get into some domain in the space of the parameters α, β, and γ which is bounded by a fold bifurcation surface of multiplicity-two limit cycles.

The corresponding maximal one-parameter family of multiplicity-two limit cycles cannot be cyclic, otherwise there will be at least one point corresponding to the limit cycle of multiplicity three (or even higher) in the parameter space. Extending the bifurcation curve of multiplicity-three limit cycles through this point and parameterizing the corresponding maximal one-parameter family of multiplicity-three limit cycles by the field rotation parameter, γ, according to Theorem 4, we will obtain a monotonic curve which, by the Wintner–Perko termination principle (Theorem 3), terminates either at the point A_1 or on some separatrix cycle surrounding this point. Since we know at least the cyclicity of the singular point which on our assumption is equal to one in this case, we have obtained a contradiction with the termination principle.

If the maximal one-parameter family of multiplicity-two limit cycles is not cyclic, using the same principle (Theorem 3), this again contradicts the cyclicity of A_1 not admitting the multiplicity of limit cycles higher than one. Moreover, it also follows from the termination principle that either an ordinary (small) separatrix loop or a big loop, or an eight-loop cannot have the multiplicity (cyclicity) higher than one in this case. Therefore, according to the same principle, there are no more than one limit cycle in the exterior domain surrounding all three finite singularities, A_1, S, and A_2.

Thus, taking into account all other possibilities for limit cycle bifurcations (see [4, 5, 27, 38]), we conclude that system (39) (and (38) as well) cannot have either a multiplicity-three limit cycle or more than two limit cycles in any configuration. The theorem is proved.

Acknowledgements This work was supported by the German Academic Exchange Service (DAAD). The author is also very grateful to the Center for Technologies in Robotics and Mechatronics Components of the Innopolis University for hospitality during his stay in January–April 2020.

References

1. Agarwal, A., Ananthkrishnan, N.: Bifurcation analysis for onset and cessation of surge in axial flow compressors. Int. J. Turbo Jet-Eng. **17**, 207–217 (2000)
2. Bautin, N.N., Leontovich, E.A.: Methods and Examples of the Qualitative Analysis of Dynamical Systems in a Plane. Nauka, Moscow (1990). (in Russian)
3. Bessa, W.M., Dutra, M.S., Kreuzer, E.: Depth control of remotely operated underwater vehicles using an adaptive fuzzy sliding mode controller. Robot. Auton. Syst. **56**, 670–677 (2008)
4. Broer, H.W., Naudot, V., Roussarie, R., Saleh, K.: Dynamics of a predator-prey model with non-monotonic response function. Discrete Contin. Dyn. Syst. Ser. A **18**, 221–251 (2007)
5. Broer, H.W., Gaiko, V.A.: Global qualitative analysis of a quartic ecological model. Nonlinear Anal. **72**, 628–634 (2010)
6. Chang-Jian, C.: Nonlinear analysis of a rub-impact rotor supported by turbulent couple stress fluid film journal bearings under quadratic damping. Nonlinear Dyn. **56**, 297–314 (2009)
7. De Maesschalck, P., Dumortier, F.: Classical Liénard equations of degree $n \geq 6$ can have $[(n-1)/2]+2$ limit cycles. J. Diff. Equat. **250**, 2162–2176 (2011)
8. Dumortier, F., Panazzolo, D., Roussarie, R.: More limit cycles than expected in Liénard equations. Proc. Amer. Math. Soc. **135**, 1895–1904 (2007)
9. Gaiko, V.A.: Global Bifurcation Theory and Hilbert's Sixteenth Problem. Kluwer, Boston (2003)
10. Gaiko, V.A.: Limit cycles of Liénard-type dynamical systems. Cubo **10**, 115–132 (2008)
11. Gaiko, V.A.: On the geometry of polynomial dynamical systems. J. Math. Sci. **157**, 400–412 (2009)
12. Gaiko, V.A.: The geometry of limit cycle bifurcations in polynomial dynamical systems. Discrete Contin. Dyn. Syst. Suppl. **5**, 447–456 (2011)
13. Gaiko, V.A.: On limit cycles surrounding a singular point. Differ. Equ. Dyn. Syst. **20**, 329–337 (2012)
14. Gaiko, V.A.: The applied geometry of a general Liénard polynomial system. Appl. Math. Letters **25**, 2327–2331 (2012)
15. Gaiko, V.A.: Limit cycle bifurcations of a general Liénard system with polynomial restoring and damping functions. Int. J. Dyn. Syst. Differ. Equ. **4**, 242–254 (2012)
16. Gaiko, V.A.: Limit cycle bifurcations of a special Liénard polynomial system. Adv. Dyn. Syst. Appl. **9**, 109–123 (2014)
17. Gaiko, V.A.: Maximum number and distribution of limit cycles in the general Liénard polynomial system. Adv. Dyn. Syst. Appl. **10**, 177–188 (2015)
18. Gaiko, V.A.: Global qualitative analysis of a Holling-type system. Int. J. Dyn. Syst. Differ. Equ. **6**, 161–172 (2016)
19. Gaiko, V.A.: Global bifurcation analysis of the Kukles cubic system. Int. J. Dyn. Syst. Differ. Equ. **8**, 326–336 (2018)

20. Gaiko, V.A., Vuik, C.: Global dynamics in the Leslie-Gower model with the Allee effect. Int. J. Bifurcat. Chaos **28**, 1850151 (2018)
21. Gaiko, V.A.: Limit cycles of a Topp system. In: Proceedings of 9th International Science and Conference on Physics and Control. Innopolis, Russia, pp. 96–99 (2019)
22. Goel, P.: Insulin resistance or hypersecretion? The βIG picture revisited. J. Theor. Biol. **384**, 131–139 (2015)
23. González-Olivares, E., Mena-Lorca, J., Rojas-Palma, A., Flores, J.D.: Dynamical complexities in the Leslie-Gower predator-prey model as consequences of the Allee effect on prey. Appl. Math. Model. **35**, 366–381 (2011)
24. Han, M.A., Tian, Y., Yu, P.: Small-amplitude limit cycles of polynomial Liénard systems. Sci. China Math. **56**, 1543–1556 (2013)
25. Laalej, H., Lang, Z.Q., Daley, S., Zazas, I., Billings, S.A., Tomlinson, G.R.: Application of non-linear damping to vibration isolation: an experimental study. Nonlin. Dyn. **69**, 409–421 (2012)
26. Lamontagne, Y., Coutu, C., Rousseau, C.: Bifurcation analysis of a predator-prey system with generalized Holling type III functional response. J. Dyn. Diff. Equat. **20**, 535–571 (2008)
27. Li, Y., Xiao, D.: Bifurcations of a predator-prey system of Holling and Leslie types. Chaos Solit. Fract. **34**, 606–620 (2007)
28. Lins, A., de Melo, W., Pugh, C.C.: On Liénard's equation. Lecture Notes in Mathematics, vol. 597. Springer, Berlin, pp. 335–357 (1977)
29. Lloyd, N.G.: Liénard systems with several limit cycles. Math. Proc. Cambridge Philos. Soc. **102**, 565–572 (1987)
30. Owens, D.B., Capone, F.J., Hall, R.M., Brandon, J.M., Chambers, J.R.: Transonic free-to-roll analysis of abrupt wing stall on military aircraft. J. Aircraft **41**, 474–484 (2004)
31. Perko, L.: Differential Equations and Dynamical Systems. Springer, New York (2002)
32. Rychkov, G.S.: The maximal number of limit cycles of the system $\dot{y} = -x,\ \dot{x} = y - \sum_{i=0}^{2} a_i\, x^{2i+1}$ is equal to two. Differ. Equ. **11**, 301–302 (1975)
33. Savin, S., Khusainov, R., Klimchik, A.: Control of actuators with linearized variable stiffness. IFAC-PapersOnLine **52**, 713–718 (2019)
34. Slight, T.J., Romeira, B., Liquan, W., Figueiredo, J.M.L., Wasige, E., Ironside, C.N.A.: Liénard oscillator resonant tunnelling diode-laser diode hybrid integrated circuit: model and experiment. IEEE J. Quantum Electron. **44**, 1158–1163 (2008)
35. Shiriaev, A., Robertsson, A., Perram, J., Sandberg, A.: Periodic motion planning for virtually constrained Euler-Lagrange systems. Syst. Control Lett. **55**, 900–907 (2006)
36. Smale, S.: Mathematical problems for the next century. Math. Intelligencer **20**, 7–15 (1998)
37. Topp, B., Promislow, K., Devries, G., Miuraa, R.M., Finegood, D.T.: A model of β-cell mass, insulin, and glucose kinetics: pathways to diabetes. J. Theor. Biol. **206**, 605–619 (2000)
38. Zhu, H., Campbell, S.A., Wolkowicz, G.S.K.: Bifurcation analysis of a predator-prey system with nonmonotonic functional response. SIAM J. Appl. Math. **63**, 636–682 (2002)

Chapter 7
Topological Shooting of Solutions for Fickian Diffusion into Core-Shell Geometry

T. G. de Jong and A. E. Sterk

Abstract King et al. (2019) introduced a model for Fickian diffusion into core-shell geometry. The purpose of this model is to study diffusion of oxygen through protective shells encapsulating pancreatic Langerhan islets. These core-shells are of interest for the preparation of artificial pancreas to treat diabetes. In this paper we prove the existence of viable core-shell solutions for King's model using a topological shooting method. The governing equations of the diffusion model can be reduced to a 2-dimensional non-autonomous first order ordinary differential equation. Solutions which correspond to viable core-shell diffusion are required to satisfy global constraints and boundary conditions in both the core and the encapsulating shell. These boundary conditions each give rise to one free parameter. We call solutions satisfying the core boundary condition core solutions. We identify two parameter spaces corresponding to core solution families. The viable core-shell solutions are on the boundary of these two core solution families. Using analytically obtained bounds we apply the intermediate value theorem to prove the existence of core-shell solutions. In addition, we obtain rigorous approximations for the boundary conditions of the viable diffusion core-shell solution.

7.1 Introduction

Type 1 diabetes is a chronic disease that is characterized by the autoimmune system destroying the insulin producing pancreatic langerhan islets. Treatment requires monitoring and maintaining insulin levels. Insulin has to be entered externally into

---

T. G. de Jong
Media Analytics and Computing Laboratory, School of Information Science and Engineering, Xiamen University, Xiamen, 361005, China
e-mail: t.g.de.jong.math@gmail.com

A. E. Sterk (✉)
Bernoulli Institute, University of Groningen, PO Box 407, 9700 AK Groningen, The Netherlands
e-mail: a.e.sterk@rug.nl

A. K. Abramian et al. (eds.), *Nonlinear Dynamics of Discrete and Continuous Systems*, Advanced Structured Materials 139,
https://doi.org/10.1007/978-3-030-53006-8_7

the body by means of injections or an insulin pump. Transplantation of pancreatic Langerhan islets might provide a new medical solution for type 1 diabetes patients. The transplanted islets need to be protected from the host immune system while being able to absorb nutrients and secrete insulin. Encapsultation by an alginate membrane seems promising since alginates can be fabricated to selectively diffuse or block certain molecules [7]. In addition, alginates are relatively inert when exposed to mammalian cells [1].

King et al. [9] proposed a mathematical model for Fickian diffusion into core-shell geometry. Specifically, their model describes the diffusion of oxygen through protective shells encapsulating a core of donor cells with the aim of avoiding hypoxia within the core. These core-shells are of interest for the preparation of an artificial pancreas to treat type 1 diabetes. The governing equations of the diffusion model are separated into a core and a shell. Under the assumption of radial symmetry, both diffusion models are described by a 2-dimensional system of 1st-order ODEs. Viable solutions of this model are solutions for which the oxygen levels within the core are above the hypoxia threshold. The existence of such solutions was shown by King et al. using non-rigorous numerics. In addition, their results suggest that these solutions are meagre in the phase space.

The aim of this paper is to prove rigorous results for the diffusion model of King et al. For biologically realistic parameter values taken from [2, 3, 10] we prove that there exists a solution which corresponds to viable diffusion of oxygen in the core-shell. More specifically, the boundary conditions imposed by the model gives rise to one free parameter on each boundary. We call solutions satisfying the core boundary condition core solutions. A topological shooting method is used to shoot these core solutions to the boundary conditions of the shell. For general theory on topological shooting we refer to [8] and for examples of topological shooting we refer to [5, 12]. Furthermore, a non-rigorous approximation of the viable solution is computed numerically to check our rigorous work.

7.2 Model

We briefly review the reaction-diffusion transport model of King et al. [9]. The core-shell is comprised of a core which consists of pancreatic Langerhan islets and a shell which consists of an protective alginate membrane. Based on experimental observations it is assumed that the islets and alginate encapsulation shape are described by a radial variable, see Fig. 7.1.

Assuming Fickian diffusion and Michaelis-Menten consumption a PDE can be derived for the concentration within the core and shell. The assumption that the concentration is described by a radially symmetric steady state solution reduces the concentration equations to an ODE. In [9] the equations are derived for the dimensionless variables. The dimensionless independent radial variable is denoted by ρ. The dependent variables are given by the dimensionless oxygen concentration

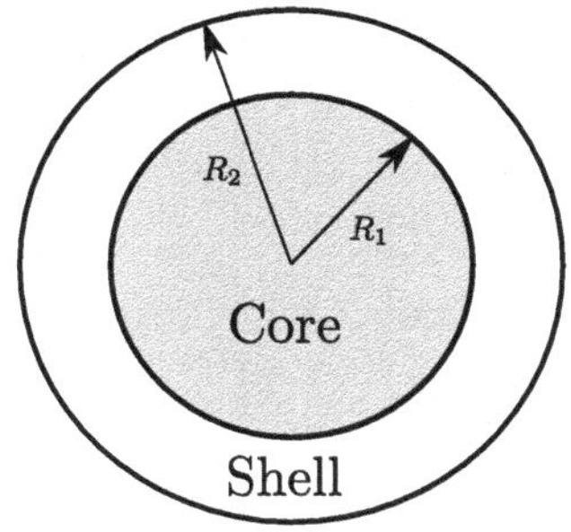

Fig. 7.1 Core-shell geometry. The donor cells are represented by the core and the protective alginate is represented by the shell. The core and shell are assumed to be spherical with outer-radius R_1 and R_2, respectively

for the interior and exterior which are denoted by χ_i and χ_e, respectively. Following [9] the resulting differential equations are

$$\begin{aligned} \frac{2}{\rho}\chi_i' + \chi_i'' - \frac{\nu\chi_i}{\kappa + \chi_i} &= 0, \text{ for } 0 < \rho < \rho_1, \\ \frac{2}{\rho}\chi_e' + \chi_e'' &= 0, \text{ for } \rho_1 < \rho < 1, \end{aligned} \tag{7.1}$$

where $\rho_1 = \frac{R_1}{R_2}$ with R_1, R_2 as given in Fig. 7.1, ν and κ are positive (dimensionless) parameters for Michaelis-Menten consumption and the prime denotes the derivative with respect to ρ. The boundary conditions are given by

$$\begin{aligned} &\lim_{\rho\to 0} \chi_i(\rho) = a > 0, \qquad \lim_{\rho\to 0} \chi_i'(\rho) = 0, \qquad \lim_{\rho\to 1} \chi_e(\rho) = 1, \\ &\lim_{\rho\to \rho_1} \chi_i(\rho) = \lim_{\rho\to \rho_1} \chi_e(\rho), \quad \eta \lim_{\rho\to \rho_1} \chi_i'(\rho) = \lim_{\rho\to \rho_1} \chi_e'(\rho), \end{aligned} \tag{7.2}$$

where a is a free parameter and η is the diffusion coefficient quotient of the interior divided by the exterior. Oxygen deprivation of the cells inside the core occurs if χ_i is below or equal to $\chi_* := 4.1 \times 10^{-3}$. In addition, χ_i cannot exceed the dimensionless concentration at the shell's external boundary. Consequently, we obtain the following global condition:

$$\chi_* < \chi_i(\rho) < 1 \qquad \forall \rho \in (0, \rho_1). \tag{7.3}$$

The aim of this paper is to prove that given parameters the ν, κ, η, and ρ_1 there exists a solution of the boundary value problem (7.1), (7.2) satisfying (7.3).

Definition 7.2.1 (*Viable core-shell solution*) A solution pair (χ_i, χ_e) of the ODE (7.1) is called a viable core-shell solution if it satisfies boundary conditions (7.2) and global condition (7.3).

In [9] viable core-shell solutions are computed numerically.

7.3 Main Theorem

Our main result concerns the existence of core-shell solutions for biologically realistic parameters:

Theorem 7.3.1 (Existence viable core-shell solution) *For* $\rho_1 = 1/2$, $\eta = 22/13$, $\nu = 9$, $\kappa = 5 \cdot 10^{-3}$ *the ODE* (7.1) *has a viable core-shell solution* (χ_i, χ_e) *as given by Definition 7.2.1. Furthermore,* χ_i *satisfies*

$$\lim_{\rho \to 0} \chi_i(\rho) \in \left[\frac{39}{100}, \frac{42}{100}\right], \quad \lim_{\rho \to \rho_1} \chi_i(\rho) \in \left[\frac{312}{425}, \frac{159}{200}\right].$$

Appendix provides the experimental support for the parameter values in Theorem 7.3.1.

In the remainder of this section Theorem 7.3.1 will be proven. We first reduce the governing equations (7.1) and define a dynamical system with a reduced phase space. Then, we define and prove the existence of so-called core solutions. These core solutions will be used to shoot the viable core-shell solution, Definition 7.2.1. Finally, we formulate the shooting method and give analytical bounds on core solutions which will allow for the implementation of the shooting method.

7.3.1 Reduction Governing Equations

The solution χ_e in (7.1) with boundary condition $\lim_{\rho \to 1} \chi_e(\rho) = 1$ from (7.2) can be solved analytically:

$$\chi_e(\rho) = 1 - b + \frac{b}{\rho}, \tag{7.4}$$

where b is a free parameter. Note that Definition 7.2.1 imposes a condition on the sign of b:

Proposition 7.3.1 *If* (χ_i, χ_e) *is a viable core-shell solution as given by Definition 7.2.1, then* $b < 0$ *in* (7.4).

Proof Observe that $\lim_{\rho \to \rho_1} \chi_i(\rho) = \lim_{\rho \to \rho_1} \chi_e(\rho)$ in (7.2) and the upper bound in (7.3) can only be satisfied if $b < 0$. □

The boundary value problem corresponding to (7.1) and (7.2) can be reformulated as a boundary value problem involving only the variable χ_i. For notational convenience denote χ_i by x. In preparation of the dynamical systems analysis we formulate the differential equations corresponding to x given in (7.1) as the first order ODE

$$x' = y,$$
$$y' = \frac{\nu x}{\kappa + x} - \frac{2}{\rho} y, \tag{7.5}$$

with phase space given by

$$M_0 = \{(x, y) \in \mathbb{R}_{>0} \times \mathbb{R}\}.$$

In the reduced setting the boundary conditions (7.2) take the form

$$\lim_{\rho \to 0} x(\rho) = a > 0, \qquad \lim_{\rho \to 0} y(\rho) = 0, \tag{7.6}$$

$$x(\rho_1) = 1 + \frac{b(1-\rho_1)}{\rho_1}, \qquad y(\rho_1) = -\frac{b\eta}{\rho_1^2}. \tag{7.7}$$

The boundary condition (7.7) was obtained by substituting (7.4) into (7.2). The boundary conditions (7.6), (7.7) give the free parameters a and b.

If (x, χ_e) is a viable core-shell solution then $(x, y) = (x, x')$ is a solution of (7.5) satisfying (7.6), (7.7) and

$$\chi_* < x(\rho) < 1, \quad \forall \rho \in (0, \rho_1). \tag{7.8}$$

Hence, our analysis will be restricted to (7.5). For notational convenience we will denote a solution (x, y) of (7.5) by $\mathbf{x}$. We introduce an equivalent definition for Definition 7.2.1.

Definition 7.3.1 (*Reduced core-shell solution*) A solution $\mathbf{x}$ of (7.5) is called a reduced core-shell solution if it satisfies boundary conditions (7.6), (7.7) and global condition (7.8).

7.3.2 Reduced Phase Space

We will consider (7.5) with phase space

$$M = \{(x, y) \in \mathbb{R}_{>0} \times \mathbb{R}_{>0}\} \subset M_0. \tag{7.9}$$

In Sect. 7.3.4 we will show that the reduced core-shell solution, Definition 7.3.1, exists in M. From an analysis perspective it is easier to work in M since solution curves in M cannot change sign. In addition, we have that:

Lemma 7.3.1 *Let* $\mathbf{x}$ *be a solution of* (7.5) *satisfying* $\mathbf{x}(\rho_0) \in M$ *with* $\rho_0 > 0$. *Then* $\mathbf{x}(\rho) \in M$ *for all* $\rho \in (\rho_0, \infty)$.

Proof Let $\rho_s > \rho_0$ be arbitrary. For any C^2-function $x : [\rho_0, \rho_s] :\to \mathbb{R}$ we define its *defect* as

$$Px = x'' - f(\rho, x, x') \quad \text{where} \quad f(\rho, x, x') = \frac{\nu x}{\kappa + x} - \frac{2}{\rho}x'.$$

Note that f is increasing in x and locally Lipschitz in both x and x'.

Assume that $\mathbf{x} = (x, y)$ is a solution of (7.5) such that $x(\rho_0) > 0$ and $x'(\rho_0) = y(\rho_0) > 0$. Define the linear function

$$w(\rho) = \varepsilon(\rho - \rho_0) + x(\rho_0),$$

where

$$0 < \varepsilon \leq \min\left\{x'(\rho_0), \frac{\rho_0}{2} \cdot \frac{\nu x(\rho_0)}{\kappa + x(\rho_0)}\right\}.$$

Then, for $\rho \in [\rho_0, \rho_s]$ we have that

$$f(\rho, w, w') \geq \frac{\nu w(\rho_0)}{\kappa + w(\rho_0)} - \frac{2\varepsilon}{\rho_0} \geq 0.$$

In conclusion, we have the following inequalities:

$$w(\rho_0) \leq x(\rho_0), \quad w'(\rho_0) \leq x'(\rho_0), \quad Pw \leq 0 = Px \quad \forall \rho \in [\rho_0, \rho_s].$$

The Comparison Theorem [14, Theorem 11.XVI] implies that

$$x(\rho) \geq w(\rho) > 0 \quad \text{and} \quad y(\rho) = x'(\rho) \geq w'(\rho) = \varepsilon > 0 \quad \forall \rho \in [\rho_0, \rho_s].$$

This shows that $\mathbf{x}(\rho) \in M$ for all $\rho \in [\rho_0, \rho_s]$. Since $\rho_s > \rho_0$ is arbitrary, the proof is complete. □

7.3.3 Core Solutions

We will apply topological shooting to solutions that satisfy the local conditions corresponding to $\rho \to 0$ in (7.6). Consequently, we introduce:

Definition 7.3.2 (*Core solution*) A solution $\mathbf{x}$ of (7.5) is called a core solution if it satisfies (7.6). A core solution $\mathbf{x} = (x, y)$ satisfying $\lim_{\rho \to 0} x(\rho) = a > 0$ is denoted by $\mathbf{x}_a$.

We will show that the range of $\mathbf{x}_a$ corresponds to an invariant manifold. Observe that the vector field corresponding to (7.6) is not defined for $\rho \to 0$. We will introduce variables such that the new governing equations are autonomous and have equilibria corresponding to the limits in (7.6).

7.3.3.1 Change of Variables

Define the new independent variable t by setting $\rho = e^t$ so that the limit $\rho \to 0$ corresponds to the limit $t \to -\infty$. Denote by the sup-dot the derivative with respect to t. If $\hat{x}(t) = x(e^t)$, $\hat{y}(t) = e^t y(e^t)$, then

$$\begin{aligned}\dot{\hat{x}}(t) &= e^t y(e^t) \\ &= \hat{y}(t), \\ \dot{\hat{y}}(t) &= e^t y(e^t) + e^{2t} y'(e^t) \\ &= -\hat{y} + \frac{\nu \hat{x} e^{2t}}{\kappa + \hat{x}}.\end{aligned}$$

We introduce the variable $\hat{z} = e^{2t}$. Hence, the first order ODE (7.5) can be written as an autonomous system:

$$\begin{aligned}\dot{\hat{x}} &= \hat{y}, \\ \dot{\hat{y}} &= -\hat{y} + \frac{\nu \hat{x} \hat{z}}{\kappa + \hat{x}}, \\ \dot{\hat{z}} &= 2\hat{z}.\end{aligned} \tag{7.10}$$

It follows from (7.6), (7.7) and M in (7.9) that the new phase space becomes

$$N := \{(\hat{x}, \hat{y}, \hat{z}) \in \mathbb{R}_{>0} \times \mathbb{R}_{>0} \times (0, e^{2\rho_1})\}.$$

7.3.3.2 Formulation in Terms of Unstable Manifolds

Observe that the limit in (7.6) corresponds to

$$\lim_{t \to -\infty} (\hat{x}, \hat{y}, \hat{z})(t) = (a, 0, 0).$$

We have that $q(a) := (a, 0, 0) \notin N$. Therefore, we consider the closure of N as phase space which will be denoted by $\overline{N}$. The Jacobian matrix of (7.10) evaluated at $q(a)$ has eigenvalues $\lambda_1 = 0$, $\lambda_2 = -1$, and $\lambda_3 = 2$. The eigenvector corresponding to λ_1 and λ_3 will be of importance:

$$\mathbf{v}_1 = (1, 0, 0)^T, \quad \mathbf{v}_3 = \left(\frac{a\nu}{6(a+\kappa)}, \frac{a\nu}{3(a+\kappa)}, 1 \right)^T. \tag{7.11}$$

Denote the unstable, stable and center manifold corresponding to $q(a)$ by $W^u(q(a))$, $W^s(q(a))$ and $W^c(q(a))$. The invariant manifolds $W^u(q(a))$, $W^s(q(a))$, $W^c(q(a))$ are one dimensional. Denote by $\mathcal{W}^c$ the union of all center manifolds.

Lemma 7.3.2 *We have that* $\mathcal{W}^c = \{(\hat{x}, 0, 0) : \hat{x} > 0\}$.

Proof Observe that $W^c(q(a))$ in the $\hat{z}$-direction is 0 since otherwise $W^c(q(a))$ would exhibit dynamics corresponding to $W^u(q(a))$. Then, $W^c(q(a))$ must also equal zero in the $\hat{y}$-direction since otherwise $W^c(q(a))$ would exhibit dynamics corresponding to the stable manifold $W^s(q(a))$. Hence, we have that $W^c(q(a)) \subset \{(\hat{x}, 0, 0) \in \mathbb{R}^3 : \hat{x} > 0\}$. Consequently, we obtain that $\mathcal{W}^c = \{(\hat{x}, 0, 0) \in \mathbb{R}^3 : \hat{x} > 0\}$. □

The invariant manifold $W^u(q(a))$ will be used to construct core solutions. We define $\Psi : N \to M \times \mathbb{R}_{>0}$ given by

$$\Psi(\hat{x}, \hat{y}, \hat{z}) = \left((\hat{x}, \hat{y}/\sqrt{\hat{z}}), \sqrt{\hat{z}}\right).$$

Lemma 7.3.3 *There exists a local unstable manifold* $W^u_{\text{loc}}(q(a))$ *such that* $W^u_{\text{loc}}(q(a)) \cap N$ *is non-empty and connected.*

Proof The components of $\mathbf{v}_3$ have the same sign. Hence, any $W^u_{\text{loc}}(q(a))$ intersects N. It follows from the smoothness of $W^u_{\text{loc}}(q(a))$ at $q(a)$ that sufficiently close to $q(a)$ the manifold $W^u_{\text{loc}}(q(a))$ is connected. □

Take $W^u_{\text{loc}}(q(a))$ as in Lemma 7.3.3. Denote the invariant manifold generated from $W^u_{\text{loc}}(q(a)) \cap N$ using the forward flow by $W^u_+(q(a))$.

Lemma 7.3.4 $W^u_+(q(a))$ *is non-empty and connected. Furthermore, we have that* $W^u_+(q(a)) \subset N$.

Proof From Lemma 7.3.2 it follows that $W^u_+(q(a))$ is non-empty. From Lemma 7.3.1 and Lemma 7.3.2 it follows that $W^u_+(q(a))$ is connected. The manifold $W^u_+(q(a))$ is generated from $W^u_{\text{loc}}(q(a)) \cap N$. Hence, using Lemma 7.3.1 we get that $W^u_+(q(a)) \subset N$. □

Theorem 7.3.2 $\{(\mathbf{x}(\rho), \rho) : \rho \in (0, \rho_1)\} = \Psi(W^u_+(q(a))$ *if and only if* $\mathbf{x} = \mathbf{x}_a$ *is a core solution as given in Definition 7.3.2.*

Proof ($\Rightarrow$) Observe that we only need to prove that $\mathbf{x} = (x, y)$ satisfies (7.6). It follows from Lemma 7.3.4 that $W^u_+(q(a))$ is a non-empty, connected invariant manifold. Hence, there exists a solution $(\hat{x}, \hat{y}, \hat{z})$ with range $W^u_+(q(a))$. Observe that $\lim_{\rho\to 0} x(\rho) = \lim_{t\to-\infty} \hat{x}(t) = 0$. We have that

$$\lim_{\rho\to 0} y(\rho) = \lim_{t\to 0} \frac{\hat{y}(t)}{\sqrt{\hat{z}(t)}}.$$

From $\mathbf{v}_3$ it follows that $\frac{a\nu}{3(a+\kappa)}\hat{y}(t) \sim \hat{z}(t)$ for $t \ll 1$. Hence, we have that

$$\lim_{t\to 0} \frac{\hat{y}(t)}{\sqrt{\hat{z}(t)}} = 0.$$

($\Leftarrow$) Since $\mathbf{x}$ is a core solution and since $t \to -\infty$ is equivalent to $\rho \to 0$ one of the following must be true:

(i) $\Psi(W^c_{\text{loc}}(q(a)) \cap N) \subset \{(\mathbf{x}(\rho), \rho) : \rho \in (0, \rho_1)\}$,
(ii) $\Psi(W^u_{\text{loc}}(q(a)) \cap N) \subset \{(\mathbf{x}(\rho), \rho) : \rho \in (0, \rho_1)\}$.

From Lemma 7.3.2 it follows that (i) is not true. Hence, (ii) is true. The global version of $W^u_{\text{loc}}(q(a))) \cap N$ is given by $W^u_+(q(a))$. From Lemma 7.3.4 we get that $W^u_+(q(a)) \subset N$. Consequently, we obtain that $\{(\mathbf{x}(\rho), \rho) : \rho \in (0, \rho_1)\} = \Psi(W^u_+ (q(a)))$. □

Denote by $\mathcal{W}^{cu}$ the center unstable manifold induced by $\mathcal{W}^c$.

Lemma 7.3.5 *There exists a smooth map* $f : \mathbb{R}_{>0} \times [0, e^{2\rho_1}) \to \mathbb{R}^2$ *satisfying* $(f(a, 0), 0) = q(a)$ *and such that*

$$\mathcal{W}^{cu} = \{(f(a, z), z) \in \overline{N} : (a, z) \in \mathbb{R}_{>0} \times [0, e^{2\rho_1})\}.$$

Proof To avoid working with clopen sets we first extend the phase space for (7.10) to $\hat{N} := \{(\hat{x}, \hat{y}, \hat{z}) \in \mathbb{R}_{>0} \times \mathbb{R} \times (-e^{2\rho_1}, e^{2\rho_1})\}$. We will show that:

Claim: There exists a neighbourhood $U \subset \mathbb{R}^3$ of $q(a)$, a neighbourhood $V \subset \mathbb{R}^2$ of $(a, 0)$, and a smooth $f_{\text{loc}} : V \to \mathbb{R}^2$ satisfying

$$\mathcal{W}^{cu} \cap U = \{(f_{\text{loc}}(a, z), z) \in \hat{N} : (a, z) \in V\},$$
$$(f_{\text{loc}}(a, 0), 0) = q(a) \quad \forall (a, 0) \in V.$$

From the claim it follows that the parametrization exists in a local neighbourhood of $q(a)$. The local parametrization can be extended using transitivity of the flow and the property that solutions of the system (7.10) satisfying $\hat{z} \neq 0$ can be smoothly parametrized in the $\hat{z}$-variable. Which proves the lemma. To show the claim observe that the tangent space of $W^{cu}(q(a))$ at $q(a)$ is $\text{span}(\mathbf{v}_1, \mathbf{v}_3)$. The $\hat{x}$-component of $\mathbf{v}_1$ is non-zero and the $\hat{z}$-component of $\mathbf{v}_3$ is non-zero. Hence, using the implicit function theorem the claim follows. □

Corollary 7.3.1 *For all* $a \in \mathbb{R}_{>0}$ *there exists a unique core solution* $\mathbf{x}_a$ *specified by Definition 7.3.2. In addition, the map* $G : a \mapsto \mathbf{x}_a$ *is continuous in* a.

Proof Uniqueness of $\mathbf{x}_a$ follows from Theorem 7.3.2. Consider $\mathcal{W}^{cu} \cap N$ foliated by solutions curves, applying the transformation Ψ and using Lemma 7.3.5 gives that the map $G : a \mapsto \mathbf{x}_a$ is continuous in a. □

7.3.4 Shooting Method

We consider a closed connected $U \subset M$. We use the boundary conditions at $\rho = \rho_1$ given by (7.7) to divide U into three subsets given by

$$\begin{aligned} A(U) &:= \left\{(x, y) \in U \ : \ x < 1 - \frac{y\rho_1(1-\rho_1)}{\eta}\right\}, \\ B(U) &:= \left\{(x, y) \in U \ : \ x > 1 - \frac{y\rho_1(1-\rho_1)}{\eta}\right\}, \\ X(U) &:= U \backslash (A(U) \cup B(U)) . \end{aligned} \tag{7.12}$$

We then arrive at the shooting lemma:

Lemma 7.3.6 *Let $\mathbf{x}_a = (x_a, y_a)$ be a core solution as given by Definition 7.3.2. If there exist $a_1, a_2 \in [\chi_*, 1]$ with $a_1 < a_2$, $U \subset M$ such that*

$$\mathbf{x}_{a_1}(\rho_1) \in A(U), \quad \mathbf{x}_{a_2}(\rho_1) \in B(U), \tag{7.13}$$

$$\mathbf{x}_a(\rho_1) \in U, \tag{7.14}$$

then there exists an $a_* \in (a_1, a_2)$ such that $\mathbf{x}_{a_*}(\rho_1) \in X(U)$ and $\mathbf{x}_{a_*}$ is a reduced core-shell solution as specified by Definition 7.3.1.

The conditions in Lemma 7.3.6 are visualised in Fig. 7.2. In the next section Lemma 7.3.6 will be applied by computing explicit bounds on a domain.

Proof Theorem 7.3.2 implies that $G : a \mapsto \mathbf{x}_a$ is continuous. Hence, it follows from applying the intermediate value theorem and using (7.13), (7.14) that there exists an

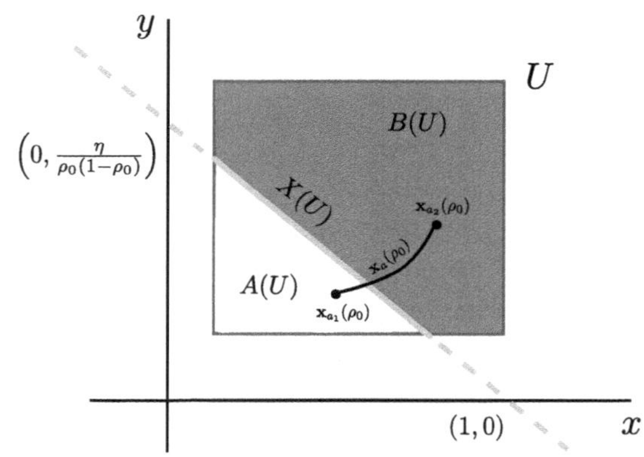

Fig. 7.2 Conditions of Lemma 7.3.6 for an example set. U is given by the rectangle which is subdivided into: $A(U)$, white area, $B(U)$, dark grey, and $X(U)$, light grey line. In addition, we have continued $X(U)$ beyond U with a dashed line to show the intersection points with axes. Observe that the core solutions $\mathbf{x}_a$ with $a \in [a_1, a_2]$ satisfy the conditions (7.13), (7.14) of Lemma 7.3.6

$a_* \in (\chi_*, 1)$ such that $\mathbf{x}_{a_*}(\rho_1) \in X(U)$. Then, $\mathbf{x}_{a_*}$ satisfies the boundary condition (7.7). Observe that $\lim_{\rho\to 0} x_a(\rho) \geq \chi_*$ for all $a \in (a_1, a_2)$. Since $\mathbf{x}_{a_*}(\rho_1) \in X(U)$, it follows that $x_{a_*}(\rho_1) < 1$. Then, using the increasing monotonicity on x_a from Lemma 7.3.1 it follows that $\mathbf{x}_{a_*}$ satisfies the global condition (7.8). Consequently, $\mathbf{x}_{a_*}$ is a reduced core-shell solution. □

We will compute analytical bounds on $\mathbf{x}_a = (x_a, y_a)$. For the parameters of Theorem 7.3.1 these bounds allow us to apply Lemma 7.3.6. We define

$$\mathcal{B}(a, \rho) := \left\{(x, y) \in \mathbb{R}^2 \ : \ \frac{\nu a \rho^2}{6(\kappa + a)} \leq x - a \leq \frac{\nu\rho^2}{6}, \frac{\nu a \rho}{3(\kappa + a)} \leq y \leq \frac{\nu\rho}{3}\right\}.$$

Lemma 7.3.7 *We have that* $\mathbf{x}_a(\rho) \in \mathcal{B}(a, \rho)$.

Proof Consider the term $\frac{x_a}{\kappa + x_a}$, using Lemma 7.3.1 we obtain an under bound for $\frac{x_a}{\kappa + x_a}$ and using the positivity of x_a and κ we obtain an upper bound:

$$\frac{a}{\kappa + a} \leq \frac{x_a}{\kappa + x_a} \leq 1. \tag{7.15}$$

From the y-equation in (7.5) it follows that

$$(y_a\rho^2)' = \frac{\nu x_a \rho^2}{\kappa + x_a}. \tag{7.16}$$

Applying the bounds (7.15) to (7.16) we obtain the bounds:

$$\frac{\nu a \rho}{3(\kappa + a)} \leq y_a \leq \frac{\nu\rho}{3}. \tag{7.17}$$

From the x-equation in (7.5) and using (7.17) we obtain:

$$\frac{\nu a \rho^2}{6(\kappa + a)} \leq x_a - a \leq \frac{\nu\rho^2}{6}. \tag{7.18}$$

□

Proof (Theorem 7.3.1) To apply Lemma 7.3.6 we will show the following:

Claim. Given the parameters of Theorem 7.3.1 there exist $a_1, a_2 \in [\chi_*, 1]$ such that

$$\chi_* < x \quad \forall (x, y) \in U. \tag{7.19}$$
$$\mathcal{B}(a_1, 1/2) \subset A(U), \quad \mathcal{B}(a_2, 1/2) \subset B(U), \tag{7.20}$$
$$\mathcal{B}(a, 1/2) \subset U, \quad \forall a \in [a_1, a_2], \tag{7.21}$$

It follows from Lemma 7.3.7 that the Claim proves Theorem 7.3.1. We continue with a proof of the claim. Take $a_1 = \frac{39}{100}, a_2 = \frac{42}{100}$ and $U = \left[\frac{312}{425}, \frac{159}{200}\right] \times \left[\frac{117}{170}, \frac{3}{4}\right]$. By choice of U we have that (7.19) is satisfied. The property (7.20) follows directly from the definition of $\mathcal{B}$. Finally, (7.21) can be shown by applying $\mathcal{B}$ on $[a_1, a_2] \times [1/2, 1/2]$ and using interval arithmetic to obtain bounds on the resulting domain. For readers unfamiliar with interval arithmetic we refer to [13]. □

7.4 Numerical Validation

The numerically computed viable core-shell solution corresponding to Theorem 7.3.1 is displayed in Fig. 7.3. Recall $A(U)$, $B(U)$, $X(U)$ defined in (7.12) and visualized in Fig. 7.2. Shooting core solutions, Definition 7.3.2, we approximated $a \in X(U)$ by using that a is on the boundary of $A(U)$ and $B(U)$. This gives the χ_i-solution. The numerical method is similar to those applied in [6, 8]. The connecting χ_e-solution was computed using (7.4).

The numerical solution satisfies $\lim_{\rho \to 0} \chi_i(\rho) \approx 0.4087 \in \left[\frac{39}{100}, \frac{42}{100}\right]$ and $\lim_{\rho \to 1/2} \chi_i(\rho) \approx 0.7802 \in \left[\frac{312}{425}, \frac{159}{200}\right]$ which is in accordance with the bounds in Theorem 7.3.1.

7.5 Conclusion

In this paper we considered a mathematical model for Fickian diffusion into core-shell geometry proposed by King et al. [9]. In particular, we proved the existence of a solution which corresponds to viable diffusion of oxygen in the core-shell for biologically realistic parameters. The main ingredient of the proof is an implementation of a fully analytical topological shooting method.

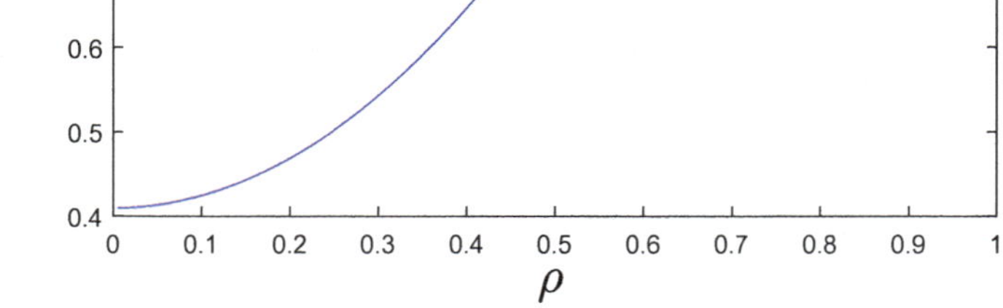

Fig. 7.3 Numerically computed viable core-shell solution (χ_i, χ_e)

The analytically computed bounds from Lemma 7.3.7 give a good approximation of solutions for application of the shooting lemma, Lemma 7.3.6, since κ in Theorem 7.3.1 is relatively small compared to the bounds on $\lim_{\rho\to 0} \chi_i(\rho)$ in Theorem 7.3.1. For κ outside this regime rigorous numerics could be used to approximate core solutions for application of the shooting lemma.

The model by King et al. [9] only considers dynamics of the oxygen concentration inside the core-shell. Biologically, the consumption of glucose and production of insulin are important to obtain an accurate model for an artificial pancreas. Buchwald [4] presents a local glucose-and oxygen concentration-based insulin secretion model for pancreatic islets. The model by Buchwald could be considered in the core-shell setting. This new model would be an extension of the model by King et al. The governing equations will be higher dimensional and will have more parameters. Hence, we expect that this will lead to more complex dynamics. Both the analytical and numerical investigation of such a model is an interesting topic for future research.

Appendix: Parameters

We show that the parameters in Theorem 7.3.1 correspond to the parameters in an experimental setting. From the experimental results in [10] we observe that the following radii are viable: $R_1 = 135\,\mu\text{m}$, $R_2 = 270\,\mu\text{m}$. Consequently, we have that $\rho = R_1/R_2 = 1/2$. For the diffusion coefficient of the Langerhan islets we have that $D_i = 1.3 \cdot 10^{-9}\,\text{m}^2/\text{s}$ [2] and for the diffusion coefficient of the alginate (in vivo, intraperitoneal) we have that $D_e = 2.2 \cdot 10^{-9}\,\text{m}^2/\text{s}$ [11]. Hence, we obtain that $\eta = \frac{22}{13}$. The dimensionless Michaelis-Menten parameters are determined from

$$\nu = \frac{V_m R_2^2}{D_i C_2}, \qquad \kappa = \frac{K_m}{C_2}, \tag{7.22}$$

with V_m the Michaelis-Menten maximum consumption rate, K_m the Michaelis constant and C_2 the ambient oxygen concentration. For a derivation of (7.22) see [9]. We consider the following parameters used by [3]: $V_m = 3.4 \cdot 10^{-2}\,\text{mol/m}^3\text{s}$, $K_m = 1.0 \cdot 10^{-3}\,\text{mol} \cdot \text{m}^{-3}$, $C_2 = 0.20\,\text{mol/m}^3$. Hence, we obtain that $\nu \approx 9$ and $\kappa = 5 \cdot 10^{-3}$.

References

1. Augst, A.D., Kong, H.J., Mooney, D.J.: Alginate hydrogels as biomaterials. Macromol. Biosci. **8**(6), 623–633 (2006)
2. Avgoustiniatos, E.S., Dionne, K.E., Wilson, D.F., Yarmush, M.L.: Measurements of the effective diffusion coefficient of oxygen in pancreatic islets. Ind. Eng. Chem. Res. **46**(19), 6157–6163 (2007)

3. Buchwald, P.: FEM-based oxygen consumption and cell viability models for avascular pancreatic islets. Theor. Biol. Med. Model, **6**(5) (2009)
4. Buchwald, P.: A local glucose-and oxygen concentration-based insulin secretion model for pancreatic islets. Theor. Biol. Med. Model **8**(20) (2011)
5. De Jong, T.G., Sterk, A.E., Broer, H.W.: Fungal tip growth arising through a codimension-1 bifurcation. Int. J. Bifurc. Chaos (2020)
6. De Jong, T.G., Sterk, A.E., Guo, F.: Numerical method to compute hypha tip growth for data driven validation. IEEE Access **7**, 53766–53776 (2019)
7. Goosen, M.F., O'Shea, G.M., Gharapetian, H.M., Chou, S., Sun, A.M.: Optimization of microencapsulation parameters: semipermeable microcapsules as a bioartificial pancreas. Biotechnol. Bioeng. **2**(27) (1985)
8. Hastings, S.P., McLeod, J.B.: Classical Methods in Ordinary Differential Equations. AMS, Rhode Island (2012)
9. King, C.C., Brown, A.A., Sargin, I., Bratlie, K.M., Beckman, S.P.: Modelling of reaction-diffusion transport into core-shell geometry. J. Theor. Biol. **260**, 204–208 (2019)
10. Ma, M., Chiu, A., Sahay, G., Doloff, J.C., Dholakia, N., Thakrar, R., Cohen, J., Vegas, A., Chen, D., Bratlie, K.M., Dang, T., York, R.L., Hollister-Lock, J., Weir, G.C., Anderson, D.G.: Core-shell hydrogel microcapsules for improved islets encapsulation. Adv. Healthc. Mater. **2**(5) (2013)
11. Najdahmadi, A., Lakey, J.: Diffusion coefficient of alginate microcapsules used in pancreatic islet transplantation, a method to cure type 1 diabetes. In: Proceedings of SPIE (2018)
12. Peletier, L.A., Troy, W.C.: A topological shooting method and the existence of kinks of the extended Fisher-Kolmogorov equation. Topol. Methods Nonlinear Anal. **6**(2), 331–355 (1995)
13. Tucker, W.: Validated Numerics: A Short Introduction to Rigorous Computations. Princeton University Press (2011)
14. Walter, W.: Ordinary Differential Equations. Springer (1998)

Chapter 8
The Dynamic Interactions and Control of Long Slender Continua and Discrete Inertial Components in Vertical Transportation Systems

Stefan Kaczmarczyk

Abstract Dynamic phenomena such as transient and steady-state resonant vibrations in vertical transportation systems deployed to move goods and passengers in the modern built environment affect the performance of the entire installation. In extreme high rise structures traction drive elevator systems comprise long slender continua such as ropes and cables with discrete mass elements that exhibit low-frequency modes and nonlinear modal interactions. This results in the need to predict and control their non-linear stationary and non-stationary dynamic responses. The underlying causes of these dynamic responses/vibrations are varied. They include low frequency sway motions of the host structure induced by high winds and seismic activities. Consequently, conditions for external, parametric and autoparametric resonances can readily arise during the operation of such installations. In this context, a general approach to model the dynamic behaviour of a typical vertical transportation system is demonstrated. Subsequently, a mathematical model is developed which is solved numerically to predict the non-stationary/nonlinear dynamic responses. An active control strategy is then proposed to minimize the effects of adverse dynamic responses of the system.

8.1 Introduction

The design and operation of high-performance systems for vertical transportation in the modern built environment present many technical challenges due to adverse dynamic responses that arise due to various sources of excitation present in these systems. In the modern high-rise built environment excitations induced by winds and earthquake/ground motion can result in large responses of buildings and civil structures [1, 2]. The dynamic responses at low frequencies and large amplitudes then arise that induce complex resonance interactions affecting the performance of

S. Kaczmarczyk (✉)
The University of Northampton, University Drive, Northampton NN1 5PH, UK
e-mail: stefan.kaczmarczyk@northampton.ac.uk

A. K. Abramian et al. (eds.), *Nonlinear Dynamics of Discrete and Continuous Systems*,
Advanced Structured Materials 139,
https://doi.org/10.1007/978-3-030-53006-8_8

vertical transportation systems (VTS) deployed in tall structures [3–5]. Passive and active control strategies can then be applied to mitigate their effects [6, 7].

In this work a mathematical model to predict and to analyse the resonance behaviour of the tall host structure—VTS is presented. The VTS is equipped with a nonlinear damper/actuator 'tie-down' device. The performance of the installation is studied by numerical simulation. It is shown that the characteristics of the tie-down device can be adjusted to minimize the effects of adverse dynamic responses of the system. The active stiffness strategy is then proposed to minimize the effects of adverse dynamic responses of suspension and compensating ropes in VTS.

8.2 Mathematical Model

A VT system may be considered as an assemblage of axially moving elastic one-dimensional long slender continua (LSC) divided into $p = 1, 2, \ldots, P$ sections of *slowly* varying length [8, 9], constrained by discrete elements such as rigid-body masses and rotating inertia elements. Its response can be described by a system of nonlinear partial differential equations of the following form

$$\rho_s(x_p)\mathbf{U}^p_{,tt} + \mathbf{C}^p[\mathbf{U}^p_{,t}] + \mathbf{L}^p[\mathbf{U}^p] = \mathbf{N}^p[\mathbf{U}] + \mathbf{F}^p\big(x_p, t, \theta_p\big),$$
$$x_p \in \big\{0 < x_p < L_p(\tau)\big\},\ 0 \le t < \infty, \tag{8.1}$$

where the boundary conditions are given as

$$\mathbf{B}^p_1\big(\mathbf{U}^p\big) = 0 \text{ at } x_p = 0,\ \mathbf{B}^p_2\big(\mathbf{U}^p\big) = 0 \text{ at } x_p = L_p(\tau) \tag{8.2}$$

where x_p denotes the spatial co-ordinate, $\mathbf{U}^p(x_p, t) = \big[U^p_1(x_p, t), U^p_2(x_p, t), U^p_3(x_p, t)\big]$ is a local (component) dynamic displacement vector representing motion of the component p in the lateral and longitudinal directions, $()_{,t}$ designates partial derivatives with respect to time t, $\tau = \varepsilon t$ represents the slow time scale, where ε is a small parameter [9], and $\mathbf{C}^p$ and $\mathbf{L}^p$ are local linear operators. Furthermore, $\mathbf{N}^p$ is an operator acting upon the global displacement vector $\mathbf{U}$, and representing non-linear couplings and inter-component constraints in the system. $\mathbf{F}^p$ is a forcing function with harmonic terms of frequency $\dot{\theta}_p = \Omega_p$, where the overdot indicates total differentiation with respect to time. The local (component) mass distribution function is defined as

$$\rho_p\big(x_p\big) = m_p + \sum_{i=1}^{N_M} M_i\delta\big(x_p - L_p\big) \tag{8.3}$$

In the model given by Eq. 8.1 the Lagrangian coordinates or Eulerian coordinates may be applied as the spatial coordinate x_p. If the Lagrangian formulation is applied, then it is convenient to refer the dynamic elastic deformations of LSC to a moving frame associated with the overall axial transport motion of the system [9]. Otherwise, a fixed (inertial) frame is used to describe the deformations. In order to discretize the continuous slowly varying nonlinear system Eq. 8.1 the following expansion can be used

$$U_k^p(x_p, t) = \sum_{n=1}^{N_p} Y_n^k(x_p; L_p(\tau))q_n^p(t) \tag{8.4}$$

where $Y_n^k(x_p; L_p(\tau))$ is the nth eigenfunction of the corresponding linear system and $q_n^p(t)$ represent the nth modal coordinate. This expansion leads to the following first-order ordinary differential equation (ODE) system given as

$$\dot{\mathbf{y}}(t) = \mathbf{A}(t, \tau)\mathbf{y}(t) + \tilde{\mathbf{N}}(\tau, \mathbf{y}) + \tilde{\mathbf{F}}(t, \tau) \tag{8.5}$$

where $\mathbf{y}$ is the system state vector, $\mathbf{A}$ is a slowly varying linear coefficient matrix, $\tilde{\mathbf{N}}$ is a vector function which represents the non-linear coupling terms, and $\tilde{\mathbf{F}}$ is the external excitation vector. This system cannot be solved exactly. An approximate solution can be sought using asymptotic (perturbation) methods and/or numerical techniques. Alternatively, in some cases, the system of partial differential equations Eq. 8.1 with the boundary conditions given by Eq. 8.2 can be treated directly without discretization and perturbations methods (such as the method of multiple scales) can be applied to investigate the non-stationary behaviour of the system [10–13].

8.3 Vertical Transportation System—Traction Drive Elevator

In the modern high-rise built environment high-speed high-capacity traction drive elevator (lift) systems are used. A diagram which illustrates the dynamic model of a high-rise lift system is shown in Fig. 8.1. The modulus of elasticity, cross-sectional effective area and mass per unit length of the ropes are denoted as E_1, A_1, m_1 and E_2, A_2, m_2 for the compensating cables and the suspension ropes, respectively. The compensating cables are of length L_1 at the car side and the suspension ropes are of length L_2 at the counterweight side, respectively. The length of the suspension rope at the car side and the compensating rope at the counterweight side are denoted as L_3 and L_4, respectively. The lengths of suspension ropes and compensating cables are varying with the position of the car in the shaft (denoted by l_{car}). The masses and dynamic displacements of the car, counterweight and the compensating sheave assembly are represented by M_{car}, M_{cwt} and M_{comp}, q_1, q_2 and q_3, respectively. The compensating sheave rotational motion is represented by the angular coordinate θ and the second moment of inertia is I_{comp}. The compensating sheave assembly (CSA) is equipped with a damper/tie-down device. The building structure sway deformations due to ground motions $s_v(t)$, $s_w(t)$ are represented by the shape function

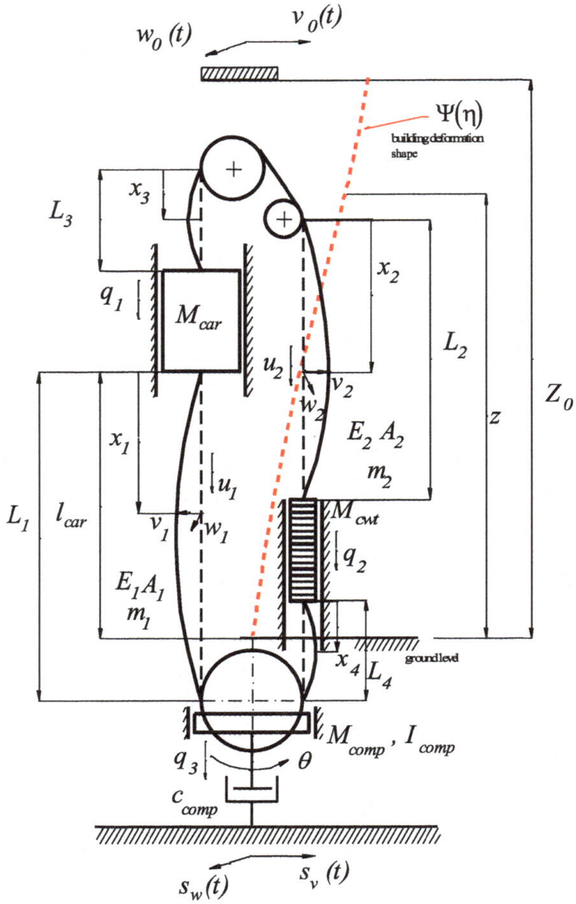

Fig. 8.1 Model of a high-rise VTS

$\Psi(\eta) = 3\eta^2 - 2\eta^3$, $\eta = z/Z_0$. Consider that the ground motions are harmonic of frequency Ω_1, Ω_2 in the in-plane direction and out-of-plane direction, respectively. The deformations then result in harmonic motions $v_0(t)$ and $w_0(t)$ at the top of the building structure, in the in-plane direction and out-of-plane direction, respectively.

The natural frequencies of the system change with the position of the car. An adverse situation arises when the building sways at its fundamental natural frequency which in turn is tuned to the natural frequency of the VTS, thus leading to resonance conditions. The resonance phenomena can be captured by the development of a suitable dynamic model. The model based on the formulation given in Eq. 8.1 is represented by Eq. 8.6, where V, a represent the speed and acceleration/deceleration of the car, $\bar{v}_i(x_i, t)$, $\bar{w}_i(x_i, t)$, $i = 1,2,\ldots, 4$, represent the dynamic displacements of the ropes, T_i, denote the rope quasi-static tension terms.

$$m_i\bar{v}_{itt} - \left\{T_i - m_i\left[V^2 + (g - a_i)x_i\right] + E_iA_ie_i\right\}\bar{v}_{ixx} + m_ig\bar{v}_{ix} + 2m_iV\bar{v}_{ixt} = F_i^v\left[t, L_i(t)\right],$$

$$
\begin{aligned}
& m_i \bar{w}_{itt} - \left\{T_i - m_i\left[V^2 + (g - a_i)x_i\right] + E_i A_i e_i\right\}\bar{w}_{ixx} + m_i g \bar{w}_{ix} + 2m_i V \bar{w}_{ixt} = F_i^w\left[t, L_i(t)\right],\\
& M_{car}\ddot{q}_1 - E_1A_1e_1 + E_2A_2e_3 = 0,\\
& M_{cwt}\ddot{q}_2 - E_1A_1e_4 + E_2A_2e_2 = 0,\\
& M_{comp}\ddot{q}_3 + E_1A_1e_1 + E_1A_1e_4 + F_d = 0,\\
& I_{comp}\ddot{\theta} - RE_1A_1e_1 + RE_1A_1e_4 = 0,
\end{aligned}
\tag{8.6}
$$

where $m_3 = m_2$, $m_4 = m_1$, $a_1 = a$, $a_2 = -a_1$, $a_3 = a_1$, $a_4 = a_2$, $E_3A_3 = E_2A_2$, $E_4A_4 = E_1A_1$ and e_i denote the quasi-static axial strains in the ropes and are given as

$$
\begin{aligned}
e_1 =& \frac{1}{L_1(t)}\left[u_1(L_i,t) - q_1(t) + \frac{1}{2}\int_0^{L_1}\left(\bar{v}_{1x}^2 + \bar{w}_{1x}^2\right)dx_1 + \frac{\Psi_1^2}{2L_1(t)}\left(v_0^2 + w_0^2\right)\right],\\
e_2 =& \frac{1}{L_2(t)}\left[q_2(t) + \frac{1}{2}\int_0^{L_2}\left(\bar{v}_{2x}^2 + \bar{w}_{2x}^2\right)dx_2 + \frac{(\Psi_h - \Psi_2)^2}{2L_2(t)}\left(v_0^2 + w_0^2\right)\right],\\
e_3 =& \frac{1}{L_3(t)}\left[q_1(t) + \frac{1}{2}\int_0^{L_3}\left(\bar{v}_{3x}^2 + \bar{w}_{3x}^2\right)dx_3 + \frac{(\Psi_{car} - \Psi_{mach})^2}{2L_3(t)}\left(v_0^2 + w_0^2\right)\right],\\
e_4 =& \frac{1}{L_4(t)}\left[u_4(L_i,t) - q_2(t) + \frac{1}{2}\int_0^{L4}\left(\bar{v}_{4x}^2 + \bar{w}_{4x}^2\right)dx_4 + \frac{\Psi_{cwt}^2}{2L_4(t)}\left(v_0^2 + w_0^2\right)\right],
\end{aligned}
\tag{8.7}
$$

where the constraint $2q_3 - u_1 - u_4 = 0$ needs to be applied and the force F_d given as

$$
F_d = c_{comp}\dot{q}_3|\dot{q}_3|^{\alpha-1},\ 0 < \alpha \leq 1 \tag{8.8}
$$

is the damping force provided by the hydraulic tie-down of the damping coefficient c_{comp}.

8.4 The Dynamic Behaviour and Numerical Results

The dynamic responses of the system can be determined by solving the nonlinear set of partial differential equations (PDE) given by Eq. 8.6. In this study the dynamic interactions when the frequency of the building is tuned to the natural frequencies of the VT system are investigated.

Figures 8.2, 8.3 and 8.4 show the variation of the natural frequencies of a VTS comprising a car of mass 5500 kg which carries rated load of 3000 kg. The travel

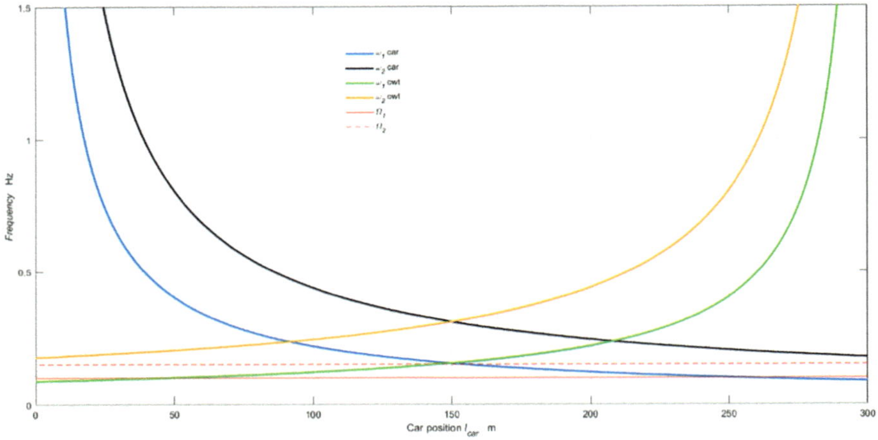

Fig. 8.2 The natural frequencies—compensating cable lateral modes

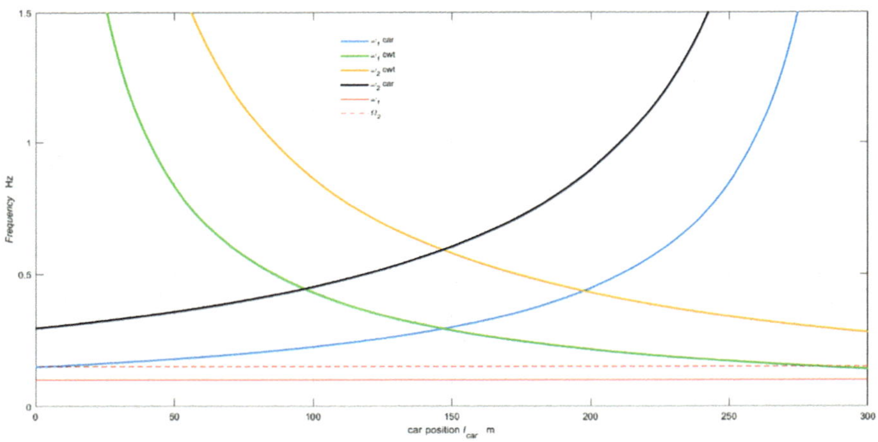

Fig. 8.3 The natural frequencies—suspension rope lateral modes

height is 300 m and the installation is equipped with compensating ropes with a synthetic fiber core (SFC) of diameter 36 mm and mass per unit length $m_{cr} =$ 4.76 kg/m each. The CSA mass is 4500 kg. The car and counterweight (balanced at 50%) are suspended on 9-stranded steel core ropes of diameter 19 mm and mass per unit length $m_{sr} = 1.54$ kg/m each. The horizontal (bending mode) natural frequencies of the building structure are given as $\Omega_1 = 0.1$ Hz in the in-plane direction and $\Omega_2 = 0.15$ Hz in the out-of-plane direction, respectively. The frequency curves are plotted against the position of the car in the shaft (measured from the bottom landing level), with the in-plane and out-of-plane excitation frequencies represented by red solid/dashed horizontal lines, respectively. It is evident that in this arrangement primary resonance interactions within the suspension/compensating system involve

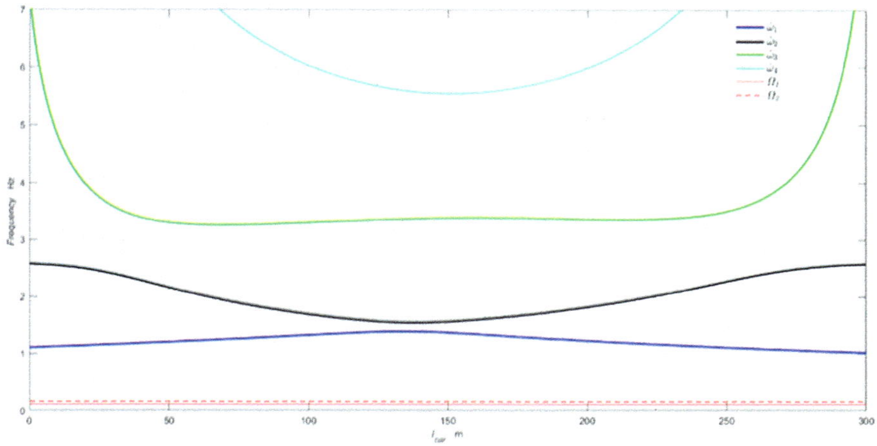

Fig. 8.4 The natural frequencies—vertical modes

the lateral modes of the ropes. On the other hand the frequencies of vertical mode are much higher than the frequencies of the building.

Following the methodology outlined above the PDE system Eq. 8.6 is discretized by using the Galerkin method so that the resulting set of nonlinear ordinary differential equations (ODE) can be simulated numerically. The simulated dynamic responses of the system for the scenario when the car is stationary at the level corresponding to the resonance length of $L_4 = 257$ m of the compensating cables at the counterweight side ($l_{car} = 45$ m, see Fig. 8.2) are presented in Figs. 8.5, 8.6, 8.7, where the damping force given by Eq. 8.8 is determined for $c_{comp} = 4.0 \times 10^5$ N(m/s)$^{-0.3}$ ($\alpha = 0.3$). Figure 8.5 shows the in-plane and out-of plane dynamic displacements of

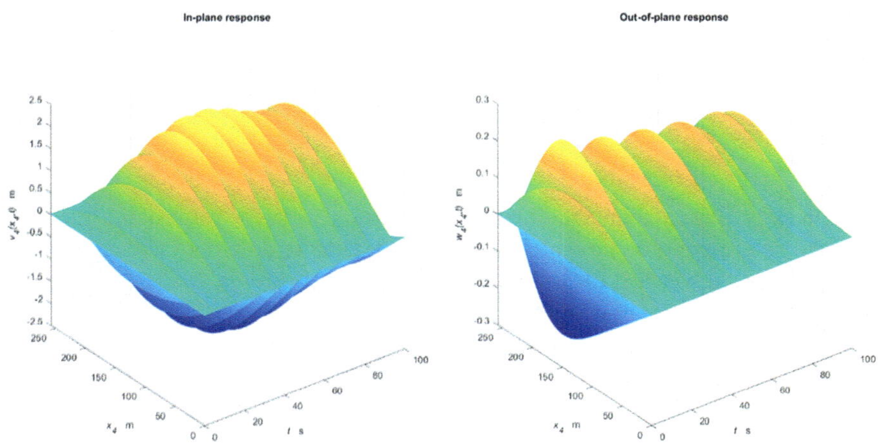

Fig. 8.5 The response of compensating cables at the counterweight side

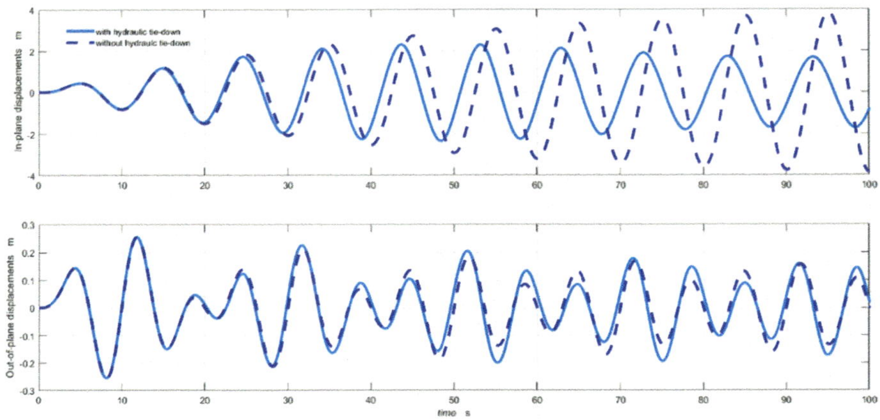

Fig. 8.6 The maximum lateral displacements of compensating cables with the hydraulic tie-down (solid lines) and with no tie-down applied (dashed lines)

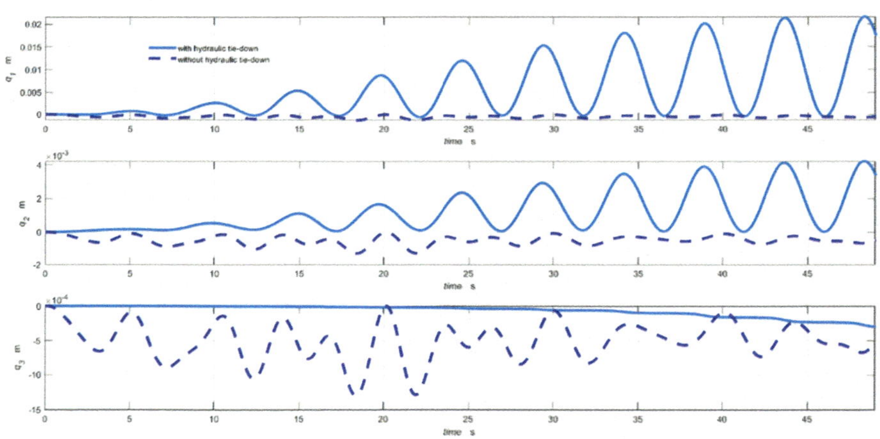

Fig. 8.7 Vertical displacements of the car, counterweight and CSA, with the hydraulic tie-down (solid lines) and with no tie-down applied (dashed lines)

the compensating cables at the counterweight side. The time records of the maximum lateral displacements of the cables (with the hydraulic tie-down and with no tie-down applied) and the vertical displacements are presented in Figs. 8.6 and 8.7, respectively. It is evident that the application of the hydraulic damper (tie-down) at the CSA results in the reduction of motions of the cables. The vertical response of the CSA is almost completely damped out whilst the vertical motions of the car are amplified.

8.5 Active Control Strategy

The application of a passive hydraulic tie-down can be effective in reducing the resonance motions of the cables and vertical motions of the CSA. However, in practice this does not fully mitigate the effects of the resonance conditions. The resonance frequencies of the ropes can be shifted/changed using different masses of the CSA. The mass of the CSA can be increased or decreased in order to shift the resonance conditions. However, the dynamic conditions present in the building structure are such that even small changes in the natural frequencies of the structure might result in large changes of the resonance conditions. Thus, the potential effects of the application of passive control techniques and the resonance shifting strategy to achieve enough reduction the dynamic responses are limited.

The active stiffness strategy can be sought to minimize the effects of adverse dynamic responses of suspension and compensating ropes in VTS [7]. To implement this strategy a servo-actuator is installed within the CSA tie-down system to control its vertical motion (q_3). The motion of the CSA is then dictated by a suitable feedback control law. The following multimode feedback control law can be applied

$$q_3 \equiv u_{comp}(t) = a_u \frac{\sum_{n=1}^{N} q_n \dot{q}_n}{\sum_{n=1}^{N} \alpha_n^2 q_n^2} \tag{8.9}$$

where a_u is the control factor q_n represent the modes of the rope/cable system and α_n are the mode weighting coefficients. This law is implemented in the numerical simulation to demonstrate its effectiveness in reducing the resonance responses of the compensating ropes when the car position corresponds to the distance $l_{car} = 45$ m. The active control is more effective than the passive damper/tie-down and results in a substantial reduction of the rope displacements, as demonstrated by the plots shown in Figs. 8.8 and 8.9. The control motion of the CSA (shown in black line in Fig. 8.9) is generated by using the control factor a_u=0.5. The FFT (Fast Fourier Transform) spectrum of the control signal is shown in Fig. 8.10. The control law accommodates the in-plane as well as the out-plane modes to avoid the modal spillover. The dominant frequency of the signal is 0.2 Hz which is twice the frequency of the fundamental resonance frequency of the cables.

8.6 Conclusion

Dynamic interactions that take place in VTS operating in high-rise structures result in adverse behaviour of their components compromising the structural integrity and safety of the installation. The application of passive hydraulic tie-down system at the CSA can mitigate the effects of fundamental resonances that occur in the compensating/suspension cable systems. The study presented in this paper demonstrates that the active stiffness control is more effective. The case study demonstrates that when

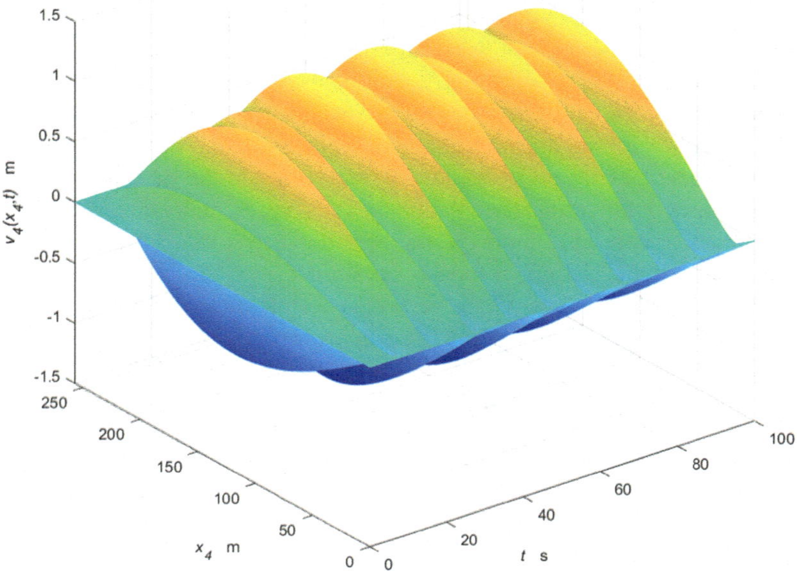

Fig. 8.8 The response of compensating cables at the counterweight side with active control strategy applied

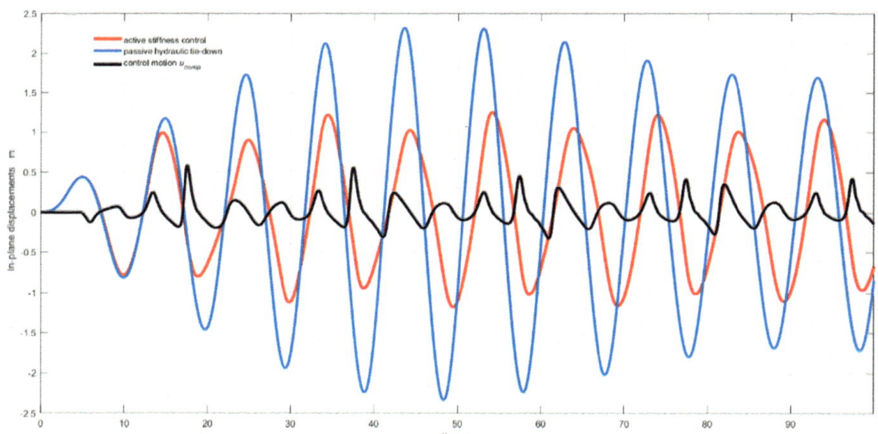

Fig. 8.9 The maximum lateral displacements of compensating ropes with active stiffness and passive hydraulic damper/tie-down with the control motion shown

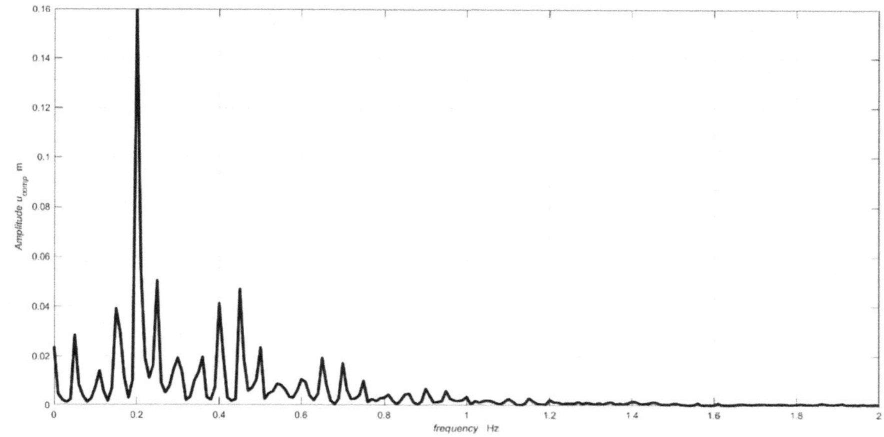

Fig. 8.10 FFT spectrum of the control motion

the proposed active control algorithm is used the response can be reduced by about 50% (in comparison with the response levels when the passive tie-down is applied).

References

1. Kijewski-Correa, T., Pirinia, D.: Dynamic behavior of tall buildings under wind: insights from full-scale monitoring. Struct. Des. Tall Spec. **16**(4), 471–486 (2007)
2. Hu, R.P., Xu, Y.L., Zhao, X.: Long-period ground motion simulation and its impact on seismic response of high-rise buildings. J. Earthq. Eng. **22**(7), 1–31 (2018)
3. Kaczmarczyk, S.: The dynamics of vertical transportation systems: from deep mine operations to modern high-rise applications. In: Awrejcewicz, J., Kazmierczak, M., Mrozowski, J., Olejnik, P. (eds.) Dynamical Systems: Mechatronics and Life Sciences, pp. 249–260. Lodz University of Technology, Lodz, Poland (2015)
4. Kaczmarczyk, S., Iwankiewicz, R.: Gaussian and non-Gaussian stochastic response of slender continua with time-varying length deployed in tall structures. Int. J. Mech. Sci. **134**, 500–510 (2017)
5. Crespo, R.S., Kaczmarczyk, S., Picton, P.D., Su, H.: Modelling and simulation of a stationary high-rise elevator system to predict the dynamic interactions between its components. Int. J. Mech. Sci. **137**, 24–45 (2018)
6. Kaczmarczyk, S.: The prediction and control of dynamic interactions between tall buildings and high-rise vertical transportation systems subject to seismic excitations. In: The Proceedings of the 25th International Congress on Sound and Vibration (ICSV 25), Hiroshima, Japan, July 08–12 (2018)
7. Kaczmarczyk, S., Picton, P.D.: The prediction of nonlinear responses and active stiffness control of moving slender continua subjected to dynamic loadings in vertical host structures. Int. J. Acoust. Vib. **18**(1), 39–44 (2013)
8. Mitropolskii, Y.A.: Problems of the Asymptotic Theory of Nonstationary Vibrations. Israel Program for Scientific Translations Ltd., Jerusalem (1965)
9. Kaczmarczyk, S.: The passage through resonance in a catenary – vertcal cable hoisting system with slowly varying length. J. Sound Vib. **208**(2), 243–269 (1997)

10. Terumichi, Y., Ohtsuka, M., Yoshizawa, M., Fukawa, Y., Tsujioka, Y.: Nonstationary vibrations of a string with time varying length and a mass-spring system at the lower end. Nonlinear Dyn. **12**, 39–55 (1997)
11. Sandilo, S.H., Van Horssen, W.T.: On variable length induced vibrations of a vertical string. J. Sound Vib. **333**(11), 2432–2449 (2013)
12. Sandilo, S.H., Van Horssen, W.T.: On a cascade of autoresonances in an elevator cable system. Nonlinear Dyn. **80**, 1613–1630 (2015)
13. Gaiko, N.V., Van Horssen, W.T.: Resonances and vibrations in an elevator cable system due to boundary sway. J. Sound Vib. **424**, 272–292 (2018)

Chapter 9
Free Generalized van der Pol Oscillators: Overview of the Properties of Oscillatory Responses

Ivana Kovacic

Abstract This work is concerned with generalized van der Pol oscillators, the damping-like force of which depends nonlinearly on the displacement and velocity with the powers that can be any positive real numbers, while the restoring force is either linear or purely nonlinear. The cases of small and large values of the damping parameter are considered. In the former case, an overview of contributions related to the amplitude and frequency of free limit cycle oscillations of different forms of generalized van der Pol oscillators are given and then the most general case examined. In the latter case, the jumps, outer curves and period of relaxation oscillations are found.

9.1 Introduction

The standard (classical) van der Pol oscillator

$$\ddot{x} + x = \varepsilon\left(1 - x^2\right)\dot{x}, \tag{9.1}$$

represents one of archetypical oscillators. It is named after Balthasar van der Pol (1889–1959), a Dutch physicist, whose achievements and life have attracted the attention of many researchers both from the viewpoint of his scientific contributions and biography [1–4].

Balthasar van der Pol entered the University of Utrecht, where he graduated *cum laude* in Physics. He then studied under John Ambrose Fleming, who was an inventor of a diode, and John Joseph Thomson, who discovered the electron. He was a friend and colleague with Edward Appleton, who was the Nobel Prize laureate for his discovery of a certain layer of the ionosphere. Balthasar van der Pol was assistant to Hendrik Antoon Lorentz, who shared the 1902 Nobel Prize in Physics in recognition

I. Kovacic (✉)
Faculty of Technical Sciences, Centre of Excellence for Vibro-Acoustic Systems and Signal Processing, University of Novi Sad, Novi Sad, Serbia
e-mail: ivanakov@uns.ac.rs

A. K. Abramian et al. (eds.), *Nonlinear Dynamics of Discrete and Continuous Systems*, Advanced Structured Materials 139,
https://doi.org/10.1007/978-3-030-53006-8_9

of the research on the influence of magnetism upon radiation phenomena. Balthasar van der Pol worked for Philips Company and also had an academic career at the Technical University, Delft. He held a temporary professorship at the University of California, Berkeley and the Victor Emanuel Professorship at Cornell in Ithaca, New York.

Balthasar van der Pol pioneered the fields of radio and telecommunications [1]. However, his scientific work did not cover only radio and electrical engineering, but also pure and applied mathematics, which included number theory, special functions, operational calculus and nonlinear differential equations.

He was a theoretician and an experimentalist. While conducting experiments with oscillations in a vacuum tube triode circuit, he concluded that all initial conditions converged to the same periodic orbit of finite amplitude. He proposed a nonlinear differential equation (9.1) as a nondimensional mathematical model for the behaviour observed experimentally [5]. The nonlinear 'damping-like' force that appears on the right-hand side of Eq. (9.1) dissipates energy for large displacements as the expression in the parentheses is negative; it feeds energy for small displacements since this expression is then positive. This behaviour gives rise to self-sustaining/self-exciting oscillations. For small values of the 'damping coefficient' ε ($0 < \varepsilon \ll 1$), this behaviour is characterized by the appearance of a stable limit cycle with the steady-state amplitude $|a_{LC,s}| = 2$ (note that the index 's' stans for the 'standard van der Pol oscillator' and this abbreviation will be used through the whole manuscript) and the angular frequency approximately equal to unity (Fig. 9.1).

While investigating the case $\varepsilon \gg 1$, van der Pol discovered the importance of what has become known as relaxation oscillations [6]—the motion consisting of very slow asymptotic behaviour along outer curves followed by a sudden discontinuous jump. The jump-down points $x_{\text{jd},s}$ are located at $x_{\text{jd},s} = 1$, from which the amplitude jumps to $x_{\text{d},s} = -2$. Then the motion proceeds along the outer curve and undergoes a jump-up from $x_{\text{ju},s} = -1$ to $x_{\text{u},s} = 2$ (Fig. 9.2). Ginoux pointed out in [7] that around the same time when van der Pol published the paper [6] in English, he also published

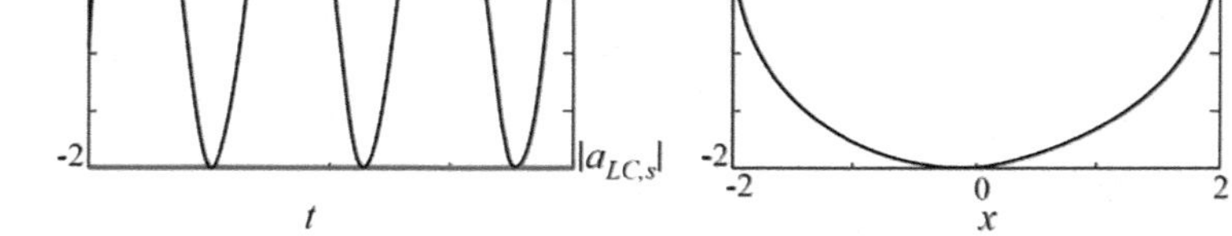

Fig. 9.1 Characteristic behaviour of the van der Pol oscillator (1) for $0 < \varepsilon \ll 1$: **a** oscillations; **b** phase trajectory

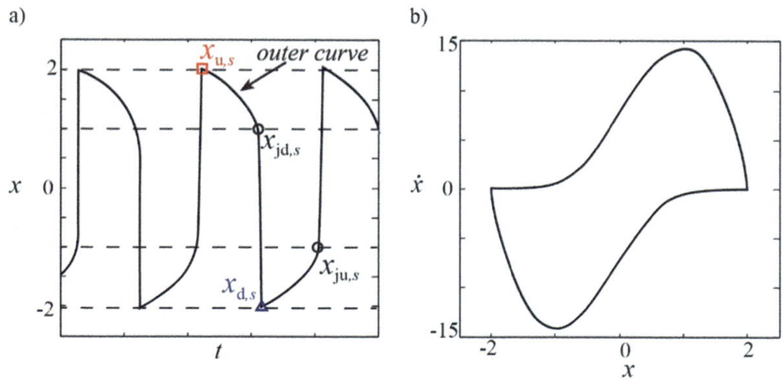

Fig. 9.2 Characteristic behaviour of the van der Pol oscillator (1) for $\varepsilon \gg 1$: **a** oscillations; **b** phase trajectory

three more contributions in Dutch and German. They were all introducing relaxation oscillations, while 'their conclusions differ in the choice of the devices exemplifying the phenomenon of relaxation oscillations'. A few years later, van der Pol and van der Mark modelled the electric activity of the heart by using relaxation oscillations [8].

In this work, the generalized van der Pol oscillator governed by the following nondimensional equation of motion is considered:

$$\ddot{x} + \operatorname{sgn}(x)|x|^{\alpha} = \varepsilon f(x, \dot{x}),$$
$$f(x, \dot{x}) = \left(1 - |x|^{\beta}\right)|\dot{x}|^{\gamma}\operatorname{sgn}(\dot{x}), \tag{9.2a,b}$$

where $\alpha > 0$, $\beta > 0$, $\gamma \geq 0$ and $\varepsilon > 0$. Here, the nonlinearity appears in both terms of the 'damping-like' force given by Eq. (9.2b) as well as in the restoring force, which is given by the second term on the left side of Eq. (9.2a); the sign and absolute value functions are used in Eqs. (9.2a,b) to assure that these forces have the properties of odd and even functions as in the standard van der Pol oscillator modelled by Eq. (9.1). The aim is to show how the properties of oscillatory responses of this generalized van der Pol oscillator differ with respect to the one described above for the standard van der Pol oscillator. This work also contains a literature survey on previous achievements related to these characteristics of different generalized van der Pol type oscillators.

9.2 Small Values of the Damping Coefficient: Limit Cycle

Minorsky [9] examined the generalized van der Pol Eq. (9.2a,b) for $\alpha = \gamma = 1$ and $\beta = 2n$, where n is a positive integer ($n \geq 1$). He used the stroboscopic method, obtained the steady-state amplitude, concluding that this amplitude is smaller than $|a_{LC,s}| = 2$. In addition, he indicated that for $0 < n < 1$, this amplitude is higher than $|a_{LC,s}| = 2$. Moremedi et al. [10] used a perturbation scheme to conclude that for $\alpha = \gamma = 1$ and $\beta = 2n$, where n is a positive integer, one has $|a_{LC}| \to 1$, when $n \to \infty$. They also concluded that the effect of increasing n is to increase the period of the limit cycle oscillations. Obi [11] analysed the model (9.2a,b) with $\gamma = 1$, but only for the case when the powers $\alpha = 2\text{n} + 1$ and $\beta = 2\text{n} + 2$ ($n \geq 1$) are an odd and even number, respectively. He gave an approximate value of the amplitude of the limit cycle as $|a_{LC}| = (3n + 4)^{\frac{1}{2n+2}}$. By applying the harmonic balance method and the averaging method, Mickens and Oyedeji [12] found that the oscillatory response of a cubic van der Pol oscillator with $\alpha = 3$, $\beta = 2$ and $\gamma = 1$ is characterised by $|a_{LC}| = 2$ and the angular frequency $\omega_{LC} = \sqrt{3}$. This work was the motivation for the subsequent papers [13, 14], where the same oscillator was considered, and its limit cycle described analytically in terms of Jacobi elliptic functions. The approach of elliptic averaging gave $|a_{LC}| = 1.9098$ and the period $T_{LC} = 3.8833$, with an error of less than 1‰ with respect to the corresponding numerical result. The elliptic balancing used in [14] produced the result that depends on ε, which gave more accurate approximate solutions. By developing an elliptic perturbation method, improved accuracy is achieved in [15] even for higher values of ε. Mickens [16] adjusted the averaging method to derive the expression for the limit cycle of the generalized van der Pol oscillator with $\alpha = 1$, $\beta = 2$ and $\gamma = 1/3$, showing that its steady-state amplitude is lower than that of the standard van der Pol oscillator: $|a_{LC}| = 1.82574$, while the frequency stays the same at the level of approximation used. By applying an iterative technique, the same author [17] showed that when $\alpha = 1/3$ and $\beta = 2$, the amplitude of the limit cycle stays the same, but the frequency decreases for about 15% with respect to the standard van der Pol oscillator. Oyedeji [18] considered the quadratic van der Pol oscillator ($\alpha = \beta = 2$ and $\gamma = 1$) and used the first order harmonic balance method to calculate the limit cycle amplitude $a_{LC,s} = 2$ and the frequency $\omega_{LC} = \sqrt{16/(3\pi)} \approx 1.30294$. Waluya and van Horssen constructed asymptotic results on long time-scales t for the periods of the generalized van der Pol Eqs. (9.2a,b) with $\beta = 2$, $\gamma = 1$ and $\alpha = (2\,m + 1)/(2n + 1)$, $m, n \in \text{N}$ [19]. First, they showed how approximations of first integrals can be obtained and, then, how the existence, stability, and the period of time-periodic solutions can be determined from them. In [20], a more general class of oscillators with α being any positive real number is dealt with. Approximate expressions for the period is obtained for $\alpha \gg 1$, $\alpha \ll 1$ and $\alpha \to 1$. Kovacic [21] applied the averaging method for purely nonlinear systems to determine the amplitude of the limit cycle for $\alpha > 0$, $\beta > 0$, $\gamma = 1$ [21], while Kovacic and Mickens [22] generalized this case to $\gamma \geq 0$. Some of their results are summarised below. They also showed how to calculate the time needed to reach the limit cycle.

9.2.1 General Case

In order to determine the properties of the response of the oscillators governed by Eqs. (9.2a,b) in a general case $\alpha > 0$, $\beta > 0$, $\gamma \geq 0$ and $0 < \varepsilon \ll 1$, the averaging method for purely nonlinear systems is used [21]. When $\varepsilon = 0$, Eq. (9.2a) corresponds to conservative purely nonlinear oscillators. Their energy integral can be used to derive the exact value of their frequency:

$$\omega(a) = c\sqrt{|a|^{\alpha-1}}, \quad c = \sqrt{\frac{\pi(\alpha+1)}{2}}\frac{\Gamma\left(\frac{\alpha+3}{2(\alpha+1)}\right)}{\Gamma\left(\frac{1}{\alpha+1}\right)}, \tag{9.3a,b}$$

where Γ is the Euler gamma function and a is the amplitude of motion.

The first approximation to motion of the perturbed systems (9.2a,b) can then be assumed as

$$x = a\cos\psi, \quad \dot{x} = -a\omega\sin\psi, \tag{9.4a,b}$$

where

$$\psi = \int_0^t \omega(a)dt + \theta(t), \tag{9.5}$$

while the frequency ω depends on the amplitude a and the power α was defined by Eqs. (9.3a,b)

Differentiating Eq. (9.4a) with respect to time, one follows

$$\dot{x} = \dot{a}\cos\psi - a\omega\sin\psi - a\dot{\theta}\sin\psi, \tag{9.6}$$

which, owing to Eq. (9.4b), imposes the following constraint:

$$\dot{a}\cos\psi - a\dot{\theta}\sin\psi = 0. \tag{9.7}$$

Substituting the second time derivative of Eqs. (9.4b) together with Eq. (9.4a) into Eq. (9.2a,b), one can derive:

$$-\dot{a}\omega\sin\psi - a\frac{d\omega}{da}\dot{a}\sin\psi - a\omega\dot{\theta}\cos\psi - a\omega^2\cos\psi + \operatorname{sgn}(a\cos\psi)|a\cos\psi|^{\alpha} = -\varepsilon\left(1 - |a\cos\psi|^{\beta}\right)|-a\omega\sin\psi|^{\gamma}\operatorname{sgn}(-a\omega\sin\psi). \tag{9.8}$$

It should be noted that the last term on the left-hand side of Eq. (9.8) can be approximated by the first term from the corresponding Fourier series expansion [23]

$$\operatorname{sgn}(a\cos\psi)|a\cos\psi|^{\alpha}\approx|a|^{\alpha}b_{1_\alpha}\cos\psi,\quad b_{1_\alpha}=\frac{2}{\sqrt{\pi}}\frac{\Gamma\left(1+\frac{\alpha}{2}\right)}{\Gamma\left(\frac{3+\alpha}{2}\right)}.\tag{9.9a,b}$$

Now, this term can be cancelled by the term in front of it

$$-a\omega^2\cos\psi+|a|^{\alpha}b_{1_\alpha}\cos\psi=0,\tag{9.10}$$

assuming that $\omega^2\approx b_{1_\alpha}|a|^{\alpha-1}\approx c|a|^{\alpha-1}$, as given by Eqs. (9.3a,b)

Next, based on Eqs. (9.3a,b), the second term on the left-hand side of Eq. (9.8) can be expressed as:

$$\frac{d\omega}{da}=\frac{\alpha-1}{2a}\omega.\tag{9.11}$$

Substituting Eqs. (9.10) and (9.11) into Eqs. (9.8) and combining it with Eq. (9.7), one can derive

$$\begin{aligned}&\dot{a}\left(1+\frac{\alpha-1}{2}\sin^2\psi\right)=\\&\varepsilon\left(1-|a\cos\psi|^{\beta}\right)|-a\omega\sin\psi|^{\gamma}\operatorname{sgn}(-a\omega\sin\psi)\sin\psi,\end{aligned}\tag{9.12}$$

$$\begin{aligned}&a\dot{\theta}+\dot{a}\frac{\alpha-1}{2}\sin\psi\cos\psi=\\&\varepsilon\left(1-|a\cos\psi|^{\beta}\right)|-a\omega\sin\psi|^{\gamma}\operatorname{sgn}(-a\omega\sin\psi)\cos\psi.\end{aligned}\tag{9.13}$$

Averaging Eqs. (9.12) and (9.13), the following first-order differential equations for the amplitude a and the phase shift θ are obtained:

$$\dot{a}=-\frac{2\varepsilon}{\pi c(\alpha+3)|a|^{\frac{\alpha-1}{2}}}\int_0^{2\pi}\left(1|a\cos\psi|^{\beta}\right)|-a\omega\sin\psi|^{\gamma}\operatorname{sgn}(-a\omega\sin\psi)\sin\psi d\psi,\tag{9.14}$$

$$\begin{aligned}&a\dot{\theta}=\\&-\frac{\varepsilon}{2\pi c|a|^{\frac{\alpha-1}{2}}}\int_0^{2\pi}\left(1-|a\cos\psi|^{\beta}\right)|-a\omega\sin\psi|^{\gamma}\operatorname{sgn}(-a\omega\sin\psi)\cos\psi d\psi.\end{aligned}\tag{9.15}$$

Solving the integrals on the right-hand sides, yields:

$$\dot{a}=\frac{4\varepsilon c^{\gamma-1}}{\pi(\alpha+3)}\frac{|a|^{\gamma\frac{\alpha+1}{2}-\frac{\alpha-1}{2}}\Gamma\left(1+\frac{\gamma}{2}\right)\left[\sqrt{\pi}\Gamma\left(\frac{3+\beta+\gamma}{2}\right)-|a|^{\beta}\Gamma\left(\frac{1+\beta}{2}\right)\Gamma\left(\frac{3+\gamma}{2}\right)\right]}{\Gamma\left(\frac{3+\gamma}{2}\right)\Gamma\left(\frac{3+\beta+\gamma}{2}\right)},\tag{9.16}$$

$$\dot{\theta} = 0. \tag{9.17}$$

Based on Eq. (9.17) one concludes that in all generalized van der Pol type oscillators modelled by Eqs. (9.2a,b), the phase shift is constant to terms of order ε. The amplitude of the limit cycle a_{LC} corresponds to $\dot{a} = 0$ and is calculated to be

$$|a_{LC}| = \left[\frac{\sqrt{\pi}\Gamma\left(\frac{3+\beta+\gamma}{2}\right)}{\Gamma\left(\frac{1+\beta}{2}\right)\Gamma\left(\frac{3+\gamma}{2}\right)} \right]^{1/\beta}, \tag{9.18}$$

while Eqs. (9.3a) define the corresponding frequency

$$\omega_{LC} = c\sqrt{|a_{LC}|^{\alpha-1}}. \tag{9.19}$$

The expression (9.18) is also obtained in [21] for the generalized van der Pol with $\gamma = 1$ and indicates that the first approximation for the amplitude of the limit cycle depends on the parameters appearing in the model of the 'damping-like' force.

Equation (9.18) is used to plot how the amplitude of the limit cycle changes with the parameter β for two different values of the power γ (Fig. 9.3a, b).

In addition, numerically obtained amplitudes of the limit cycle are also presented in this figure for three different values of the parameter α corresponding to the linear $\alpha = 1$, under-linear $\alpha = 2/3$ and over-linear restoring forces $\alpha = 2$. It is seen that the analytical and numerical result agree reasonably well for the whole range of the powers considered.

As the power of the restoring force α increases, the amplitude of the limit cycle decreases. As the power of the geometric term in the 'damping' force β increases, the amplitude of the limit cycle decreases as well.

9.2.2 *Special Case: $\gamma = 1$*

If the velocity term in the 'damping' force is linear as in the standard van der Pol oscillator, the amplitude of the limit cycle is [21]:

$$|a_{LC}| = \left[\frac{\sqrt{\pi}\Gamma\left(\frac{4+\beta}{2}\right)}{\Gamma\left(\frac{1+\beta}{2}\right)} \right]^{1/\beta}. \tag{9.20}$$

The way how this amplitude changes with the parameter β is plotted in Fig. 9.4. Numerically obtained amplitudes of the limit cycle calculated for different values of the parameter α are also shown.

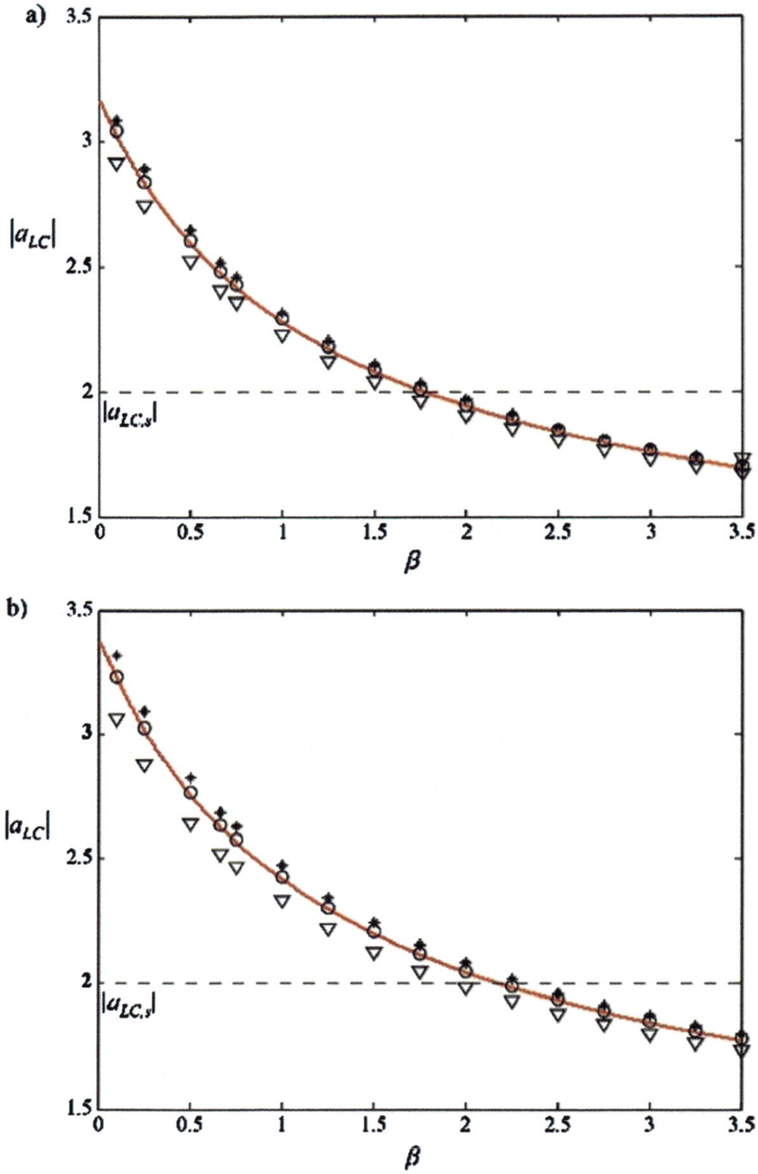

Fig. 9.3 Amplitude of the limit cycle obtained analytically Eq. (9.18) (solid line) and numerically for $\varepsilon = 0.1$, $\alpha = 2/3$ (stars), $\alpha = 1$ (circles) and $\alpha = 2$ (triangles): **a** $\gamma = 0.8$; **b** $\gamma = 1.2$

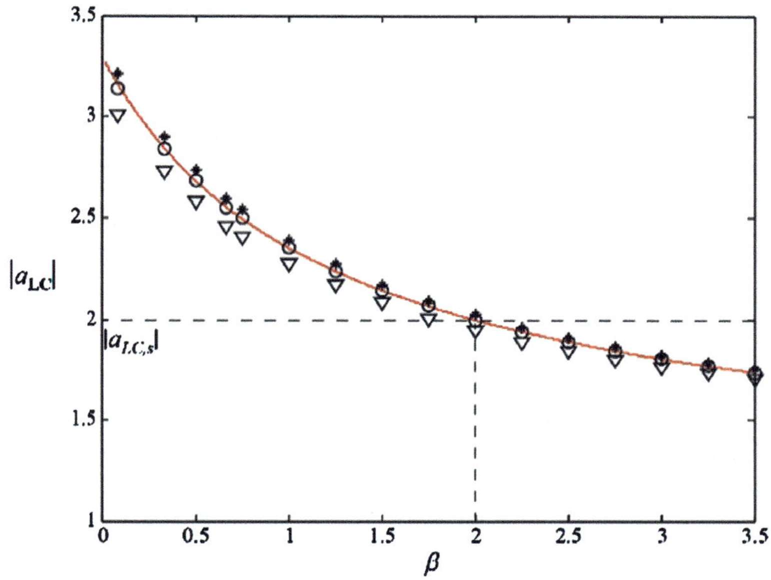

Fig. 9.4 Amplitude of the limit cycle for $\gamma = 1$, $\varepsilon = 0.1$ obtained analytically Eq. (9.20) (solid line) and numerically: $\alpha = 2/3$ (stars), $\alpha = 1$ (circles) and $\alpha = 2$ (triangles)

Equation (9.20) implies that when $\beta \to 0$, $|a_{LC}| \to 2\sqrt{e}$ as well as when $\beta \to \infty$, $|a_{LC}| \to 1$, which agrees with the results found in [10]. When $\beta = 2$, as is in the standard van der Pol oscillator, the amplitude is obtained $|a_{LC}| = 2$, which is also seen in Fig. 9.1.

Using Eqs. (9.19) and (9.20), the frequency of the limit cycle oscillations is plotted in Fig. 9.5. For the under-linear case ($\alpha < 1$), this frequency is lower than $\omega_{LC,s}$ and increases with β; for the over-linear case ($\alpha > 2$), this frequency is higher than $\omega_{LC,s}$ and decreases with β.

9.2.3 *Special Case: $\gamma = 0$*

When the parameter γ is equal to zero, Eq. (9.18) yields the following amplitude for the limit cycle

$$|a_{LC}| = (1 + \beta)^{1/\beta}. \tag{9.21}$$

Thus, when $\beta \to 0$, one has $|a_{LC}| \to e$. For the case when $\beta = 2$, one can calculate $|a_{LC}| = \sqrt{3}$.

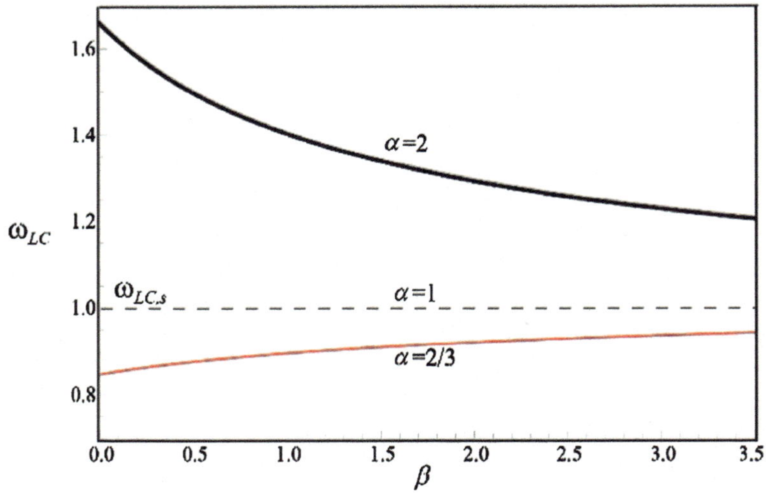

Fig. 9.5 Frequency of limit cycle oscillations for $\gamma = 1$ and different values of α, Eqs. (9.19), (9.20), (9.3a,b)

9.3 Large Values of the Damping Coefficient: Relaxation Oscillations

As illustrated in Fig. 9.2, the oscillatory response of the standard van der Pol oscillator (1) corresponding to $\varepsilon \gg 1$ has the form of relaxation oscillations. Besides the characteristic coordinates labelled in Fig. 9.2, their period has also attracted the interest of researchers. Based on geometrical considerations, van der Pol wrote in [6] that 'the period T, instead of being 2π (as was the case when $\varepsilon \ll 1$) increases with increase of ε, and when $\varepsilon \gg 1$ becomes equal to approximately ε itself'. Later on, he improved this conclusion to $T = 1.61\varepsilon$ [8]. This expression was found not to be accurate for larger values of ε, such as, for example $\varepsilon = 10$, as the experiments had showed $T \approx 20$. Haag [24, 25] and Dorodnitsyn [26] provided more accurate approximations, giving the expressions whose first term coincided with $T_s = 1.61\varepsilon$, while additional ones were either power or log-forms of ε. Stoker's approximation for the period [27] was even more accurate, with the error of 0.8% for $\varepsilon = 5$ and 0.1% for $\varepsilon = 10$. The interested reader is referred to [7] for a rich historical review of the discovery and investigations of relaxation oscillations.

The aim of this section is to analyse relaxation oscillations of the generalized van der Pol oscillator (9.2a,b) and to determine the analytical expressions for the coordinates of jump points, outer curves, and the period. To that end, a perturbation approach with slow and fist time scales will be used. First, time is scaled by setting $t_1 = \varepsilon^{\vartheta} t$, where ϑ is to be determined. Equations (9.2a,b) turns into

$$\varepsilon^{2\vartheta} \frac{d^2 x}{dt_1^2} + \mathrm{sgn}(x)|x|^{\alpha} - \varepsilon^{1+\gamma\vartheta}\left(1 - |x|^{\beta}\right)\left|\frac{dx}{dt_1}\right|^{\gamma} \mathrm{sgn}\left(\frac{dx}{dt_1}\right) = 0. \tag{9.22}$$

Selecting $\vartheta = -1/\gamma$, the third and the second term are of the same order, while the first term becomes of the order $(1/\varepsilon)^{2/\gamma}$. This term can be neglected as being small for γ being around unity. Thus, for the remaining procedure, it will be selected that $\gamma = 1$. As a result, one has $\vartheta = -1$ now. The slow time scale is defined as $t_1 = t/\varepsilon$, and Eq. (9.22) turns into

$$\frac{1}{\varepsilon^2}\frac{d^2x}{dt_1^2} + \operatorname{sgn}(x)|x|^\alpha - \left(1 - |x|^\beta\right)\frac{dx}{dt_1} = 0. \tag{9.23}$$

The first term can be neglected as being small, which yields

$$\frac{dx}{dt_1} = \frac{\operatorname{sgn}(x)|x|^\alpha}{1 - |x|^\beta}. \tag{9.24}$$

Jumps occur when dx/dt_1 is infinite, i.e. when the nominator in Eq. (9.24) is zero, giving the jump-down x_{jd} and jump-up x_{ju} values of the coordinates: $x_{\mathrm{jd}} = 1$ and $x_{\mathrm{ju}} = -1$. These values are the same as the one in the standard van der Pol oscillator (see Fig. 9.2) and are obtained as independent of the values of the powers α and β.

In order to define the characteristic amplitudes x_{d} and x_{u} labelled in Fig. 9.2 for a generalized van der Pol oscillator, the fast time scale is introduced as $t_2 = \varepsilon t$, resulting in

$$\frac{d^2x}{dt_2^2} + \frac{1}{\varepsilon^2}\operatorname{sgn}(x)|x|^\alpha - \left(1 - |x|^\beta\right)\frac{dx}{dt_2} = 0. \tag{9.25}$$

By neglecting the term of $O\left(1/\varepsilon^2\right)$, the following first integral can be derived

$$\frac{dx}{dt_2} - x + \frac{x|x|^\beta}{\beta + 1} = C, \tag{9.26}$$

where C is a constant. Its value can be calculated by considering jumps for which $dx/dt_2 = 0$. When $x_{\mathrm{jd}} = 1$, one can calculate $C = -\beta/(\beta + 1)$, giving the following implicit equation for the amplitude x_{d}:

$$x_{\mathrm{d}} - \frac{x_{\mathrm{d}}|x_{\mathrm{d}}|^\beta}{\beta + 1} - \frac{\beta}{\beta + 1} = 0. \tag{9.27}$$

In a similar way, the value of C corresponding to $x_{\mathrm{ju}} = -1$ can be obtained, resulting in the following implicit equation for x_{u}:

$$x_{\mathrm{u}} - \frac{x_{\mathrm{u}}|x_{\mathrm{u}}|^\beta}{\beta + 1} + \frac{\beta}{\beta + 1} = 0. \tag{9.28}$$

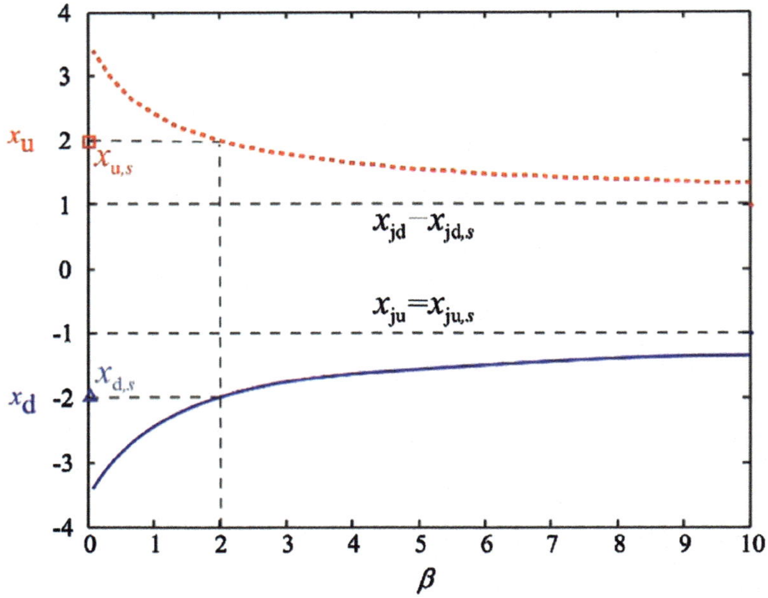

Fig. 9.6 Change of the coordinates x_d and x_u with the power β, Eqs. (9.27) and (9.28)

Equations (9.27) and (9.28) imply that x_d and x_u depend on the damping power β, but do not depend on the power of the restoring force α.

The solutions of Eqs. (9.27) and (9.28) are plotted in Fig. 9.6 as a function of the power β. As β increases infinitely, the values of x_d and x_u approach 1 and -1, respectively, i.e. x_jd and x_ju. This means that outer curves will be flatter as β increases.

Integration of Eq. (9.26) can give analytical expressions for the outer curves. Thus, for $\alpha = 1$, one can derive [21]:

$$\ln|x| - \frac{|x|^\beta}{\beta} = t_1 + D, \tag{9.29}$$

where D is a constant.

When $\alpha > 1$ and $\beta = \alpha - 1$, one can obtain

$$\frac{|x|^{1-\alpha}}{1-\alpha} - \ln|x| = t_1 + D. \tag{9.30}$$

If $\alpha \neq 1$and$\beta \neq \alpha - 1$, the integration gives

$$\frac{|x|^{1-\alpha}}{1-\alpha} - \frac{|x|^{\beta-\alpha+1}}{\beta-\alpha+1} = t_1 + D. \tag{9.31}$$

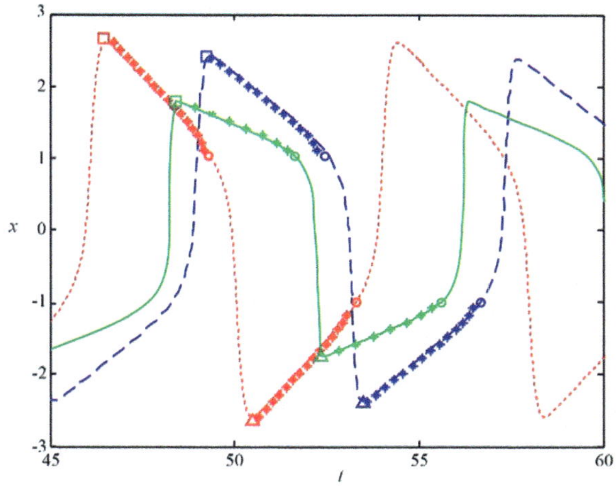

Fig. 9.7 Relaxation oscillations of the generalized van der Pol oscillators modelled by Eqs. (9.2a,b) for $\gamma = 1$, $\varepsilon = 10$: numerical solution for $\alpha = 5/3$, $\beta = 2/3$ (red dotted line); numerical solution for $\alpha = 2$, $\beta = 1$ (blue dashed line), numerical solution for $\alpha = 5/2$, $\beta = 3/2$ (green solid line); outer curves defined by Eq. (9.30) are depicted by stars; jump-up points x_{ju} and jump-down points x_{jd} by circles; points x_{d} by triangles and points x_{u} by squares

To validate the analytical results obtained, their comparison with the numerical results from direct integration of the equation of motion is carried out for different values of the powers α and β. Figure 9.7 shows this comparison for the outer curves, and the characteristic coordinates: x_{u}, x_{d}, x_{jd} and x_{ju}.

For the sake of easier visual comparison, the legend used for the characteristics coordinates corresponds to the one used in Fig. 9.2 for the standard van der Pol oscillator. It is seen that the analytical results obtained are in good agreement with the numerical results. These figures illustrate the effects of the powers α and β influence on the relaxation oscillations, including their amplitude, i.e. the coordinates x_{d} and x_{u}. They also give insight into the time spent moving along the outer curves from x_{u} to x_{jd}, which corresponds to the first half of the period. It is seen that this time is affected by the powers α and β. In order to obtain the analytical expression for the half of the period, one can utilize the previously derived analytical expression for the outer curves and the coordinate x_{u}, Eq. (9.27). Thus, the half-period on the slow time scale t_1 corresponding to $\alpha = 1$ is

$$\frac{T_1}{2} = \left[\ln|x| - \frac{|x|^{\beta}}{\beta}\right]_{x_{\mathrm{u}}}^{x_{\mathrm{jd}}=1} = -\frac{1}{\beta} - \ln x_{\mathrm{u}} + \frac{x_{\mathrm{u}}^{\beta}}{\beta}, \tag{9.32}$$

where the absolute value of x_{u} has been omitted as $0 < x_{\mathrm{u}} < 1$. On the original time scale t, the period of relaxation oscillations T is

$$T = 2\left[-\frac{1}{\beta} - \ln x_{\mathrm{u}} + \frac{x_{\mathrm{u}}^{\beta}}{\beta}\right]\varepsilon. \tag{9.33}$$

When $\alpha > 1$, $\beta = \alpha - 1$, this period is

$$T = 2\left[\frac{1}{1-\alpha} + \ln x_u - \frac{x_u^{1-\alpha}}{1-\alpha}\right]\varepsilon. \quad (9.34)$$

If $\alpha \neq 1$, $\beta \neq \alpha - 1$, this period is defined by

$$T = 2\left[\frac{\beta}{(1-\alpha)(\beta-\alpha+1)} - \frac{x_u^{1-\alpha}}{1-\alpha} + \frac{x_u^{\beta-\alpha+1}}{\beta-\alpha+1}\right]\varepsilon. \quad (9.35)$$

Figure 9.8 shows how the ratio of the period of relaxation oscillations and the 'damping' coefficient ε changes with the power β. The cases defined by Eqs. (9.33) and (9.35) are plotted in Fig. 9.8a and c, respectively. They illustrate that the ratio T/ε increases as β increases. Figure 9.8b is plotted based on Eq. (9.34) and shows different trends of T/ε with respect to the value $\beta^* \approx 1.84$, which corresponds to $T^* \approx 0.637\varepsilon$. For $\beta < \beta^*$, the ratio considered increases with the increase of β, and then decreases.

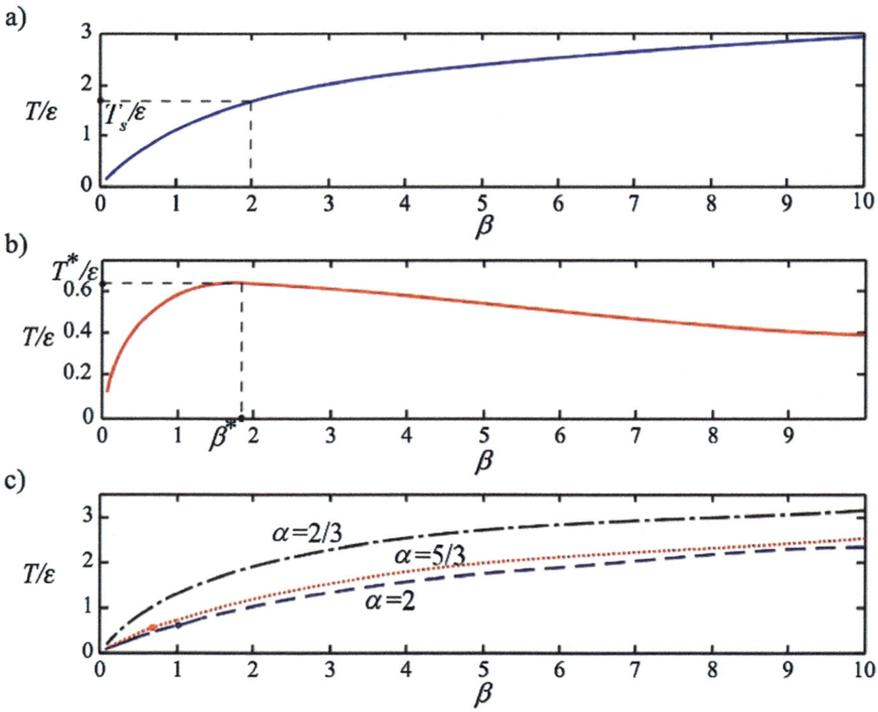

Fig. 9.8 Ratio of the period of relaxation oscillations T and the 'damping' coefficient ε versus the damping power β: **a** Eq. (9.33); **b** Eq. (9.34); **c** Eq. (9.35), $\alpha = 2/3$ (black dashed-dotted line), $\alpha = 5/3$ (red dotted line), $\alpha = 2$ (blue dashed line), discontinuity of the curves due to the condition $\beta \neq \alpha - 1$ is depicted by circles

9.4 Conclusions

This work has first given a tribute to Balthasar van der Pol and his contribution related to the standard equation named after him. Two main cases and their properties have been pointed out: (i) the case of small values of the 'damping' parameter with the amplitude and frequency of free limit cycle oscillations, (ii) the case of large values of the 'damping' parameter and the resulting relaxation oscillations. Then, generalized van der Pol oscillators, have been investigated in both cases. Their restoring force and the 'damping-like' force are of power-form. The results have been compared with those for the standard van der Pol oscillator. In the former case, the method of averaging has been used. The expressions for the amplitude and frequency of the limit cycle have been derived, and also simplified for certain special cases related to certain system parameters. In the latter case, the expressions for jumps and outer curves have been obtained by using a perturbation technique for distinguishable combinations of the system parameters. The resulting period of relaxation oscillations has been obtained for these three combinations and the differences between them have been pointed out.

Acknowledgements This research has been supported by the Ministry of Education, Science and Technological Development through the project no. 451-03-68/2020-14/200156.

References

1. Cartwright, M.L.: Balthasar van der Pol. J. Lond. Math. Soc. **35**, 367–376 (1960)
2. Bremmer, H.: The scientific work of Balthasar van der Pol. Philips Tech. Rev. **22**, 36–52 (1960/1961)
3. Matviishin, Ya.A.: The investigations of B. van der Pol in the theory of nonlinear oscillations (in Russian). Application of asymptotic methods in the theory of nonlinear differential equations (Russian). Akad. Nauk Ukrain SSR Kiev 70–77 (1987)
4. Stumpers, F.L.H.M: Balthasar van dr Pol's work on nonlinear circuits. IRE Trans. CT-**7** 366–367 (1960)
5. van der Pol, B.: A theory of the amplitude of free and forced triode vibrations. Radio Rev. **1**, 701–710, 754–762 (1920)
6. van der Pol, B.: On 'Relaxation Oscillations'. Phil. Mag. **2**, 978–992 (1926)
7. Ginoux, J.M.: van der Pol and the history of relaxation oscillations: towards the emergence of a concept. Chaos **22**, 023120 (37 pages) (2012)
8. van der Pol, B., van der Mark, J.: The heartbeat considered as a relaxation oscillation, and an electrical model of the heart. Lond. Edinb. Dublin Phil. Mag. J. Sci. Ser. 7, **6**, 763–775 (1928)
9. Minorsky, N.: Nonlinear Oscillations. Princeton, van Nostrand (1962)
10. Moremedi, G.M., Mason, D.P., Gorringe, V.M.: On the limit cycle of a generalized van der Pol equation. Int. J. Nonlinear Mech. **28**, 237–250 (1993)
11. Obi, C.: Analytical theory of nonlinear oscillations IV: the periodic oscillations of the equation $\ddot{x} - \epsilon(1 - x^{2n+2})\dot{x} + x^{2n+1} = \epsilon a \cos \omega t$, $a > 0$, $\omega > 0$ independent of ε. SIAM J. Appl. Math. **21**, 345–357 (1976)
12. Mickens, R.E., Oyedeji, K.: Construction of approximate analytical solutions to a new class of nonlinear oscillator equation. J. Sound Vib. **102**, 579–582 (1985)

13. Bravo Yuste, S., Diaz Bejarano, J.: Construction of approximate analytical solutions to a new class of nonlinear oscillator equation. J. Sound Vib. **110**, 347–350 (1986)
14. Garcia-Margallo, J., Diaz Bejarano, J.: A generalization of the method of the harmonic balance. J. Sound Vib. **116**, 1591–1595 (1987)
15. Chen, S.H., Cheung, Y.K.: An elliptic perturbation method for certain strongly nonlinear oscillators. J. Sound Vib. **192**, 453–464 (1996)
16. Mickens, R.E.: Fractional van der Pol equations. J. Sound Vib. **259**, 457–460 (2003)
17. Mickens, R.E.: Iteration method solutions for conservative and limit-cycle $x^{1/3}$ force oscillators. J. Sound Vib. **292**, 964–968 (2006)
18. Oyedeji, K.: An analysis of a nonlinear elastic force van der Pol oscillator equation. J. Sound Vib. **281**, 417–422 (2005)
19. Waluya, S.B., van Horssen, W.T.: On the periodic solutions of a generalized nonlinear van der Pol oscillator. J. Sound Vib. **268**, 209–215 (2003)
20. Andrianov, I.V., van Horssen, W.T.: Analytical approximations of the period of a generalized nonlinear van der Pol oscillator. J. Sound Vib. **295**, 1099–1104 (2006)
21. Kovacic, I.: A limit cycle and relaxation oscillations in a generalized van der Pol oscillator. Commun. Nonlinear Sci. **16**, 1640–1649 (2011)
22. Kovacic, I., Mickens, R.E.: A generalized van der Pol type oscillator: investigation of the properties of its limit cycle. Math. Comput. Model. **55**, 645–653 (2012)
23. Kovacic, I.: The method of multiple scales for forced oscillators with some real-power nonlinearities in the stiffness and damping force. Chaos Soliton Frac. **44**, 891–901 (2011)
24. Haag, J.: Asymptotic study of relaxation oscillations. Annales de l'Ecole Normale Superieure III **60**, 35–64, 65–111, 289 (errata) (1943)
25. Haag, J.: Concrete examples of asymptotic studies of relaxation oscillations. Annales de l'Ecole Normale Superieure III **61**, 73–117 (errata) (1944)
26. Dorodnitsyn, A.A.: Asymptotic solution of the van der Pol equation. Prikladnaya Mathematika i Mekhanika **XI**, 313–328 (1947)
27. Stoker, J.J.: Nonlinear Vibrations in Mechanical and Electrical Systems. Wiley-Interscience (1992)

Chapter 10
Theoretical Determination of the Five Physical Constants of the Toupin-Mindlin Gradient Elasticity for Polycrystalline Materials

Victor I. Malyi

Abstract A polycrystal is considered as a micro-inhomogeneous elastic medium, which consists of homogeneous anisotropic crystallites with random orientation. It is shown that the elastic energy density of the polycrystalline medium depends non-locally on the non-uniform field of average deformations. In the case of smooth fields of average deformations it could be considered that the energy density locally depends not only on the values of deformations, but also on the values of their derivatives. This dependence in the isotropic case is characterized by five gradient modules, in addition to the two elastic moduli of the classical theory of elasticity. For the first time, explicit expressions are obtained for the gradient modules of polycrystalline materials. This makes it possible to solve problems of the Toupin-Mindlin strain gradient theory for polycrystalline materials using real, rather than artificial parameters.

10.1 Introduction

Since the release of the classic monograph of Cosserat brothers (1909), activity has continued in the direction of generalizations of the theory of elasticity, which expand the possibilities for describing the effects accompanying elastic deformations of materials with an internal structure [1–11]. A good description of these theories can be found in comprehensive reviews, see, i.g., [1, 2]. It was expected that the results of such work will lead to a significant change in the values of the concentration coefficients and the types of stress singularities [1, 3, 12–14]. As a problem that impedes the use of generalized theories for solving real problems, there is a lack of quantitative data on the physical constants of these theories [1, 2]. As it is mentioned in [2], Generalized elasticity theories even for isotropic materials contain many additional constants that are difficult or impossible to determine experimentally. Theoretically, gradient moduli were previously determined [8] for a micro-inhomogeneous medium

V. I. Malyi (✉)
HSE Tikhonov Moscow Institute of Electronics and Mathematics,
34 Tallinskaya Ulitsa, 123458 Moscow, Russia
e-mail: vict-maly@mtu-net.ru

A. K. Abramian et al. (eds.), *Nonlinear Dynamics of Discrete and Continuous Systems*,
Advanced Structured Materials 139,
https://doi.org/10.1007/978-3-030-53006-8_10

composed of homogeneous and isotropic grains, whose Lamé's constants are random variables. Recently, gradient modules have been defined for two-phase composites in the case of a low concentration of inclusions in a homogeneous matrix [15, 16]. It is shown that it is possible to estimate the values of the gradient modules and the corresponding characteristic lengths for the crystal lattice of metals by the methods of the density functional theory (DFT) [11].

In this paper closed-form expressions for five physical constants of the Toupin-Mindlin theory of isotropic gradient elasticity for polycrystalline materials were obtained.

10.2 Description of the State of a Micro-Inhomogeneous Elastic Medium

We consider a polycrystalline medium as micro-inhomogeneous, but statistically homogeneous and statistically isotropic. Tensor of its elastic modules is written in the form

$$\lambda_{iklm}(\mathbf{r}) = \langle\lambda_{iklm}\rangle + \Delta_{iklm}(\mathbf{r}),$$

where the angle brackets denote averaging over the statistical ensemble of micro-inhomogeneous media. In this case, for random deviations we have $\langle\Delta_{iklm}(\mathbf{r})\rangle = 0$ and the Voigt average tensor of elastic moduli $\langle\lambda_{iklm}\rangle$ due to macroscopic homogeneity and isotropy of the medium must have the following structure

$$\langle\lambda_{iklm}\rangle = \lambda_V\delta_{ik}\delta_{lm} + \mu_V(\delta_{il}\delta_{km} + \delta_{im}\delta_{kl}). \tag{10.1}$$

Here δ_{pq} is the Kronecker delta, λ_V, μ_V are the Lamé's constants, corresponding to the Voigt average tensor.

Static displacements $u_i(\mathbf{r})$ that occur in an infinite domain under the action of distributed forces $F_i(\mathbf{r})$ are described by the equations

$$\langle\lambda_{iklm}\rangle\, u_{l,mk}(\mathbf{r}) + \left(\Delta_{iklm}(\mathbf{r})\, u_{l,m}(\mathbf{r})\right)_{,k} + F_i(\mathbf{r}) = 0 \tag{10.2}$$

with boundary conditions $u_i(\mathbf{r}) \to 0$ at $\mathbf{r} \to \infty$. It is possible to create any displacement field by selecting the appropriate forces $F_i(\mathbf{r})$.

Solution of the problem (10.2) satisfies the integral equation

$$u_i(\mathbf{r}) = u_i^0(\mathbf{r}) + \int dV^1 g_{ij}(\mathbf{r}-\mathbf{r}^1)\left(\Delta_{jklm}(\mathbf{r}^1)\, u_{l,m}(\mathbf{r}^1)\right)_{,k},$$

where

$$u_i^0(\mathbf{r}) = \int dV^1 g_{ij}(\mathbf{r}-\mathbf{r}^1) F_j\left(\mathbf{r}^1\right) \tag{10.3}$$

are displacements of a homogeneous medium in the field of forces $F_i(\mathbf{r})$ and $g_{ij}(\mathbf{r})$ is the Green's function of a homogeneous medium with Lamé's constants λ_V, μ_V. Considering $\Delta_{iklm}(\mathbf{r})$ as a perturbation, in the 2nd approximation we obtain the expression for displacements

$$u_i(\mathbf{r}) = u_i^0(\mathbf{r}) + u_i^1(\mathbf{r}) + u_i^2(\mathbf{r}) + O(\Delta^3), \tag{10.4}$$

where

$$u_i^1(\mathbf{r}) = \int dV^1 g_{ij}(\mathbf{r} - \mathbf{r}^1) \left(\Delta_{jklm}(\mathbf{r}^1)\, u_{l,m}^0(\mathbf{r}^1)\right)_{,k}, \tag{10.5}$$

$$u_i^2(\mathbf{r}) = \int dV^1 g_{ij}(\mathbf{r} - \mathbf{r}^1) \left(\Delta_{jklm}(\mathbf{r}^1)\, u_{l,m}^1(\mathbf{r}^1)\right)_{,k}. \tag{10.6}$$

The expressions (10.3), (10.5), (10.6) for each of the terms of the sum (10.4) are integrals of the product of deterministic quantities $F_i(\mathbf{r})$, $g_{ij}(\mathbf{r})$, $u_i^0(\mathbf{r})$ by random variable $\Delta_{iklm}(\mathbf{r})$. This allows us to express the density of elastic energy

$$\begin{aligned} A[\mathbf{r}, u_i] &= \tfrac{1}{2}\lambda_{iklm} u_{i,k} u_{l,m} \\ &= \tfrac{1}{2}\left(\langle\lambda_{iklm}\rangle + \Delta_{iklm}\right)\left(u_{i,k}^0 + u_{i,k}^1 + u_{i,k}^2 + \cdots\right)\left(u_{l,m}^0 + u_{l,m}^1 + u_{l,m}^2 + \cdots\right) \end{aligned}$$

in terms of the same values. Since, when averaging, all the terms of the first order of $\Delta_{iklm}(\mathbf{r})$ turn to zero, for the average density of elastic energy we obtain

$$\begin{aligned} \langle A[\mathbf{r}, u_i]\rangle &= \tfrac{1}{2}\langle\lambda_{iklm}\rangle u_{i,k}^0 u_{l,m}^0 + \langle\lambda_{iklm}\rangle u_{i,k}^0 \left\langle u_{l,m}^2\right\rangle \\ &\quad + u_{i,k}^0 \left\langle \Delta_{iklm} u_{l,m}^1\right\rangle + \tfrac{1}{2}\langle\lambda_{iklm}\rangle \left\langle u_{i,k}^1 u_{l,m}^1\right\rangle + O(\Delta^3) \end{aligned} \tag{10.7}$$

10.3 Macroscopic Description of a Micro-Inhomogeneous Medium

According to (10.3)–(10.6), the displacements $u_i(\mathbf{r})$ of a micro-inhomogeneous medium and the corresponding deformations $\varepsilon_{ij}(\mathbf{r})$ are random functions that are deterministically expressed through random elastic modules. Therefore, measurements at the point $\mathbf{r}_0$ of local values of functions $u_i(\mathbf{r})$ and $\varepsilon_{ij}(\mathbf{r})$ for different experimental samples lead to different results, $u_i^{(r)}(\mathbf{r}_0)$ and $\varepsilon_{ij}^{(r)}(\mathbf{r}_0)$, depending on the random realization of the modules $\Delta_{iklm}^{(r)}(\mathbf{r})$ in a particular sample. Due to this scatter of values, the result of the experiment in the macroscopic sense is considered to be the average value of the results, $\langle u_i(\mathbf{r}_0)\rangle$ and $\langle\varepsilon_{ij}(\mathbf{r}_0)\rangle$, for the entire set of samples [17, 18]. At the same time, the number of samples must be so large that its further increase does not affect the result of averaging. A rather representative set of samples in this sense is called the ensemble of realizations of a micro-inhomogeneous medium [17, 18].

We will proceed from the fact that in macroscopic experiments the directly observable quantities are only external forces $F_i(\mathbf{r})$, the average values of the displacements $\langle u_i(\mathbf{r})\rangle$ and the corresponding deformations $\langle \varepsilon_{ij}(\mathbf{r})\rangle$. In this case, the average density of elastic energy $\langle A[\mathbf{r}, u_i]\rangle$ is determined by the work of external forces $F_i(\mathbf{r})$ at average displacements $\langle u_i(\mathbf{r})\rangle$. On the other hand, elastic energy (10.7) is formally expressed through the terms $u_i^0(\mathbf{r})$, $u_i^1(\mathbf{r})$ and $\langle u_i^2(\mathbf{r})\rangle$ of the displacement fields (10.4) and the corresponding expression for the average values

$$\langle u_i(\mathbf{r})\rangle = u_i^0(\mathbf{r}) + \langle u_i^2(\mathbf{r})\rangle + O(\Delta^3). \tag{10.8}$$

The influence of external forces $F_i(\mathbf{r})$ on the components of the displacements $u_i^0(\mathbf{r})$ is elementarily described by the relation (10.3), but their influence on the functions $u_i^1(\mathbf{r})$ and $\langle u_i^2(\mathbf{r})\rangle$ is difficult to analyze using the formulas (10.5), (10.6).

All macroscopic properties of an elastic medium are contained in the expression for its average elastic energy $\langle A[\mathbf{r}, u_i]\rangle$. Therefore, for theoretical modeling of experimental situations, it is desirable to explicitly express the functional $\langle A[\mathbf{r}, u_i]\rangle$ through the field of average displacements $\langle u_i(\mathbf{r})\rangle$. For this, it is necessary to exclude from the expressions (10.5) and (10.7) the functions $u_i^0(\mathbf{r})$ and $\langle u_i^2(\mathbf{r})\rangle$. This can be done by change of variables in expressions (10.5), (10.7)

$$u_i^0(\mathbf{r}) = \langle u_i(\mathbf{r})\rangle - \langle u_i^2(\mathbf{r})\rangle + O(\Delta^3), \tag{10.9}$$

which is a direct consequence of equality (10.8). As a result, the general expression of the average elastic energy density (10.7) up to the third order of smallness is reduced to the functional depending only from the average displacements

$$\begin{aligned}\langle A[\mathbf{r}, u_i]\rangle &= \tfrac{1}{2}\langle \lambda_{iklm}\rangle \langle u_{i,k}(\mathbf{r})\rangle \langle u_{l,m}(\mathbf{r})\rangle \\ &\quad + \langle u_{i,k}(\mathbf{r})\rangle \langle \Delta_{iklm} U_{l,m}^1\rangle + \tfrac{1}{2}\langle \lambda_{iklm}\rangle \langle U_{i,k}^1 U_{l,m}^1\rangle + O(\Delta^3),\end{aligned} \tag{10.10}$$

where

$$U_i^1(\mathbf{r}) = \int dV^1 g_{ij}(\mathbf{r}-\mathbf{r}^1)\left(\Delta_{jklm}(\mathbf{r}^1)\langle u_{l,m}(\mathbf{r}^1)\rangle\right)_{,k}. \tag{10.11}$$

When calculating the second and third terms of the sum (10.10), the average of the pair products of the modules

$$\langle \Delta_{jrpq}(\mathbf{r})\Delta_{styz}(\mathbf{r}^1)\rangle = f_{styz}^{jrpq}(\mathbf{r}, \mathbf{r}^1)$$

appears, which is the correlation function of the field of these modules. Due to the macroscopic homogeneity of the properties of a polycrystal, its correlation function can depend only on the difference of the arguments [17, 18]

$$f_{styz}^{jrpq}(\mathbf{r}-\mathbf{r}^1) = \langle \Delta_{jrpq}(\mathbf{r})\Delta_{styz}(\mathbf{r}^1)\rangle. \tag{10.12}$$

Taking into account (10.11), (10.12) expression (10.10) can be reduced to the following form:

$$
\begin{aligned}
\langle A[\mathbf{r}, u_i]\rangle = {} & \tfrac{1}{2}\langle \lambda_{iklm}\rangle \langle u_{i,k}(\mathbf{r})\rangle \langle u_{l,m}(\mathbf{r})\rangle \\
& + \langle u_{l,m}(\mathbf{r})\rangle \int dV^1 g_{ij,kr}(\mathbf{r}-\mathbf{r}^1) f^{iklm}_{jrpq}(\mathbf{r}-\mathbf{r}^1)\langle u_{p,q}(\mathbf{r}^1)\rangle \\
& + \tfrac{1}{2}\langle \lambda_{iklm}\rangle \iint dV^1 dV^2 g_{ij,kr}(\mathbf{r}-\mathbf{r}^1)\, g_{ls,mt}(\mathbf{r}-\mathbf{r}^2) \\
& \cdot f^{jrpq}_{styz}(\mathbf{r}^1-\mathbf{r}^2)\langle u_{p,q}(\mathbf{r}^1)\rangle \langle u_{y,z}(\mathbf{r}^2)\rangle .
\end{aligned}
\tag{10.13}
$$

Although, in general, formula (10.13) for the energy density was obtained by methods traditional for the mechanics of micro-inhomogeneous media, the use of substitution (10.9) was nontrivial. As a result, the expression for the density of elastic energy $\langle A[\mathbf{r}, u_i]\rangle$ was freed from the cumbersome expressions for the corrections (10.3), (10.5), (10.6) to the displacements (10.4).

According to (10.13), for a micro-inhomogeneous elastic medium, the general structure of the relationship between the average density of elastic energy and average displacements corresponds to the non-local theory of elasticity [8, 9, 19, 20]. In this case, in accordance with the results of [8], the specific properties of the nonlocality of the energy functional (10.13) are completely determined by the correlation function of the micro-inhomogeneities of the elastic moduli $f^{jrpq}_{styz}(\mathbf{r}-\mathbf{r}^1)$. In particular, the characteristic length of a decrease in the correlation function $f^{jrpq}_{styz}(\mathbf{r})$ determines the characteristic distances at which non-local effects occur for a medium with an energy density (10.13). This characteristic length for polycrystals always exceeds the grain size, since within each individual crystallite the properties are constant, i.e. completely correlated.

Although a single sample of a polycrystalline body is described by the local classical theory of elasticity of micro-inhomogeneous bodies, the averaged macroscopic properties of a representative ensemble of polycrystalline samples turn out to be homogeneous in a macroscopic sense. Therefore, the transition from the consideration of a particular sample to the description of the averaged macroscopic properties of an ensemble of samples can be called homogenization. In this case, it turns out that macroscopic properties are described by a nonlocal theory of elasticity with an energy density (10.13).

It is necessary to distinguish the homogenization of a continuous microinhomogeneous medium (see also [8]) from the continualization of a discrete elastic medium, such as a crystal lattice [9, 19, 20, 22, 23]. Formally, both approaches lead to theories with a nonlocal dependence of the density of elastic energy on the displacement field, but from the physical point of view, the result of the lattice continualization differs significantly from the homogenization of a continuous micro-inhomogeneous medium. First, non-local effects for the lattice are include into the theory from the very beginning due the long-range interatomic forces. Secondly, the characteristic lengths of nonlocal lattice effects correspond in order of magnitude to interatomic distances, i.e. their influence is many orders of magnitude less than in the case of polycrystals. And thirdly, a correct description of the lattice behavior requires consideration of its anisotropy, which further complicates the already cumbersome description of non-local effects. For example, the local Toupin - Mindlin gradient theory obtained as a result of homogenization requires the introduction of five additional physical constants (see [8] and Sect. 6 below). But after the continualization of even a simple cubic lattice, eleven additional physical constants appear [21].

10.4 Accounting for the Random Texture of Crystallites Constituting a Polycrystal

Correlation function (10.12) in the general case of materials with a texture, when the orientations of neighboring grains cannot be considered independent, can only be determined experimentally. But for an isotropic polycrystal in the absence of texture, when the orientations of neighboring grains are random and independent, the average product of random variables $\left\langle \Delta_{jrpq}(\mathbf{r})\Delta_{styz}(\mathbf{r}^1)\right\rangle$ vanishes if the points $\mathbf{r}$, $\mathbf{r}^1$ belong to different grains. If these points $\mathbf{r}$, $\mathbf{r}^1$ belong to the same grain, then $\Delta_{iklm}(\mathbf{r}^1) = \Delta_{iklm}(\mathbf{r})$ because of the homogeneity of the polycrystal grains, and then we get $\left\langle \Delta_{jrpq}(\mathbf{r})\Delta_{styz}(\mathbf{r}^1)\right\rangle = f_{styz}^{jrpq}(0)$. Therefore, if we define a scalar function $w(\left|\mathbf{r} - \mathbf{r}^1\right|)$ as the probability of finding points $\mathbf{r}$, $\mathbf{r}^1$ in a single grain, then for an isotropic polycrystal in the absence of texture, the correlation function of micro-inhomogeneities takes the form

$$f_{styz}^{jrpq}(\mathbf{r} - \mathbf{r}^1) = w(\left|\mathbf{r} - \mathbf{r}^1\right|)\, f_{styz}^{jrpq}(0). \tag{10.14}$$

As a result, the general expression of the average density of the elastic energy (10.13) for an isotropic polycrystal in the absence of texture is reduced to the functional

$$\begin{aligned}
\langle A[\mathbf{r}, u_i]\rangle &= \tfrac{1}{2}\,\langle \lambda_{iklm}\rangle \left\langle u_{i,k}(\mathbf{r})\right\rangle \left\langle u_{l,m}(\mathbf{r})\right\rangle \\
&+ \left\langle u_{l,m}(\mathbf{r})\right\rangle f_{jrpq}^{iklm}(0) \int dV^1 g_{ij,kr}(\mathbf{r} - \mathbf{r}^1)\, w(\left|\mathbf{r} - \mathbf{r}^1\right|) \left\langle u_{p,q}(\mathbf{r}^1)\right\rangle \\
&+ \tfrac{1}{2}\,\langle \lambda_{iklm}\rangle\, f_{styz}^{jrpq}(0) \iint dV^1 dV^2 g_{ij,kr}(\mathbf{r} - \mathbf{r}^1)\, g_{ls,mt}(\mathbf{r} - \mathbf{r}^2) \\
&\quad \cdot w(\left|\mathbf{r}^1 - \mathbf{r}^2\right|) \left\langle u_{p,q}(\mathbf{r}^1)\right\rangle \left\langle u_{y,z}(\mathbf{r}^2)\right\rangle .
\end{aligned} \tag{10.15}$$

10.5 Gradient Modules for Micro-Inhomogeneous Media

Due to the rapid decay of the probability function $w(r)$ at distances exceeding the average grain size R, a small neighborhood of a point $\mathbf{r}$ essential for integration in (10.15). In the case of sufficiently smooth distortion fields, they can be approximated in this region by linear expressions

$$\left\langle u_{p,q}(\mathbf{r}^1)\right\rangle = \left\langle u_{p,q}(\mathbf{r})\right\rangle + \left\langle u_{p,qv}(\mathbf{r})\right\rangle \left(r_v^1 - r_v\right) + O\left(\left|\mathbf{r}^1 - \mathbf{r}\right|^2\right). \qquad (10.16)$$

In this case, the functional (10.15) becomes a quadratic form of the expansion coefficients (10.16):

$$\langle A[\mathbf{r}, u_i]\rangle = \tfrac{1}{2}\Lambda_{iklm}\left\langle u_{i,k}\left(\mathbf{r}\right)\right\rangle\left\langle u_{l,m}\left(\mathbf{r}\right)\right\rangle + \tfrac{1}{2}B_{yzw}^{pqv}\left\langle u_{p,qv}\left(\mathbf{r}\right)\right\rangle\left\langle u_{y,zw}\left(\mathbf{r}\right)\right\rangle. \qquad (10.17)$$

The coefficients Λ_{iklm} of the first term of the quadratic form (10.17) coincide with the effective elastic moduli of the polycrystal, defined in [17]:

$$\begin{aligned}\Lambda_{iklm} &= \langle\lambda_{iklm}\rangle - \frac{1}{15}\frac{3\lambda_V + 8\mu_V}{\mu_V\left(\lambda_V + 2\mu_V\right)} f_{ikpq}^{lmpq}(0) \\ &+ \frac{1}{15}\frac{\lambda_V + \mu_V}{\mu_V\left(\lambda_V + 2\mu_V\right)} f_{ikpp}^{lmjj}(0).\end{aligned} \qquad (10.18)$$

The modulus tensor B_{yzw}^{pqv} of the gradient theory of elasticity of a polycrystalline medium appearing in (10.17) is determined by explicit expressions:

$$B_{yzw}^{pqv} = f_{styz}^{jrpq}\left(0\right) L^2 \left\langle\lambda_{iklm}\right\rangle \frac{1}{4\pi} \oint\limits_{|\mathbf{k}|=1} d\Omega \left(k_k k_r G_{ij}(\mathbf{k})\right)_{,v} \left(k_m k_t G_{ls}(\mathbf{k})\right)_{,w}, \qquad (10.19)$$

$$L^2 = \int\limits_0^\infty r\, w(r)\, dr. \qquad (10.20)$$

Here $G_{ij}(\mathbf{k})$ is the Fourier transform for the Green function $g_{ij}(\mathbf{r})$, and the index after the comma denotes the derivative with respect to the corresponding component of the vector $\mathbf{k}$.

Keeping in approximation (10.16) a larger number of terms, one obtains a quadratic form containing Nth ($N > 2$) derivatives of displacements. Such expression for the density of elastic energy corresponds to the material grade N due Toupin classification [6].

10.6 Gradient Modules for Polycrystalline Materials with Cubic Symmetry of Crystallites

Further simplifications depend on the type of symmetry of the crystallites. For cubic symmetry single crystals, all elastic moduli are expressed in terms of three moduli c_{1111}, c_{1122}, c_{2323} and the result of averaging the moduli over orientations is the isotropic elastic modulus tensor (10.1) with parameter values [17, 18]

$$\lambda_V = \frac{1}{5}\left(c_{1111} + 4c_{1122} - 2c_{2323}\right), \quad \mu_V = \frac{1}{5}\left(c_{1111} - c_{1122} + 3c_{2323}\right).$$

Using in (10.18), (10.19) values of the correlation function $f^{jrpq}_{styz}(0)$ are well known, see, e.g., [18].

The effective elastic moduli of the polycrystal (10.18) are reduced to the form that coincides with the one obtained in [17]:

$$\Lambda_{iklm} = \lambda_L \delta_{ik}\delta_{lm} + \mu_L\left(\delta_{il}\delta_{km} + \delta_{im}\delta_{kl}\right),$$

where

$$\lambda_L = \lambda_V + \frac{2c^2}{375\mu_V}\frac{3\lambda_V + 8\mu_V}{\lambda_V + 2\mu_V}, \quad \mu_L = \mu_V - \frac{c^2}{125\mu_V}\frac{3\lambda_V + 8\mu_V}{\lambda_V + 2\mu_V}.$$

Here $c = c_{1111} - c_{1122} - 2c_{2323}$ is the characteristic of the anisotropy of the elastic moduli of cubic symmetry crystallites.

Performing integration in (10.19), we obtain the tensor of gradient modules for an isotropic polycrystalline in the case of cubic symmetry of crystallites:

$$\begin{aligned} B^{pqv}_{yzw} &= B_1\delta_{pq}\delta_{yz}\delta_{vw} + B_2(\delta_{py}\delta_{qz}\delta_{vw} + \delta_{pz}\delta_{qy}\delta_{vw}) \\ &+ B_3(\delta_{pq}\delta_{vy}\delta_{zw} + \delta_{pq}\delta_{vz}\delta_{yw} + \delta_{pv}\delta_{qw}\delta_{yz} + \delta_{pw}\delta_{qv}\delta_{yz}) \\ &+ B_4(\delta_{pv}\delta_{qy}\delta_{zw} + \delta_{pv}\delta_{qz}\delta_{yw} + \delta_{py}\delta_{qv}\delta_{zw} + \delta_{pz}\delta_{qv}\delta_{yw}) \\ &+ B_5(\delta_{pw}\delta_{qy}\delta_{zv} + \delta_{pw}\delta_{qz}\delta_{yv} + \delta_{py}\delta_{qw}\delta_{zv} + \delta_{pz}\delta_{qw}\delta_{yv}), \end{aligned} \tag{10.21}$$

where

$$\begin{aligned} &B_1 = -4C\left(105 - 82\kappa + 152\kappa^2\right), \quad B_2 = 4C\left(140 - 37\kappa + 212\kappa^2\right), \\ &B_3 = 2C\left(35 - 172\kappa + 32\kappa^2\right), \quad B_4 = -C\left(77 - 160\kappa + 48\kappa^2\right), \\ &B_5 = -4C\left(7 - 89\kappa + 12\kappa^2\right), \quad C = \frac{L^2c^2}{18375\mu_V}, \quad \kappa = \frac{\lambda_V + \mu_V}{\lambda_V + 2\mu_V}. \end{aligned} \tag{10.22}$$

The constant L in (10.19), (10.20) is calculated from the statistical data on the average size of polycrystal grains, which are determined experimentally by standard methods. In the absence of statistical data on the probability $w(r)$ of finding two points inside a single grain, the constant L can be estimated [8] considering the grains closing in shape to balls with a radius R. Then one obtains

$$w(r) = \begin{cases} 1 - \frac{3}{4}\frac{r}{R} + \frac{r^3}{16R^3}, & r \leqslant 2R, \\ 0, \ 2\mathrm{R} \leqslant r, \end{cases}, \quad L^2 = \int_0^\infty r\, w(r)\, dr = \frac{2R^2}{5}$$

10.7 Conclusions

In studies on the couple-stress elasticity [3, 5], it was concluded from dimensional analysis that the gradient elastic moduli should be of the order $B \sim \mu R^2$, where R is the characteristic size of micro-inhomogeneities, and μ is the macroscopic shear modulus. Relations (10.21), (10.22) show that the magnitude of the gradient modules for polycrystals should be estimated as $B \sim c^2 R^2/\mu$. It is evident a great influence of anisotropy c of the crystallites elastic moduli on the gradient modules B. It is also obvious that the real role of the macroscopic shear modulus μ is radically different from the predictions of [3, 5].

Complete and well-posed formulation of any real problem is impossible without specifying the values of all elastic constants. Therefore, because of the principal difficulties for determining the values of gradient modules [1, 2], all publications on solving problems for gradient media were characterized by non-quantitative studies of general regularities. Without a confident definition of gradient modules, it is impossible even to determine which of the large number of considered variants of gradient theories should be used in this particular case.

From the results of this work it follows that the general structure of the elastic energy density (10.17) and of the tensor (10.21) of gradient modules B^{pqv}_{yzw} corresponds to the Toupin-Mindlin strain gradient theory with five physical constants [4, 6, 7]. For polycrystalline materials, the values of gradient modules (10.21) are theoretically determined, based on the well-known values of the elastic moduli of crystallites. This makes it possible to solve problems of the Toupin-Mindlin strain gradient theory for polycrystalline materials using real, rather than artificial parameters.

Acknowledgements The author greatly appreciates discussions on different subjecs related to this paper with Professor I. Andrianov.

References

1. Maugin, G.A.: Continuum Mechanics Through the Twentieth Century. A Concise Historical Perspective. Springer, Dordrecht (2013)
2. Vasiliev, V.V., Lurie, S.A.: Correct nonlocal generalized theories of elasticity. Phys. Mesomech. **19**(3), 269–281 (2016)

3. Mindlin, R.D., Tiersten, H.F.: Effects of couple-stresses in linear elasticity. Arch. Ration. Mech. Anal. **11**(5), 415–448 (1962)
4. Toupin, R.A.: Elastic materials with couple-stresses. Arch. Ration. Mech. Anal. **11**(1), 385–414 (1962)
5. Koiter, W.T.: Couple-stresses in the theory of elasticity. Proc. Koninkl. Nederl. Akad. Wet. **67**(1), 17–44 (1964)
6. Toupin, R.A.: Theories of elasticity with couple-stresses. Arch. Ration. Mech. Anal. **17**(2), 85–112 (1964)
7. Mindlin, R.D.: Microstructure in linear elasticity. Arch. Ration. Mech. Anal. **16**(1), 51–78 (1964)
8. Malyi, V.I.: About nonlocal theory of elasticity. In: Paper Presented at IV All-Union Conference on Strength and Plasticity, Moscow. 1967; text published in: Prochnost i Plastichnost‘ (Strength and Plasticity). Trudi (Proceedings) of IV All-Union Conference on Strength and Plasticity, pp. 74–78. Nauka, Moscow (1971) (in Russian)
9. Kunin, I.A.: Model of an elastic medium of simple structure with three-dimensional dispersion. J. Appl. Math. Mech. **30**(3), 642–652 (1966)
10. Hadjesfandiari, A.R., Dargush, G.F.: Couple stress theory for solids. Int. J. Solids Struct. **48**, 2496–2510 (2011)
11. Shodja, H.M., Zaheri, A., Tehranchi, A.: Ab initio calculations of characteristic lengths of crystalline materials in first strain gradient elasticity. Mech. Mater. **61**, 73–78 (2013)
12. Hadjesfandiari, A.R., Dargush, G.F.: Fundamental solutions for isotropic size-dependent couple stress elasticity. Int. J. Solids Struct. **50**, 1253–1265 (2013)
13. Gourgiotis, P.A., Zisis, Th, Georgiadis, H.G.: On concentrated surface loads and Green’s functions in the Toupin-Mindlin theory of strain-gradient elasticity. Int. J. Solids Struct. **130–131**, 153–171 (2018)
14. Gourgiotis, P.A., Zisis, Th., Giannakopoulos, A.E., Georgiadis, H.G.: The Hertz contact problem in couple-stress elasticity. Int. J. Solids Struct. **168**, 228–237 (2019)
15. Bigoni, D., Drugan, W.J.: Analitical derivation of Cosserat moduli via homogenization of heterogeneous elastic materials. J. Appl. Mech. **74**, 741–753 (2007)
16. Bacca, M., Bigoni, D., Dal Corso, F., Veber, D.: Mindlin second-gradient elastic properties from dilute two-phase Cauchy-elastic composites. Part II: Higher-order constitutive properties and application cases. Int. J. Solids Struct. **50**, 4020–4029 (2013)
17. Lifshitz, I.M., Rosenzweig, L.N.: On the theory of the elastic properties of polycrislalls. J. Exp. Theor. Phys. **16**, 967–980 (1946). Erratum: J. Exp. Theor. Phys. **21**, 1184 (1951) (in Russian)
18. Shermergor, T.D.: The Theory of Elasticity of Microinhomogeneous Media. Nauka, Moscow (1977). (in Russian)
19. Eringen, A.C.: Linear theory of nonlocal elasticity and dispersion of plane waves. Int. J. Eng. Sci. **10**, 425–435 (1972)
20. Eringen, A.C.: Nonlocal Continuum Field Theories. Springer, New York (2002)
21. Auffray, N., Le Quang, H., He, Q.C.: Matrix representations for 3D strain-gradient elasticity. J. Mech. Phys. Solids **61**(5), 1202–1223 (2013)
22. Kunin, I.A.: Elastic Media with Microstructure. I. One-Dimensional Models. Springer, Berlin (1982)
23. Kunin, I.A.: Elastic Media with Microstructure. II. Three-Dimensional Models. Springer, Berlin (1983)

Chapter 11
A Parametrically Excited Nonlinear Wave Equation

Ferdinand Verhulst and Johan M. Tuwankotta

Abstract When considering nonlinear waves with periodic parametric forcing the geometry of the spatial domain plays a crucial part. If the spatial domain is a square we find an infinite number of 1 : 1 resonances and in addition accidental resonances. Using Galerkin projection on 2 modes in 1 : 1 resonance we find stable normal mode periodic solutions and unstable periodic solutions in general position; the location in phase-space is characterised as a triple resonance zone. In the limit case of vanishing dissipation we find neutral stability and strong recurrence of the orbits. Interaction of 1 : 1 resonances shows a selection mechanism of the 1 : 1 modes triggered off by the parametric forcing. In addition we analyse a number of prominent accidental resonances produced by the spectrum induced by our choice of a square in space.

11.1 Introduction

Consider the parametrically excited nonlinear wave equation formulated by Rand et al. [4] in the one-dimensional case, see also [1]; we will consider the equation on a square as two space dimensions often introduces new phenomena, in particular resonances.

$$u_{tt} - c^2(u_{xx} + u_{yy}) + \mu u_t + ({\omega_0}^2 + \beta \cos(\Omega t))u = \alpha u^3, \tag{11.1}$$

where $t \geq 0$ and $0 < x < \pi, 0 < y < \pi$. The boundary values are $\partial u/\partial n|_S = 0$.

The parameters μ, β are positive and small in a way to be specified.

F. Verhulst (✉)
Mathematisch Instituut, PO Box 80.010, 3508TA Utrecht, Netherlands
e-mail: F.Verhulst@uu.nl

J. M. Tuwankotta
Analysis and Geometry Group, Faculty of Mathematics and Natural Sciences, Institut Teknologi Bandung, Jl. Ganesha no 10, Bandung, Indonesia
e-mail: jmtuwankotta@itb.ac.id

A. K. Abramian et al. (eds.), *Nonlinear Dynamics of Discrete and Continuous Systems*, Advanced Structured Materials 139,
https://doi.org/10.1007/978-3-030-53006-8_11

The system of equations and conditions model the surface deflections $u(x, y, t)$ of a fluid in a square basin with parametric excitation and damping, c is the wave speed.

Resonant nonlinear waves in 2 spatial dimensions were also considered in [3, 6]. We associate with the system the eigenfunctions:

$$v_{mn}(x, y) = \cos mx \cos ny, m, n = 0, 1, 2 \ldots$$

with eigenvalues of the space-dependent operator:

$${\omega_{mn}}^2 = {\omega_0}^2 + (m^2 + n^2)c^2, \omega_{mn} = \omega_{nm} = \omega.$$

An early paper by W.T. van Horssen on the asymptotic approximation of solutions of nonlinear wave equations is [2]. The solutions of Eq. (11.1) with boundary conditions can be approximated by projection of a finite sum of eigenfunctions (Galerkin projection) followed by averaging approximation. The process results in asymptotic approximations in the mathematical sense. The procedure is summarised with references in [6] Sect. 1, we do not repeat this here.

The choice of eigenfunctions is determined by the initial values of Eq. (11.1) while keeping an eye on the resonances of the eigenvalues. It turns out that for the geometry considered here, there are an infinite number of 1 : 1 resonances. This will require our main attention. In addition we will briefly look at prominent accidental resonances.

11.2 The Two-Mode 1 : 1 Resonance

We propose a two-mode expansion with:

$$u_p(x, y, t) = u_1(t) \cos mx \cos ny + u_2(t) \cos nx \cos my, \tag{11.2}$$

$m, n = 0, 1, 2 \ldots, m \neq n$. Put $\omega_0 = 1$ and rescale $u = \sqrt{\varepsilon}\bar{u}$ (and its derivatives likewise) in Eq. (11.1) with ε a small positive parameter; we omit the bars. Substituting expansion (11.2) into Eq. (11.1) and taking inner products with the eigenfunctions we find with $\omega_{mn} = \omega, m \neq n$:

$$\begin{cases} \ddot{u}_1 + \omega^2 u_1 & = -\mu\dot{u}_1 - \beta u_1 \cos(\Omega t) + \varepsilon\alpha(\frac{9}{16}{u_1}^3 + \frac{3}{4}u_1{u_2}^2), \\ \ddot{u}_2 + \omega^2 u_2 & = -\mu\dot{u}_2 - \beta u_2 \cos(\Omega t) + \varepsilon\alpha(\frac{9}{16}{u_2}^3 + \frac{3}{4}{u_1}^2 u_2). \end{cases} \tag{11.3}$$

We choose $\Omega = 2\omega$ to study prominent Floquet resonances; rescale $\mu = \varepsilon\bar{\mu}, \beta = \varepsilon\bar{\beta}$ after which we omit the bars. System (11.3) contains the 1 : 2 Floquet resonance and in addition the 1 : 1 resonance of the Hamiltonian interaction force.

Note that because of the symmetry of system (11.3) $u_1(t) = \pm u_2(t)$ satisfies the system.

The coordinate planes $u_1, \dot{u}_1$ and $u_2, \dot{u}_2$ are invariant under the phase-flow, we start with the analysis of these normal mode planes.

11.2.1 The Invariant Normal Mode Planes

The analysis for both coordinate planes runs exactly along the same lines with symmetric results so we consider only the $u_1, \dot{u}_1$ plane. We put $u_2 = \dot{u}_2 = 0$ and introduce amplitude-phase coordinates by:

$$u_1 = r_1 \cos(\omega t + \psi_1), \dot{u}_1 = -r_1 \omega \sin(\omega t + \psi_1).$$

Deriving the equations for r_1, ψ_1 and averaging over time we find the first order averaged system:

$$\dot{r}_1 = \frac{\varepsilon}{2} r_1 (-\mu + \frac{\beta}{2\omega} \sin 2\psi_1),\ \dot{\psi}_1 = \frac{\varepsilon}{4\omega} (\beta \cos 2\psi_1 - \alpha \frac{27}{32} {r_1}^2). \tag{11.4}$$

Here and in the sequel, the solutions of first order averaged equations with appropriate initial values approximate the solutions of the original system with error $O(\varepsilon)$ on a long interval of time of order $1/\varepsilon$. A critical point corresponding with an equilibrium of system (11.4) is given by:

$$\beta \sin 2\psi_1 = 2\mu\omega,\ \beta \cos 2\psi_1 = \alpha \frac{27}{32} {r_1}^2,\ 0 < \frac{2\mu\omega}{\beta} < 1.$$

A critical point of the averaged equations corresponds under certain conditions with a periodic solution of the original equations; see theorems 11.5–11.6 in [5] (this is sometimes called the 2nd Bogoliubov theorem). We can eliminate the phase angle to find:

$${r_1}^2 = {r_0}^2 = \frac{32}{27\alpha} \sqrt{\beta^2 - 4\mu^2\omega^2}.$$

Computing eigenvalues at the critical point shows that the periodic solution is stable within the invariant coordinate plane. For the eigenvalues we have:

$$\lambda_{1,2} = -\mu \pm \sqrt{5\mu^2 - \frac{\beta^2}{\omega^2}}. \tag{11.5}$$

If $\beta > \sqrt{5}\mu\omega$ the periodic solution is complex stable in the coordinate plane, if $2\mu\omega < \beta < \sqrt{5}\mu\omega$ the periodic solution is stable with real eigenvalues. If $\beta = 2\mu\omega$ the periodic solution vanishes.

An important question is whether the periodic solution is stable or unstable in the full 4-dimensional system. For u_2, $\dot{u}_2$ near zero we should not use polar coordinates. Instead we introduce in system (11.3) the variables a, b by:

$$u_2 = a\cos\omega t + \frac{b}{\omega}\sin\omega t,\ \dot{u}_2 = -a\omega\sin\omega t + b\cos\omega t.$$

Introducing amplitude-phase variables for u_1 and a, b variables for u_2 in system (11.3) we have to average the system. To determine the stability of the normal mode periodic solution we compute the Jacobian of the averaged system for r_1, ψ_1, a, b and find the eigenvalues of the gradient of the Jacobian at the periodic $u_1(t)$ for $a = b = 0$. This means that we can leave out the quadratic and cubic expressions in a, b. For the averaged system in the variables r_1, ψ_1, a, b we find:

$$\begin{cases} \dot{r}_1 &= \frac{\varepsilon}{2} r_1 \left(-\mu + \frac{\beta}{2\omega}\sin 2\psi_1\right) + \dots, \\ \dot{\psi}_1 &= \frac{\varepsilon}{2}\left(\frac{\beta}{2\omega}\cos 2\psi_1 - \frac{27\alpha}{64\omega} {r_1}^2\right) + \dots, \\ \dot{a} &= \frac{\varepsilon}{2}\left(-\mu a + \frac{\beta}{2\omega^2} b + \frac{3\alpha}{16\omega} {r_1}^2\left((\sin 2\psi_1)\, a + \left(\frac{2 - \cos 2\psi_1}{\omega}\right) b\right)\right) + \dots, \\ \dot{b} &= \frac{\varepsilon}{2}\left(-\mu b - \frac{\beta}{2} a + \frac{3\alpha}{16} {r_1}^2\left((2 + \cos 2\psi_1)\, a - \frac{\sin 2\psi_1}{\omega} b\right)\right) + \dots . \end{cases}$$

where the dots stand for the omitted higher order terms in a, b. The gradient of the Jacobian at the periodic solution in the coordinate plane becomes when omitting the factor $\varepsilon/2$:

$$\begin{pmatrix} 0 & \frac{r_0\beta\cos 2\psi_1}{\omega} & 0 & 0 \\ -\frac{27\alpha r_0}{32\omega} & -2\mu & 0 & 0 \\ 0 & 0 & \frac{-16\mu\omega + 3\alpha {r_0}^2 \sin 2\psi_1}{16\omega} & \frac{8\beta + 3\alpha {r_0}^2(2 - \cos 2\psi_1)}{16\omega^2} \\ 0 & 0 & \frac{-8\beta + 3\alpha {r_0}^2(2 + \cos 2\psi_1)}{16} & \frac{-16\mu\omega - 3\alpha {r_0}^2 \sin 2\psi_1}{16\omega} \end{pmatrix}$$

The four eigenvalues are splitting up in two groups; the first group corresponds with the eigenvalues of Eq. (11.5), the second group produces the eigenvalues $\lambda_{3,4}$ with $\lambda_3 + \lambda_4 = -2\mu$. We find:

$$\lambda_{3,4} = -\mu \pm \sqrt{\frac{13}{108}\frac{\beta^2}{\omega^2} - \frac{88}{81}\mu^2 - \frac{128}{81}\frac{\omega^2\mu^4}{\beta^2}}.$$

$\lambda_{3,4}$ depends on the parameters $\mu, \beta, \omega, \alpha$. We conclude that the 2 periodic normal mode solutions of the $1:1$ resonances are asymptotically stable if

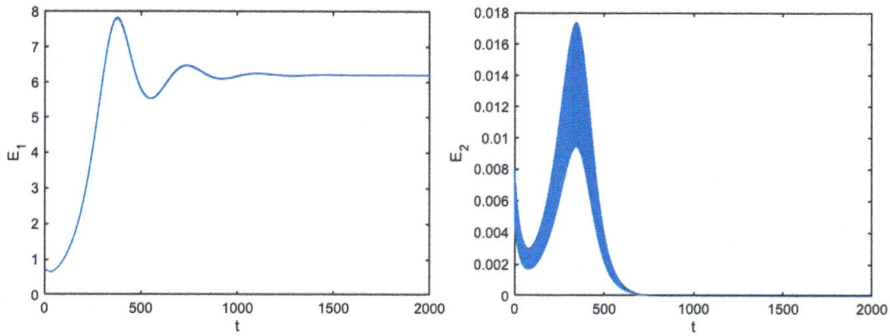

Fig. 11.1 The behaviour of the solutions of system (11.3) near the invariant u_1, $\dot{u}_1$ coordinate plane is shown by plotting $E_1(t) = 0.5(\dot{u}_1^2(t) + 6u_1^2(t))$ and $E_2(t) = 0.5(\dot{u}_2^2(t) + 6u_2^2(t))$ for the parametrically excited oscillators. The initial conditions are $u_1(0) = 0.5$, $\dot{u}_1(0) = 0$, $u_2(0) = \dot{u}_2(0) = 0.05$; $\omega^2 = 6$, $\mu = 0.01$, $\beta = 0.1$, $\alpha = 0.05$

$$\sqrt[4]{\frac{39}{2}}\,\beta \leq \omega\mu.$$

See Fig. 11.1.

11.2.2 First Order Averaging for the Orbits in General Position

Introducing amplitude-phase coordinates by:

$$u = r\cos(\omega t + \psi),\, \dot{u} = -r\omega\sin(\omega t + \psi),$$

we find by first order averaging:

$$\begin{cases} \dot{r}_1 &= \frac{\varepsilon}{2}\left(-\mu r_1 + \frac{\beta}{2\omega} r_1 \sin 2\psi_1 - \frac{3\alpha}{16\omega} r_1 {r_2}^2 \sin 2(\psi_1 - \psi_2)\right), \\ \dot{\psi}_1 &= \frac{\varepsilon}{8\omega}\left(\beta\cos 2\psi_1 - \frac{27\alpha}{16}{r_1}^2 - \frac{3\alpha}{2}{r_2}^2 - \frac{3\alpha}{4}{r_2}^2\cos 2(\psi_1 - \psi_2)\right), \\ \dot{r}_2 &= \frac{\varepsilon}{2}\left(-\mu r_2 + \frac{\beta}{2\omega} r_2 \sin 2\psi_2 + \frac{3\alpha}{16\omega} {r_1}^2 r_2 \sin 2(\psi_1 - \psi_2)\right), \\ \dot{\psi}_2 &= \frac{\varepsilon}{8\omega}\left(\beta\cos 2\psi_2 - \frac{27\alpha}{16}{r_2}^2 - \frac{3\alpha}{2}{r_1}^2 - \frac{3\alpha}{4}{r_1}^2\cos 2(\psi_1 - \psi_2)\right). \end{cases} \tag{11.6}$$

The solutions of system (11.6) approximate the exact solutions with given initial values to $O(\varepsilon)$ on the timescale $1/\varepsilon$; with some abuse of notation we kept the notation r, ψ for the approximating system.

It is important to note that the damping term (coefficient μ) is not scaled by the frequency ω, but on the other hand the parametric excitation (coefficient β) and the nonlinear interaction (coefficient α) are reduced considerably for high frequency modes (ω large). If ω is $O(1/\varepsilon)$, system (11.6) is dominated by the damping terms.

Assuming that ω is $O(1)$ with respect to ε we have for the resonant combination angle $\chi = \psi_1 - \psi_2$:

$$\dot{\chi} = \frac{\varepsilon}{8\omega}\left(\beta(\cos 2\psi_1 - \cos 2\psi_2) - \frac{3\alpha}{16}\left({r_1}^2 - {r_2}^2\right) - \frac{3\alpha}{4}\left({r_2}^2 - {r_1}^2\right)\cos 2\chi\right). \tag{11.7}$$

The resonance zones are corresponding with domains in phase-space where the three angles ψ_1, ψ_2, χ are not timelike, they are determined by the zeros of the equations for the 3 angles. From Eq. (11.7) we find for angle χ two possible resonance zones M_1, M_2 given by:

$$r_1 = r_2, \ \psi_1 = \psi_2, \text{ or } \psi_1 = \psi_2 + \pi, . \tag{11.8}$$

Dynamically most interesting is the case that we have intersection of resonance zones. For the angles ψ_1, ψ_2, using system (11.6), this leads in M_1, M_2 to the equations:

$$\beta \cos 2\psi_{1,2} = \frac{63\alpha}{16}{r_1}^2. \tag{11.9}$$

So for triple intersection of resonance zones we have the necessary condition: $r_1 = r_2, \ 0 < \frac{63\alpha}{16\beta}{r_1}^2 < 1$.

11.3 Triple Resonance for the 1 : 1 Case

We will distinguish the dynamics for the dissipative and volume-preserving cases.

11.3.1 Periodic Solutions in the Dissipative Case

Assume $\mu > 0$ and consider the resonance zones M_1, M_2 determined by Eq. (11.8). An interesting type of periodic solution may arise if $\dot{r}_{1,2} = 0$ and simultaneously $\dot{\chi} = 0$. With these assumptions we find in M_1, M_2:

$$\begin{aligned}\dot{r}_{1,2} &= \frac{\varepsilon}{2}r_{1,2}\left(-\mu + \frac{\beta}{2\omega}\sin 2\psi_{1,2}\right) = 0,\\ \dot{\psi}_{1,2} &= \frac{\varepsilon}{8\omega}\left(\beta\cos 2\psi_{1,2} - \frac{63\alpha}{16}{r_{1,2}}^2\right) = 0.\end{aligned} \tag{11.10}$$

Conditions (11.10) are satisfied if the periodic solutions are located in the triple resonance zone determined by Eq. (11.9) and moreover:

$$\sin 2\psi_{1,2} = \frac{2\mu\omega}{\beta},\ \left|\frac{2\mu\omega}{\beta}\right| \leq 1 \text{ or } \mu \leq \frac{\beta}{2\omega}, \tag{11.11}$$

which puts a bound on the size of the dissipation with respect to the other parameters. In a Galerkin projection of Eq. (11.1) with large eigenvalues ω, these periodic solutions vanish. From $\sin^2\psi + \cos^2\psi = 1$ we find for the amplitudes of the periodic solutions:

$$r_{1,2}{}^2 = \frac{16}{63\alpha}\sqrt{\beta^2 - 4\mu^2\omega^2}. \tag{11.12}$$

11.3.2 Stability in the Dissipative Case, $\mu > 0$

To establish the stability of the periodic solutions in the triple resonance zone we use theorems 11.5–11.6 from [5] (the 2nd Bogoliubov theorem). We need the Jacobian of the vector field F of the averaged system (11.6). Omitting the factor $\varepsilon/2$ we find for Jacobian ∇F:

$$\begin{pmatrix} A_1 & -\dfrac{3\alpha r_1 r_2 \sin 2\chi}{8\omega} & B_1 & \dfrac{3\alpha r_1 r_2{}^2 \cos 2\chi}{8\omega} \\ \dfrac{3\alpha r_1 r_2 \sin 2\chi}{8\omega} & A_2 & \dfrac{3\alpha r_1{}^2 r_2 \cos 2\chi}{8\omega} & B_2 \\ -\dfrac{27\alpha r_1}{32\omega} & -\dfrac{3\alpha r_2\,(2+\cos 2\chi)}{8\omega} & C_1 & -\dfrac{3\alpha r_2{}^2 \sin 2\chi}{8\omega} \\ -\dfrac{3\alpha r_1\,(2+\cos 2\chi)}{8\omega} & -\dfrac{27\alpha r_2}{32\omega} & \dfrac{3\alpha r_1{}^2 \sin 2\chi}{8\omega} & C_2 \end{pmatrix}$$

where

$$\begin{aligned} A_1 &= -\mu + \frac{\beta}{2\omega}\sin 2\psi_1 - \frac{3\alpha}{16\omega} r_2{}^2 \sin 2\chi,\ A_2 = -\mu + \frac{\beta}{2\omega}\sin 2\psi_2 + \frac{3\alpha}{16\omega} r_1{}^2 \sin 2\chi, \\ B_1 &= \frac{\beta}{\omega} r_1 \cos 2\psi_1 - \frac{3\alpha}{8\omega} r_1 r_2{}^2 \cos 2\chi,\ B_2 = \frac{\beta}{\omega} r_2 \cos 2\psi_2 - \frac{3\alpha}{8\omega} r_1{}^2 r_2 \cos 2\chi, \\ C_1 &= -\frac{\beta}{2\omega}\sin 2\psi_1 + \frac{3\alpha}{8\omega} r_2{}^2 \sin 2\chi, \text{ and } C_2 = -\frac{\beta}{2\omega}\sin 2\psi_2 - \frac{3\alpha}{8\omega} r_1{}^2 \sin 2\chi. \end{aligned}$$

Applying the Jacobian at the periodic solutions using Eqs. (11.8), (11.9), (11.11) with notation $r_1 = r_2 = r$ we find the matrix:

$$J(r) = \begin{pmatrix} 0 & 0 & \frac{57\alpha}{16\omega}r^3 & \frac{3\alpha}{8\omega}r^3 \\ 0 & 0 & \frac{3\alpha}{8\omega}r^3 & \frac{57\alpha}{16\omega}r^3 \\ -\frac{27\alpha}{32\omega}r & -\frac{9\alpha}{8\omega}r & -\mu & 0 \\ -\frac{9\alpha}{8\omega}r & -\frac{27\alpha}{32\omega}r & 0 & -\mu \end{pmatrix}$$

It is easy to see that if $r > 0$ we have $|J(r)| > 0$. The implication is from [5] that periodic solutions obtained from nontrivial equilibria of the averaged system (11.6) do exist in an ε-neighbourhood of the equilibria. Note that this also holds if $\mu = 0$.

A MATHEMATICA calculation produces the eigenvalues of matrix $J(r)$:

$$\lambda_{1,2} = -\frac{\mu}{2} \pm \frac{1}{2}\sqrt{\mu^2 + \frac{459\alpha^2}{128\omega^2}r^4},\ \lambda_{3,4} = -\frac{\mu}{2} \pm \frac{1}{2}\sqrt{\mu^2 - \frac{3969\alpha^2}{128\omega^2}r^4}.$$

The eigenvalues $\lambda_{1,2}$ are real, the plus sign results in a positive eigenvalue so we have instability of the periodic solutions. The instability is caused by the parametric excitation, it is weakened for large ω. In Fig. 11.2 we show the instability by starting near the solution where $u_1(t) = u_2(t)$.

11.3.3 Stability in the Volume-Preserving Case, $\mu = 0$

Without dissipation the flow in the time-extended phase-space is volume-preserving, the dynamics is more delicate. For the angle χ the resonance zones M_1, M_2 are unchanged. Looking for periodic solutions with constant, nontrivial amplitudes r_1, r_2 we find from system (11.6):

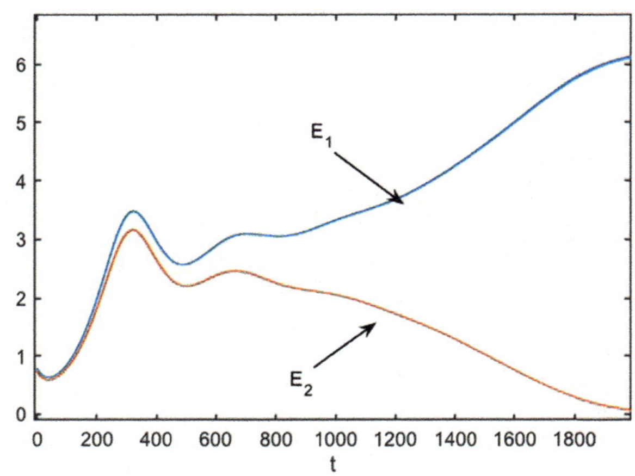

Fig. 11.2 The behaviour of the solutions of system (11.3) near the general position solution $u_1(t) = u_2(t)$ is shown by plotting $E_1(t) = 0.5(\dot{u}_1^2(t) + 6u_1^2(t))$ and $E_2(t) = 0.5(\dot{u}_2^2(t) + 6u_2^2(t))$. The initial conditions are $u_1(0) = 0.51, \dot{u}_1(0) = 0.05, u_2(0) = 0.49, \dot{u}_2(0) = 0.05$; $\omega^2 = 6, \mu = 0.01, \beta = 0.1, \alpha = 0.05$

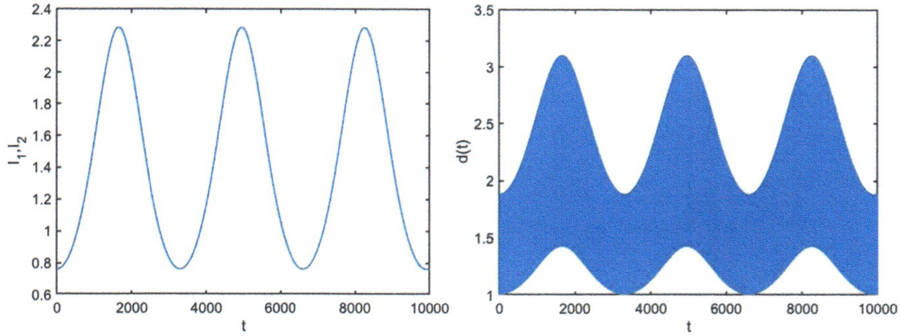

Fig. 11.3 Left we illustrate the behaviour of the solutions of system (11.3) $(m = 1, n = 2)$ by plotting $I_1(t) = I_2(t)$ in the case $\omega = \sqrt{6}, \mu = 0, \alpha = \beta = 1, \varepsilon = 0.01$ with initial conditons $u_1(0) = u_2(0) = r_0, \psi_1(0) = \psi_2(0) = 0$. Right we show the Euclidean distance $d(t)$ to the initial conditions

$$\beta \sin 2\psi_1 = \frac{3}{8}\alpha {r_2}^2 \sin 2\chi, \quad \beta \sin 2\psi_2 = -\frac{3}{8}\alpha {r_1}^2 \sin 2\chi.$$

These conditions lead in M_1, M_2 to the solutions:

$$\psi_1 = \psi_2 = 0 \text{ and } \psi_1 = 0, \, \psi_2 = \pi.$$

In system (11.6) we have $\dot{\psi}_1 = \dot{\psi}_2 = 0$ if:

$${r_1}^2 = {r_2}^2 = {r_0}^2 = \frac{16\beta}{63\alpha}.$$

Using this value of r and the eigenvalues (11.1) we obtain for the eigenvalues in the volume-preserving case:

$$\lambda_{1,2} = \pm \frac{\beta}{21\omega}\sqrt{\frac{51}{2}}, \; \lambda_{3,4} = \pm \frac{\beta}{2\omega}\sqrt{2}i. \tag{11.13}$$

We have again instability of the periodic solutions. The periodic solutions in the triple resonance zone are illustrated in Fig. 11.3. The behaviour for the cases $\psi_2 = 0, \pi$ is identical. We use the expression $I_{1,2}(t) = \frac{1}{2}(\dot{u}_{1,2}^2 + \omega^2 {u_{1,2}}^2)$. The recurrence in the volume-preserving case $\mu = 0$ is illustrated by plotting the Euclidean distance $d(t)$ to the initial values, we have:

$$d^2(t) = \sum_{i=1}^{2} (u_i(t) - u_i(0))^2 + (\dot{u}_i(t) - \dot{u}_i(0))^2.$$

With $\varepsilon = 0.01$ the typical timescale of recurrence is 3500 timesteps.

11.3.4 Interaction of 1 : 1 Resonances

As we have an infinite number of 1 : 1 resonances it is natural to study a combination of N eigenfunctions of the form:

$$u_p(x, y, t) = \sum_{i=1}^{N}(u_{1i}(t)\cos m_i x \cos n_i y + u_{2i}(t)\cos n_i x \cos m_i y), \qquad (11.14)$$

where $m_j, n_j \in \{0, 1, 2, \ldots, N\}$. We choose $m_i \neq n_i$, $i = 0, 1, \ldots, N$ and avoid accidental resonances (to be discussed in the next section) in (11.14). Substitution in wave equation (11.1) and taking inner products with the individual eigenfunctions produces a system of $2N$ second order coupled ODEs. The ε-scaling as before enables us to apply averaging; the results depend on the choice of Ω. Suppose that for one of the (m_i, n_i) combinations we have the frequency ω_i with $\Omega = 2\omega_i$. The corresponding eigenfunction will show dynamics that is different from the other modes.

We will discuss the dynamics in a particular case of $N = 2$ as this shows the essential behaviour and avoids too much notation. The results can easily be generalised for $N > 2$. Choose $m_1 \neq n_1$ with corresponding ω_1 and $\Omega = 2\omega_1$. Choose a different set $m_2 \neq n_2$ with corresponding ω_2, $\Omega \neq 2\omega_2$. We associate with m_1, n_1 the time-dependent amplitudes u_1, u_2, with m_2, n_2 the amplitudes u_3, u_4. Substituting the expansion containing 4 modes:

$$\sum_{i=1}^{2} u_1(t)\cos m_i x \cos n_i y + u_{i+1}(t)\cos n_i x \cos m_i y$$

into the wave equation (11.1) and taking inner products with the eigenfunctions we obtain after the usual ε-scaling the system:

$$\begin{cases} \ddot{u}_1 + \omega_1^2 u_1 &= -\varepsilon\mu\dot{u}_1 - \varepsilon\beta u_1\cos(2\omega_1 t) + \varepsilon\alpha P(u_1, u_2, u_3, u_4), \\ \ddot{u}_2 + \omega_1^2 u_2 &= -\varepsilon\mu\dot{u}_2 - \varepsilon\beta u_2\cos(2\omega_1 t) + \varepsilon\alpha P(u_2, u_1, u_3, u_4), \\ \ddot{u}_3 + \omega_2^2 u_3 &= -\varepsilon\mu\dot{u}_3 - \varepsilon\beta u_3\cos(2\omega_1 t) + \varepsilon\alpha P(u_3, u_1, u_2, u_4), \\ \ddot{u}_4 + \omega_2^2 u_4 &= -\varepsilon\mu\dot{u}_4 - \varepsilon\beta u_4\cos(2\omega_1 t) + \varepsilon\alpha P(u_4, u_1, u_2, u_3), \end{cases}$$

where $P(u_1, u_2, u_3, u_4) = \frac{9}{16}{u_1}^3 + \frac{3}{4}u_1{u_2}^2 + \frac{3}{4}u_1{u_3}^2 + \frac{3}{4}u_1{u_4}^2$. We assume that there are no accidental resonances as discussed in the next section. First order averaging produces in amplitude-phase coordinates for orbits in general position:

$$
\begin{cases}
\dot{r}_i &= \frac{\varepsilon}{2}\left(\left(-\mu + \frac{\beta}{2\omega_1}\sin 2\psi_i\right) r_i + (-1)^i \frac{3\alpha}{16\omega_1} {r_1}^i {r_2}^{3-i} \sin 2(\psi_1 - \psi_2)\right), \\
\dot{\psi}_i &= \frac{\varepsilon}{8}\left(\frac{\beta}{\omega_1}\cos 2\psi_i - \frac{27\alpha}{16\omega_1}{r_i}^2 - \frac{3\alpha}{2\omega_1}\left({r_{3-i}}^2 + {r_3}^3 + {r_4}^2\right)\right. \\
& \qquad \left. - \frac{3\alpha}{4\omega_1}{r_{3-i}}^2 \cos 2(\psi_1 - \psi_2)\right), \quad \text{for } i = 1, 2 \\
\dot{r}_j &= \frac{\varepsilon}{2}\left(-\mu r_j + (-1)^j \frac{3\alpha}{16\omega_2} {r_3}^{j-2} {r_4}^{5-j} \sin 2(\psi_3 - \psi_4)\right), \\
\dot{\psi}_j &= \frac{\varepsilon}{8}\left(-\frac{27\alpha}{16\omega_2}{r_j}^2 - \frac{3\alpha}{2\omega_2}\left({r_1}^2 + {r_2}^2 + {r_{7-j}}^2\right)\right. \\
& \qquad \left. - \frac{3\alpha}{4\omega_2} r_{7-j}^2 \cos 2(\psi_3 - \psi_4)\right), \quad \text{for } j = 3, 4
\end{cases}
$$

From the equations for r_3, r_4 we find:

$$
\frac{1}{2}\frac{d}{dt}({r_3}^2 + {r_4}^2) = -\frac{\varepsilon}{2}\mu({r_3}^2 + {r_4}^2),
$$

so the amplitudes r_3, r_4 will vanish with time. For the wave equation (11.1) the behaviour of the eigenfunctions corresponding with m_1, n_1 will be prominent.

We illustrate the results for an explicit case. Consider the combination $(m, n) \in \{(1, 2), (2, 1)\}$ (coefficients $u_1(t), u_2(t)$) and $\{(1, 3), (3, 1)\}$ (coefficients $u_3(t), u_4(t)$). We have $\omega_1 = \sqrt{6}$, $\omega_2 = \sqrt{11}$, the parametric excitation frequency $\Omega = 2\sqrt{6}$. We introduce as measures for the energy of the oscillators u_1, u_2 the quantity

$$
E_1 = \frac{1}{2}({\dot{u}_1}^2 + 6{u_1}^2 + {\dot{u}_2}^2 + 6{u_2}^2)
$$

and similarly for u_3, u_4 the quantity

$$
E_2 = \frac{1}{2}({\dot{u}_3}^2 + 11{u_3}^2 + {\dot{u}_4}^2 + 11{u_4}^2).
$$

The initial values of the two groups of oscillators are equal, the first group is excited, the second group is damped out; see Fig. 11.4.

11.4 Remarks on Accidental Resonances

The instability of periodic solutions in general position in the case of two modes with symmetric eigenfunctions (11.1) suggests the question whether energy can be transferred to other modes by accidental resonance. We consider a few prominent cases, the topic can be extended considerably. Choose for the eigenvalues (11.1) $\omega_0 = c = 1$.

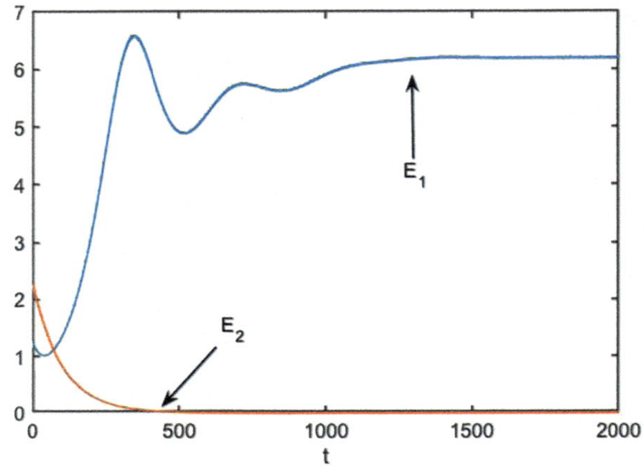

Fig. 11.4 The behaviour of the solutions of system (3.4) by plotting $E_1(t)$ for the parametrically excited oscillators u_1, u_2 and $E_2(t)$ for the oscillators u_3, u_4. The initial conditions are $u_1(0) = u_2(0) = 0.5, u_3(0) = u(_4(0) = 0.4$ with initial velocities zero; $\omega_1^2 = 6, \omega_2^2 = 11, \mu = 0.01, \beta = 0.1, \alpha = 0.05$

11.4.1 The $1 : 1 : 3$ *resonance*

Consider the 3 eigenfunctions with $(m, n) \in \{(1, 3), (3, 1), (7, 7)\}$. In this case the frequencies of the linear oscillations are given by 11, 11, and 99, producing the $1 : 1 : 3$ resonance. The eigenfunction expansion of the corresponding 3 modes is:

$$u_p(x, y, t) = u_1(t) \cos x \cos 3y + u_2(t) \cos 3x \cos y + u_3(t) \cos 7x \cos 7y. \tag{11.15}$$

Substitution of expansion (11.15) into Eq. (11.1) and taking inner products with the eigenfunctions we find with $\omega = \sqrt{11}, \omega_1 = 2\omega$:

$$\begin{cases} \ddot{u}_1 + \omega^2 u_1 &= -\varepsilon\mu\dot{u}_1 - \varepsilon\beta u_1 \cos(2\omega t) + \varepsilon\alpha P(u_1, u_2, u_3), \\ \ddot{u}_2 + \omega^2 u_2 &= -\varepsilon\mu\dot{u}_2 - \varepsilon\beta u_2 \cos(2\omega t) + \varepsilon\alpha P(u_2, u_1, u_3), \\ \ddot{u}_3 + 9\omega^2 u_3 &= -\varepsilon\mu\dot{u}_3 - \varepsilon\beta u_3 \cos(2\omega t) + \varepsilon\alpha P(u_3, u_2, u_1), \end{cases} \tag{11.16}$$

where $P(u_1, u_2, u_3) = \frac{9}{16}{u_1}^3 + \frac{3}{4}u_1{u_2}^2 + \frac{3}{4}u_1{u_3}^2$. Although we have a primary resonance it turns out that because of the symmetries of system (11.16) the $1 : 1 : 3$ resonance is not effective. First order averaging produces for the amplitude r_3 of $u_3(t)$ the equation

$$\dot{r}_3 = -\varepsilon\frac{\mu}{2}r_3,$$

so there is no interaction with the modes $u_1(t), u_2(t)$ and no quenching or transfer of energy of the first two modes. Higher order approximation will not change this picture qualitatively.

11.4.2 The 1 : 1 : 1 Resonance

Consider the 3 eigenfunctions with $(m, n) \in \{(1, 7), (7, 1), (5, 5)\}$. In this case the frequencies of the linear oscillations are the same, i.e. $\omega = \sqrt{51}$, producing the 1 : 1 : 1 resonance. The eigenfunction expansion of the corresponding 3 modes is:

$$u_p(x, y, t) = u_1(t) \cos x \cos 7y + u_2(t) \cos 7x \cos y + u_3(t) \cos 5x \cos 5y. \tag{11.17}$$

We substitute expansion (11.17) into Eq. (11.1) and we take inner products with the eigenfunctions. Put $\omega_1 = 2\sqrt{51}$ and rescale $\sqrt{51}\, t \longmapsto t$, $\varepsilon/51 \longmapsto \varepsilon$; we find the system:

$$\begin{cases} \ddot{u}_1 + u_1 &= -\varepsilon\mu\dot{u}_1 - \varepsilon\beta u_1 \cos(2t) + \varepsilon\alpha P(u_1, u_2, u_3), \\ \ddot{u}_2 + u_2 &= -\varepsilon\mu\dot{u}_2 - \varepsilon\beta u_2 \cos(2t) + \varepsilon\alpha P(u_2, u_1, u_3) \\ \ddot{u}_3 + u_3 &= -\varepsilon\mu\dot{u}_3 - \varepsilon\beta u_3 \cos(2t) + \varepsilon\alpha P(u_3, u_1, u_1), \end{cases}$$

where $P(u_1, u_2, u_3) = \frac{9}{16}{u_1}^3 + \frac{3}{4}u_1{u_2}^2 + \frac{3}{4}u_1{u_3}^2$. Because of the symmetry of the system we can recover the solutions of the preceding 1 : 1 resonances in 2 degrees-of-freedom invariant manifolds. This means that we find periodic solutions in the 3 normal mode planes and unstable periodic solutions in 3 invariant 4-dimensional manifolds when putting successively the initial conditions of one mode equal to zero.

However, we are interested in general position orbits. We can extend the averaging by adding to system (11.6) 2 equations, the angle ψ_3 and the combination angles $\psi_1 - \psi_3$ and $\psi_2 - \psi_3$. Apart from the normal mode solutions and because of the symmetry of system (11.18) we can enumerate a number of exact solutions in general position, for instance:

$$u_1(t) = u_2(t) = u_3(t). \tag{11.18}$$

We can also put $u_2(t) = -u_1(t)$, $u_3(t) = -u_1(t)$ or $u_3(t) = -u_1(t)$, $u_2(t) = -u_3(t)$. In the case of Eq. (11.18) we have the special solution from:

$$u_1(t) = u_2(t) = u_3(t) = u(t), \ , \ddot{u} + u = -\varepsilon\mu\dot{u} - \varepsilon\beta u \cos(2t) + \varepsilon\alpha\frac{33}{16}u^3.$$

With amplitude-phase coordinates as before the solutions are approximated by averaging:

$$\dot{r} = \frac{\varepsilon}{2}r(-\mu + \frac{\beta}{2}\sin 2\psi), \ \dot{\psi} = \frac{\varepsilon}{4}(\beta\cos 2\psi - \alpha\frac{99}{32}r^2). \tag{11.19}$$

In system (11.19) $r = r_1 = r_2 = r_3$ and $\psi = \psi_1 = \psi_2 = \psi_3$. A critical point of the averaged vector field is determined by:

$$\mu = \frac{\beta}{2}\sin 2\psi, \ \beta\cos 2\psi = \alpha\frac{99}{32}r^2.$$

Elimination of the phase-angle yields:

$$r^2 = \frac{32}{99\alpha}\sqrt{\beta^2 - 4\mu^2}. \tag{11.20}$$

11.4.3 The $1:1:1:1$ *Resonance*

Consider the 4 eigenfunctions with $(m, n) \in \{(3, 4), (4, 3), (0, 5), (5, 0)\}$. In this case the frequencies are $\omega = \sqrt{26}$, producing the $1:1:1:1$ resonance. The eigenfunction expansion of the corresponding 4 modes is:

$$u_p(x, y, t) = u_1(t)\cos 3x \cos 4y + u_2(t)\cos 4x \cos 3y + u_3(t)\cos 5y + u_4(t)\cos 5x.$$

The analysis by averaging and of exact solutions runs as before. This is left to the reader.

11.5 Conclusions

1. The analysis of a nonlinear wave equation with 2 spatial dimensions introduces many new problems involving resonances. The $1:1$ resonance dominates the dynamics in the case of a square domain.
2. In contrast to the case of systems without forcing, see [6], the excitation forces a strong selection of modes. This has become clear in the analysis of interaction of $1:1$ resonances.
3. An important aspect of the analysis is the choice of the parametric excitation frequency $\Omega = 2\omega$ in Eq. (11.1). In the cases of modes with $\Omega \neq 2\omega$ and $\mu > 0$ we expect reduction of these modes by damping. This becomes already clear for the $1:1:3$ resonance in Sect. 11.4.1.
4. We have omitted the analysis of detuning. Inspection of the frequencies generated by the space-dependent operator suggests a number of interesting cases. The formulation of the initial-value problem for Eq. (11.1) raises many more questions that will hopefully be discussed in later papers.

Acknowledgements The MATHEMATICA calculation to produce expressions (3.2) were performed by Roelof Bruggeman. J.M.T. acknowledges the financial support of Riset KK **B** ITB 2020.

References

1. Bakri, T., Meijer, H.G., Verhulst, F.: Emergence and bifurcation of Lyapunov manifolds in nonlinear wave equations. J. Nonlinear Sci. **19**, 571–596 (2009)

2. van Horssen, W.T.: An asymptotic theory for a class of initial-boundary value problems for weakly nonlinear wave equations. SIAM J. Appl. Math. **48**(6), 1227–1243 (1988)
3. Pals, H.: The Galerkin-averaging method for the Klein-Gordon equation in two space dimensions. Nonlinear Anal. **27**, 841–856 (1996)
4. Rand, R.H., Newman, W.I., Denardo, B.C., Newman, A.L.: Dynamics of a nonlinear parametrically excited partial differential equation. In: Proceedings of the 1995 Design Engineering Technical Conferences, vol. 3, pp. 57–68. ASME, DE-84-1 (1999)
5. Verhulst, F.: Nonlinear differential equations and dynamical systems. Springer, Berlin (2000) (rev. and extended ed.)
6. Verhulst, F.: Recurrence and resonance in the cubic Klein-Gordon equation. Acta Appl. Math. **162** (2019). https://doi.org/10.1007/s10440-019-00238-4

Chapter 12
Chaotic Dynamic of a Symmetric Tree-Shaped Wave Network

Fei Wang and Jun-Min Wang

Abstract The chaotic dynamic behavior of a symmetric tree-shaped network of wave equations described by a system of partial differential equations is considered. The nonlinearities of van der Pol type are proposed at three boundary endpoints that can cause the total energy of the system to rise and fall within certain bounds. At the interconnected point of the wave equations, the energy is injected into the system through an anti-damping velocity feedback. We show that when the parameters satisfy certain conditions, the snapback repeller is existence and the system is chaotic. Finally, we give some numerical simulations to illustrate the theoretical outcomes.

12.1 Introduction

In the past few decades, the phenomena of chaos has attracted many researchers because of its wide applications for example human brain [1], image encryption [2, 3], secure communication [4], sound encryption [5], etc. The term "chaos", literally, means confusion, disorder, disorganization, turbulence. There are several definitions to describe the chaos, Li-Yorke [6], Devaney [7], Robinson [8] and the total variation [9], and each of which reflects its physical background, respectively. There are many papers have studied the chaotic behavior for one-dimensional wave equation. For example, in [10–14], the authors proved different one-dimension wave equation is chaotic in the sense of the total variation, respectively. In [15], the author prove the one-dimension wave equation with mixed energy transports is chaotic in the sense of Li-Yorke. The same 1D wave equation as [12] is also proved to be chaotic in the sense of Devaney and Li-Yorke under some conditions in [16].

F. Wang · J.-M. Wang (✉)
School of Mathematics and Statistics, Beijing Institute of Technology,
Beijing 100081, People's Republic of China
e-mail: jmwang@bit.edu.cn

A. K. Abramian et al. (eds.), *Nonlinear Dynamics of Discrete and Continuous Systems*,
Advanced Structured Materials 139,
https://doi.org/10.1007/978-3-030-53006-8_12

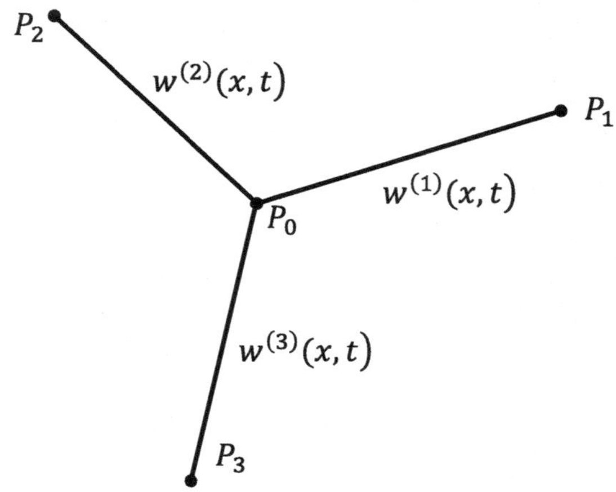

Fig. 12.1 A tree-shaped network of wave equations

In [17], the authors studied a coupled wave equation. In the coupled system, the left boundary condition is fixed and the right end is a van der Pol type nonlinear condition. By analyzing the existence of the snap-back theory, the authors proved the system is chaos.

In this paper, we consider the following wave equations (Fig. 12.1):

$$\begin{cases} w_{tt}^{(i)}(x,t) - w_{xx}^{(i)}(x,t) = 0, \quad x \in (0,1), \quad t > 0, \quad i = 1,\ 2,\ 3, \\ w^{(1)}(0,t) = w^{(2)}(0,t) = w^{(3)}(0,t), \quad t > 0, \\ w_x^{(1)}(0,t) + w_x^{(2)}(0,t) + w_x^{(3)}(0,t) = -\eta w_t^{(1)}(0,t), \quad t > 0,\ \eta > 0,\ \eta \neq 3, \\ w_x^{(i)}(1,t) = \alpha w_t^{(i)}(1,t) - \beta\left(w_t^{(i)}(1,t)\right)^3, \quad t > 0,\ 0 < \alpha < 1,\ \beta > 0, \quad i = 1,\ 2,\ 3, \\ w^{(i)}(x,0) = w_0^{(i)}(x) \in C^1([0,1]), \quad w_t^{(i)}(x,0) = w_1^{(i)}(x) \in C([0,1]), \quad i = 1,\ 2,\ 3. \end{cases} \tag{12.1}$$

The energy function $E(t)$ for system (12.1) is

$$\begin{aligned} E(t) = \frac{1}{2}\int_0^1 \Big[& \left(w_t^{(1)}(x,t)\right)^2 + \left(w_x^{(1)}(x,t)\right)^2 + \left(w_t^{(2)}(x,t)\right)^2 + \left(w_x^{(2)}(x,t)\right)^2 \\ & + \left(w_t^{(3)}(x,t)\right)^2 + \left(w_x^{(3)}(x,t)\right)^2 \Big] dx, \end{aligned} \tag{12.2}$$

and the derivative of $E(t)$ with respective to the time t yields

$$\begin{aligned} \frac{dE(t)}{dt} = & \ \eta\left(w_t^{(1)}(0,t)\right)^2 + \alpha\left(w_t^{(1)}(1,t)\right)^2 - \beta\left(w_t^{(1)}(1,t)\right)^4 + \alpha\left(w_t^{(2)}(1,t)\right)^2 \\ & -\beta\left(w_t^{(2)}(1,t)\right)^4 + \alpha\left(w_t^{(3)}(1,t)\right)^2 - \beta\left(w_t^{(3)}(1,t)\right)^4. \end{aligned} \tag{12.3}$$

Let $\Pi_1 = \eta\big(w_t^{(1)}(0,t)\big)^2$. It is found that the energy is pumping into the system through the intersection point at $x = 0$ in (12.1), and such a property is called anti-damping. Let

$$\Pi_{2i} = \alpha\big(w_t^{(i)}(1,t)\big)^2 - \beta\big(w_t^{(i)}(1,t)\big)^4,\ i = 1,\ 2,\ 3.$$

Then Π_{2i} is "self-regulating" and

$$\begin{cases} \Pi_{2i} < 0, \text{ if } |w_t^{(i)}(1,t)| > \sqrt{\dfrac{\alpha}{\beta}}, \\ \Pi_{2i} \geq 0, \text{ if } |w_t^{(i)}(1,t)| \leq \sqrt{\dfrac{\alpha}{\beta}}, \end{cases} \quad \text{for } i = 1,\ 2,\ 3.$$

As in [17], we define

$$\begin{cases} \xi_i(x,t) = \dfrac{w_x^{(i)}(x,t) + w_t^{(i)}(x,t)}{2}, \\ \chi_i(x,t) = \dfrac{w_x^{(i)}(x,t) - w_t^{(i)}(x,t)}{2}, \end{cases} \quad \text{for } i = 1,\ 2,\ 3. \tag{12.4}$$

Then system (12.1) is converted into a first-order hyperbolic system

$$\frac{\partial}{\partial t}\begin{bmatrix} \xi_1 \\ \xi_2 \\ \xi_3 \\ \chi_1 \\ \chi_2 \\ \chi_3 \end{bmatrix} = \begin{bmatrix} 1 & 0 & 0 & 0 & 0 & 0 \\ 0 & 1 & 0 & 0 & 0 & 0 \\ 0 & 0 & 1 & 0 & 0 & 0 \\ 0 & 0 & 0 & -1 & 0 & 0 \\ 0 & 0 & 0 & 0 & -1 & 0 \\ 0 & 0 & 0 & 0 & 0 & -1 \end{bmatrix} \frac{\partial}{\partial x}\begin{bmatrix} \xi_1 \\ \xi_2 \\ \xi_3 \\ \chi_1 \\ \chi_2 \\ \chi_3 \end{bmatrix}, \quad \text{for } 0 < x < 1,\ t > 0. \tag{12.5}$$

The reflection relation at $x = 0$, according to the second and third equalities of equation (12.1), is

$$\begin{bmatrix}\chi_1(0,t)\\ \chi_2(0,t)\\ \chi_3(0,t)\end{bmatrix} = \begin{bmatrix}\frac{\eta-1}{\eta-3}\xi_1(0,t)+\frac{2}{\eta-3}\xi_2(0,t)+\frac{2}{\eta-3}\xi_3(0,t)\\ \frac{2}{\eta-3}\xi_1(0,t)+\frac{\eta-1}{\eta-3}\xi_2(0,t)+\frac{2}{\eta-3}\xi_3(0,t)\\ \frac{2}{\eta-3}\xi_1(0,t)+\frac{2}{\eta-3}\xi_2(0,t)+\frac{\eta-1}{\eta-3}\xi_3(0,t)\end{bmatrix} = R_0\left(\begin{bmatrix}\xi_1(0,t)\\ \xi_2(0,t)\\ \xi_3(0,t)\end{bmatrix}\right), \quad \text{for } t>0, \;\; \eta>0, \;\; \eta\neq 3, \tag{12.6}$$

and the boundary condition at $x = 1$, according to the fourth equality of equation (12.1), is

$$\begin{bmatrix}\xi_1(1,t)\\ \xi_2(1,t)\\ \xi_3(1,t)\end{bmatrix} = \begin{bmatrix}F\big(\chi_1(1,t)\big)\\ F\big(\chi_2(1,t)\big)\\ F\big(\chi_3(1,t)\big)\end{bmatrix} = R_1\left(\begin{bmatrix}\chi_1(1,t)\\ \chi_2(1,t)\\ \chi_3(1,t)\end{bmatrix}\right), \quad \text{for } t>0, \tag{12.7}$$

where for each given $\chi_i \in \mathbb{R}$, $\xi_i = F(\chi_i)$, $i = 1, 2, 3$ is defined through the following cubic equation

$$\beta(\xi_i-\chi_i)^3+(1-\alpha)(\xi_i-\chi_i)+2\chi_i=0, \;\text{ for } i=1,\;2,\;3,\;\; 0<\alpha<1,\;\; \beta>0, \tag{12.8}$$

and the equation (12.8) has the unique real solution ξ_i [18]. Moreover, F is a continuously differentiable function in $\mathbb{R}$. The initial conditions for $\xi_1, \xi_2, \xi_3, \chi_1, \chi_2, \chi_3$ are

$$\begin{bmatrix}\xi_1(x,0)\\ \xi_2(x,0)\\ \xi_3(x,0)\\ \chi_1(x,0)\\ \chi_2(x,0)\\ \chi_3(x,0)\end{bmatrix} = \begin{bmatrix}\xi_{1,0}(x)\\ \xi_{2,0}(x)\\ \xi_{3,0}(x)\\ \chi_{1,0}(x)\\ \chi_{2,0}(x)\\ \chi_{3,0}(x)\end{bmatrix} = \begin{bmatrix}\frac{1}{2}\Big(w_x^{(1)}(x,0)+w_t^{(1)}(x,0)\Big)\\ \frac{1}{2}\Big(w_x^{(2)}(x,0)+w_t^{(2)}(x,0)\Big)\\ \frac{1}{2}\Big(w_x^{(3)}(x,0)+w_t^{(3)}(x,0)\Big)\\ \frac{1}{2}\Big(w_x^{(1)}(x,0)-w_t^{(1)}(x,0)\Big)\\ \frac{1}{2}\Big(w_x^{(2)}(x,0)-w_t^{(2)}(x,0)\Big)\\ \frac{1}{2}\Big(w_x^{(3)}(x,0)-w_t^{(3)}(x,0)\Big)\end{bmatrix}, \quad \text{for } 0\le x\le 1. \tag{12.9}$$

Using the method of characteristic, it is straight to show that system (12.5)–(12.9) has a unique solution the solution for $t = 2k + \tau, k = 0, 1, 2, ..., 0 \le \tau \le 2$ and $0 \le x \le 1$,

$$\begin{bmatrix} \xi_1(x,t) \\ \xi_2(x,t) \\ \xi_3(x,t) \end{bmatrix} = \begin{cases} (R_1 \circ R_0)^k \left(\begin{bmatrix} \xi_{1,0}(x+\tau) \\ \xi_{2,0}(x+\tau) \\ \xi_{3,0}(x+\tau) \end{bmatrix} \right), & \text{for } 0 \le \tau \le 1-x, \\ (R_1 \circ R_0)^k \circ R_1 \left(\begin{bmatrix} \chi_{1,0}(2-x-\tau) \\ \chi_{2,0}(2-x-\tau) \\ \chi_{3,0}(2-x-\tau) \end{bmatrix} \right), & \text{for } 1-x < \tau \le 2-x, \\ (R_1 \circ R_0)^{k+1} \left(\begin{bmatrix} \xi_{1,0}(x+\tau-2) \\ \xi_{2,0}(x+\tau-2) \\ \xi_{3,0}(x+\tau-2) \end{bmatrix} \right), & \text{for } 2-x < \tau \le 2, \end{cases} \tag{12.10}$$

and

$$\begin{bmatrix} \chi_1(x,t) \\ \chi_2(x,t) \\ \chi_3(x,t) \end{bmatrix} = \begin{cases} (R_0 \circ R_1)^k \left(\begin{bmatrix} \chi_{1,0}(x-\tau) \\ \chi_{2,0}(x-\tau) \\ \chi_{3,0}(x-\tau) \end{bmatrix} \right), & \text{for } 0 \le \tau \le x, \\ (R_0 \circ R_1)^k \circ R_0 \left(\begin{bmatrix} \xi_{1,0}(\tau-x) \\ \xi_{2,0}(\tau-x) \\ \xi_{3,0}(\tau-x) \end{bmatrix} \right), & \text{for } x < \tau \le 1+x, \\ (R_0 \circ R_1)^{k+1} \left(\begin{bmatrix} \chi_{1,0}(2+x-\tau) \\ \chi_{2,0}(2+x-\tau) \\ \chi_{3,0}(2+x-\tau) \end{bmatrix} \right), & \text{for } 1+x < \tau \le 2, \end{cases} \tag{12.11}$$

where $(R_0 \circ R_1)^k$ denotes the k-times iterative composition of $R_0 \circ R_1$ with itself. From the expression of the solution, we can know the dynamical behavior of the system (12.5)–(12.9) can be completely determined by two 3D maps $R_0 \circ R_1$ and $R_1 \circ R_0$, which are mutually topologically conjugate. As [19] and [20], we say the system is chaotic if the 3D map $R_0 \circ R_1$ has chaotic behavior.

The organization of the paper is as follows. In Sect. 12.2, we present basic properties of the map $R_0 \circ R_1$. In Sect. 12.3, we prove the existence of snapback repellers, which can cause the chaotic behavior of system (12.5)–(12.9). In Sect. 12.4, some numerical simulations are used to illustrate the theoretical results.

12.2 Map of $R_0 \circ R_1$

It is known that for each given $0 < \alpha < 1, \beta > 0$ and $\eta > 0, \eta \neq 3$, the maps $R_0 \circ R_1$ and $R_1 \circ R_0$ are topologically conjugate. Hence, we only study the properties of $R_0 \circ R_1$.

From (12.6) and (12.7), we have the expression of $R_0 \circ R_1$

$$\begin{bmatrix} u_1 \\ u_2 \\ u_3 \end{bmatrix} = R_0 \circ R_1 \left(\begin{bmatrix} v_1 \\ v_2 \\ v_3 \end{bmatrix} \right) = \begin{bmatrix} \frac{\eta-1}{\eta-3} F(v_1) + \frac{2}{\eta-3} F(v_2) + \frac{2}{\eta-3} F(v_3) \\ \frac{2}{\eta-3} F(v_1) + \frac{\eta-1}{\eta-3} F(v_2) + \frac{2}{\eta-3} F(v_3) \\ \frac{2}{\eta-3} F(v_1) + \frac{2}{\eta-3} F(v_2) + \frac{\eta-1}{\eta-3} F(v_3) \end{bmatrix}. \tag{12.12}$$

First, for $0 < \alpha < 1$ and $\beta > 0$, we define a new function

$$g(v) = u - v = F(v) - v, \tag{12.13}$$

then it follows from (12.8) that $g(v)$ is the unique real solution of the cubic equation [18]

$$\beta g^3(v) + (1-\alpha) g(v) + 2v = 0. \tag{12.14}$$

Moreover, we have the following lemma about $g(v)$ and $F(v)$ from [18].

Lemma 12.1 *Let g and F be given by (12.13). We have the following assertions:*

(i) $g(v)$ *is odd:* $g(-v) = -g(v)$, *and* $g(v) < 0$ *for all* $v > 0$;
(ii) $g(v)$ *is strictly decreasing, that is,*

$$g'(v) = \frac{-2}{3\beta g^2(v) + 1 - \alpha} < 0; \tag{12.15}$$

(iii) $F(v)$ *is odd;*
(iv) $F(v) = 0$ *has three zeros* $v = -v_I, 0, v_I$, *where*

$$v_I = \sqrt{\frac{1+\alpha}{\beta}}; \tag{12.16}$$

(v) $F'(v) = 1 + g'(v)$,

$$F'(0) = -\frac{1+\alpha}{1-\alpha}; \tag{12.17}$$

(vi) $F'(v) = 0$ *if* $v = \pm v_c$, *where*

$$v_c = \frac{2-\alpha}{3}\sqrt{\frac{1+\alpha}{3\beta}}. \tag{12.18}$$

We define

$$G = \frac{\eta+3}{\eta-3} \quad \text{and} \quad y = G \circ F(v) = \frac{\eta+3}{\eta-3}F(v), \quad v \in \mathbb{R}. \tag{12.19}$$

then, we have the following lemma.

Lemma 12.2 *Let* G *and* $y = G \circ F(v)$ *be given by (12.19). For* $0 < \alpha < 1$, $\beta > 0$, $\eta > 0$, $\eta \neq 3$, *and* α, β, η *be fixed, we have the following assertions:*

(1) $y = G \circ F(v)$ *has three intercepts at* $v = -v_I, 0, v_I$, *where* v_I *is given by (12.16).*
(2) (i) If $0 < \eta < 3$, *then* $y = G \circ F(v)$ *has two local extremal values:*

$$M_1 = G \circ F(v_c) = -\frac{(1+\alpha)(3+\eta)}{3(\eta-3)}\sqrt{\frac{1+\alpha}{3\beta}} > 0, \tag{12.20}$$

and

$$-M_1 = G \circ F(-v_c) = \frac{(1+\alpha)(3+\eta)}{3(\eta-3)}\sqrt{\frac{1+\alpha}{3\beta}} < 0 \tag{12.21}$$

where v_c *is given by (12.18), and* M_1 *and* $-M_1$ *are the local maximum and minimum of* $y = G \circ F(v)$ *in* $\mathbb{R}$, *respectively, (as depicted by Fig. 12.2, for the case of* $\alpha = 0.65$, $\beta = 1$ *and* $\eta = 1.6$*). The function* $y = G \circ F(v)$ *is strictly decreasing on the intervals both* $(-\infty, -v_c)$ *and* $(v_c, +\infty)$, *and strictly increasing on* $(-v_c,\ v_c)$.

(ii) If $\eta > 3$, *then* $y = G \circ F(v)$ *has two local extremal values:*

$$M_2 = G \circ F(-v_c) = \frac{(1+\alpha)(\eta+3)}{3(\eta-3)}\sqrt{\frac{1+\alpha}{3\beta}} > 0, \tag{12.22}$$

and

$$-M_2 = G \circ F(v_c) = -\frac{(1+\alpha)(\eta+3)}{3(\eta-3)}\sqrt{\frac{1+\alpha}{3\beta}} < 0, \tag{12.23}$$

where v_c *is given by (12.18), and* M_2 *and* $-M_2$ *are the local maximum and minimum of* $y = G \circ F(v)$ *in* $\mathbb{R}$, *respectively, (as depicted by Fig. 12.3, for the case of* $\alpha = 0.65$, $\beta = 1$ *and* $\eta = 4.8$*). The function* $y = G \circ F(v)$ *is strictly increasing on the intervals both* $(-\infty, -v_c)$ *and* $(v_c, +\infty)$, *and strictly decreasing on* $(-v_c,\ v_c)$.

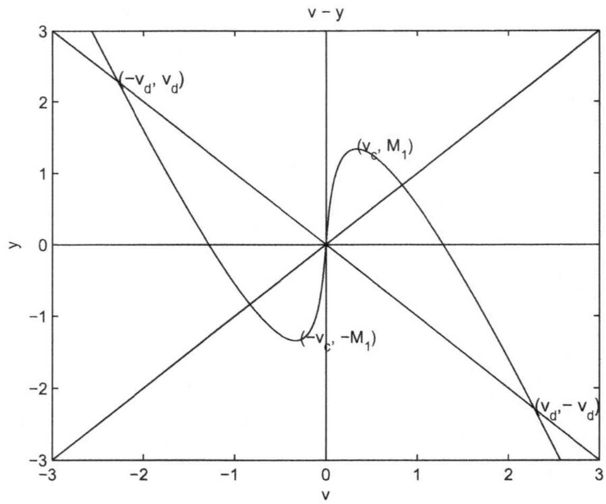

Fig. 12.2 The graph of $y = G \circ F(v)$, for the case of $\alpha = 0.65$, $\beta = 1$ and $\eta = 1.6$

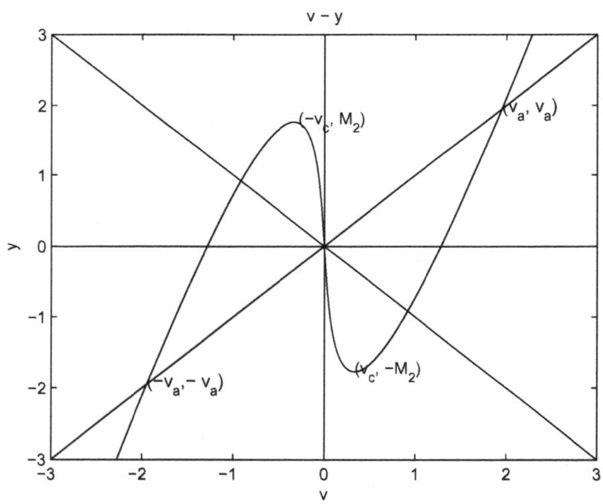

Fig. 12.3 The graph of $y = G \circ F(v)$, for the case of $\alpha = 0.65$, $\beta = 1$ and $\eta = 4.8$

(3) $y = G \circ F(v)$ *intersects the line* $y = v$ *at the points*

$$-(v_a,\ v_a),\quad (0,\ 0),\quad (v_a,\ v_a),\quad \textit{where}\quad v_a = \frac{\eta+3}{6}\sqrt{\frac{3\alpha+\eta}{3\beta}}. \tag{12.24}$$

(4) $y = G \circ F(v)$ *intersects the line* $y = -v$ *at the points*

$$(-v_d,\ v_d),\quad (0,\ 0),\quad (v_d,\ -v_d),\quad \textit{where}\quad v_d = \frac{3+\eta}{2\eta}\sqrt{\frac{3+\alpha\eta}{\beta\eta}}. \tag{12.25}$$

Proof Due to fact that G is a constant, the first assertion (1) is a direct result of the (iv) of Lemma 12.1. It follows from the fact $F(v) = g(v) + v$ in (12.13) that

$$y' = (G \circ F)'(v) = \frac{\eta+3}{\eta-3}\Big(v + g(v)\Big)' = \frac{\eta+3}{\eta-3}\Big(1 + g'(v)\Big) = 0 \tag{12.26}$$

which yields $g'(v) = -1$. By using (12.15), we get

$$g'(v) = \frac{-2}{3\beta g^2(v) + 1 - \alpha} = -1,$$

and hence $g(v) = \pm\sqrt{\frac{1+\alpha}{3\beta}}$. For $0 < \eta < 3$, there are two cases:

(a) When $g(v) = -\sqrt{\frac{1+\alpha}{3\beta}}$, substituting this into (12.14), we have

$$-\frac{1+\alpha}{3}\sqrt{\frac{1+\alpha}{3\beta}} - (1-\alpha)\sqrt{\frac{1+\alpha}{3\beta}} + 2v = 0,$$

and get the zero $v = v_c$, where v_c is given by (12.18). Thus, $F(v_c) = v_c + g(v_c)$ and

$$y = G \circ F(v_c) = \frac{\eta+3}{\eta-3}\left[\frac{2-\alpha}{3}\sqrt{\frac{1+\alpha}{3\beta}} - \sqrt{\frac{1+\alpha}{3\beta}}\right] = M_1 > 0.$$

where M_1 is given by (12.20), and y takes the local maximum at $v = v_c$ with value $M_1 > 0$.

(b) When $g(v) = \sqrt{\frac{1+\alpha}{3\beta}}$, it follows from the odd properties of g and F that $v = -v_c$ is the zero of (12.14), and $y = G \circ F(-v_c) = -M_1 < 0$, where $-M_1$ is given by (12.21), and y takes the local minimum at $v = -v_c$ with value $-M_1 < 0$.

Moreover, it is noted that F is an odd function in $\mathbb{R}$, so is for $y = G_1 \circ F(v)$. Thus, y is strictly decreasing on the intervals both $(-\infty, -v_c)$ and $(v_c, +\infty)$, and strictly increasing on $(-v_c,\ v_c)$. For $\eta > 3$, the proof is similar to $0 < \eta < 3$. This is assertion (2).

Now we show assertion (3). By $y = G \circ F(v) = v$, we have

$$\frac{\eta+3}{\eta-3}F(v) = \frac{\eta+3}{\eta-3}\Big[v + g(v)\Big] = v$$

which gives $g(v) = -\frac{6}{\eta+3}v$. Substituting $g(v)$ into (12.14), we have

$$\beta\left(\frac{-6}{\eta+3}v\right)^3-(1-\alpha)\frac{6}{\eta+3}v+2v=0,$$

which has three zeros $v=0$ and $v=\pm v_a$, where v_a is given by (12.24). The third assertion (3) is then concluded.

Same arguments to the case $y=G\circ F(v)=-v$, we have

$$\frac{\eta+3}{\eta-3}F(v)=\frac{\eta+3}{\eta-3}\Big[v+g(v)\Big]=-v,\quad g(v)=-\frac{2\eta}{\eta+3}v.$$

Substituting $g(v)$ into (12.14), v satisfies the following equation

$$\beta\left(\frac{-2\eta}{\eta+3}v\right)^3-(1-\alpha)\frac{2\eta}{\eta+3}v+2v=0,$$

which has three zeros $v=0$ and $v=\pm v_d$, where v_d is given by (12.25). The fourth assertion (4) is then concluded. □

Lemma 12.3 *Let* $0<\alpha<1$, $\beta>0$, *and* $\eta>0$, $\eta\neq 3$.

(1) If $0<\eta<3$ *and* α, η *satisfy the following inequality*

$$-\frac{(1+\alpha)(3+\eta)}{3(\eta-3)}\sqrt{\frac{1+\alpha}{3\beta}}\le\frac{3+\eta}{2\eta}\sqrt{\frac{3+\alpha\eta}{\beta\eta}},\tag{12.27}$$

then $E_1=[-v_d,v_d]$ *is the invariant set of function* $y=G\circ F(v)$.

(2) If $\eta>3$ *and* α, η *satisfy the following inequality*

$$\frac{(1+\alpha)(\eta+3)}{3(\eta-3)}\sqrt{\frac{1+\alpha}{3\beta}}\le\frac{\eta+3}{6}\sqrt{\frac{3\alpha+\eta}{3\beta}},\tag{12.28}$$

then $E_2=[-v_a,v_a]$ *is the invariant set of function* $y=G\circ F(v)$.

Proof First we consider $0<\eta<3$. From (12.18) and (12.25), we obtain

$$v_c=\frac{2-\alpha}{3}\sqrt{\frac{1+\alpha}{3\beta}}<\frac{3+\eta}{2\eta}\sqrt{\frac{3+\alpha\eta}{\beta\eta}}=v_d.$$

From the assertion (2) of Lemma 12.2, we have that y is strictly decreasing on $(-v_d,-v_c)$ and (v_c,v_d), and strictly increasing on $(-v_c,v_c)$. Thus, if $y(v_c)\le v_d$, for each $v\in[0,v_d]$, we have $y(v)\le y(v_c)\le v_d$ and $y(-v)\ge y(-v_c)\ge -v_d$. Hence E_1 is invariant set of y. Similarly, for $\eta>3$, we prove E_2 is the invariant set of y. □

Theorem 12.1 *Let* $0 < \alpha < 1$, $\beta > 0$, $\eta > 0$ *and* $\eta \neq 3$. *We have*

(1) when $1 \leq \eta < 3$, *and* α, η *satisfy (12.27), then*

$$D_1 = \left\{ (v_1, v_2, v_3) \in \mathbb{R}^3 \middle| \, |v_1| \leq v_d, |v_2| \leq v_d, |v_3| \leq v_d \right\}$$

is invariant under $R_0 \circ R_1$, *i.e.,* $(R_0 \circ R_1)(D_1) \subset D_1$.

(2) when $\eta > 3$, *and* α, η *satisfy (12.28), then*

$$D_2 = \left\{ (v_1, v_2, v_3) \in \mathbb{R}^3 \middle| \, |v_1| \leq v_a, |v_2| \leq v_a, |v_3| \leq v_a \right\}$$

is invariant under $R_0 \circ R_1$, *i.e.,* $(R_0 \circ R_1)(D_2) \subset D_2$.

Proof First, we consider $1 \leq \eta < 3$. Assume $(v_1, v_2, v_3) \in D_1$. We need to show $(u_1, u_2, u_3) \in D_1$. From Lemma 12.3, we know $[-v_d, \ v_d]$ is invariant of $y = G \circ F(v)$, and for $v \in [-v_d, \ v_d]$, we have

$$-v_d \leq \frac{\eta+3}{\eta-3} F(v) \leq v_d, \quad \text{and} \quad \frac{\eta-3}{\eta+3} v_d \leq F(v) \leq -\frac{\eta-3}{\eta+3} v_d.$$

Since $1 \leq \eta < 3$, for each $v \in [-v_d, v_d]$, we further have

$$-\frac{\eta-1}{\eta+3} v_d \leq \frac{\eta-1}{\eta-3} F(v) \leq \frac{\eta-1}{\eta+3} v_d \tag{12.29}$$

and

$$-\frac{2}{\eta+3} v_d \leq \frac{2}{\eta-3} F(v) \leq \frac{2}{\eta+3} v_d \tag{12.30}$$

Now we are in a position to show that D_1 is invariant under $R_0 \circ R_1$. For each $(v_1, v_2, v_3) \in D_1$, it follows from (12.12) that

$$u_1 = \frac{\eta-1}{\eta-3} F(v_1) + \frac{2}{\eta-3} F(v_2) + \frac{2}{\eta-3} F(v_3),$$

$$u_2 = \frac{2}{\eta-3} F(v_1) + \frac{\eta-1}{\eta-3} F(v_2) + \frac{2}{\eta-3} F(v_3),$$

and

$$u_3 = \frac{2}{\eta-3} F(v_1) + \frac{2}{\eta-3} F(v_2) + \frac{\eta-1}{\eta-3} F(v_3).$$

By (12.29) and (12.30), we have

$$\left[\frac{\eta-1}{\eta+3} + \frac{2}{\eta+3} + \frac{2}{\eta+3}\right](-v_d) \leq u_1, u_2, u_3 \leq \left[\frac{\eta-1}{\eta+3} + \frac{2}{\eta+3} + \frac{2}{\eta+3}\right] v_d$$

and $-v_d \le u_1, u_2, u_3 \le v_d$. Therefore, D_1 is invariant under $R_0 \circ R_1$. Similarly, for $\eta > 3$, we prove D_2 is the invariant set of $R_0 \circ R_1$. The proof is complete. □

Lemma 12.4 *Let* $0 < \alpha < 1$, $\beta > 0$ *and* $\eta > 0$, $\eta \neq 3$. *Then* $(0, 0, 0)^T$ *is the fixed point of the map* $R_0 \circ R_1$.

Since F is continuously differentiable, we define the Jacobian matrix of $R_0 \circ R_1$ at $(v_1, v_2, v_3)^T$ by

$$D(R_0 \circ R_1)\left(\begin{bmatrix} v_1 \\ v_2 \\ v_3 \end{bmatrix}\right) = \begin{bmatrix} \frac{\eta-1}{\eta-3}F'(v_1) & \frac{2}{\eta-3}F'(v_2) & \frac{2}{\eta-3}F'(v_3) \\ \frac{2}{\eta-3}F'(v_1) & \frac{\eta-1}{\eta-3}F'(v_2) & \frac{2}{\eta-3}F'(v_3) \\ \frac{2}{\eta-3}F'(v_1) & \frac{2}{\eta-3}F'(v_2) & \frac{\eta-1}{\eta-3}F'(v_3) \end{bmatrix}. \tag{12.31}$$

We give the following definition from [7, p. 215].

Definition 12.1 A fixed point $(z_1, z_2, z_3)^T$ of $R_0 \circ R_1$ is called a repelling point if all of the eigenvalues of $D(R_0 \circ R_1)(z_1, z_2, z_3)^T$ are greater than one in absolute value.

Theorem 12.2 *Let* $0 < \alpha < 1$, $\beta > 0$, *and* $\eta > 0$, $\eta \neq 3$, *the fixed point* $(0, 0, 0)^T$ *is a repelling fixed point of* $R_0 \circ R_1$.

Proof Due to fact that $(0, 0, 0)^T$ is a fixed point of $R_0 \circ R_1$, according the Definition 12.1, we only need to show all eigenvalues of $D(R_0 \circ R_1)(0, 0, 0)^T$ are greater than one in absolute value. From (12.17) and (12.31), we have

$$A \equiv D(R_0 \circ R_1)(0,0,0)^T = \begin{bmatrix} -\frac{1+\alpha}{1-\alpha}\frac{\eta-1}{\eta-3} & -\frac{1+\alpha}{1-\alpha}\frac{2}{\eta-3} & -\frac{1+\alpha}{1-\alpha}\frac{2}{\eta-3} \\ -\frac{1+\alpha}{1-\alpha}\frac{2}{\eta-3} & -\frac{1+\alpha}{1-\alpha}\frac{\eta-1}{\eta-3} & -\frac{1+\alpha}{1-\alpha}\frac{2}{\eta-3} \\ -\frac{1+\alpha}{1-\alpha}\frac{2}{\eta-3} & -\frac{1+\alpha}{1-\alpha}\frac{2}{\eta-3} & -\frac{1+\alpha}{1-\alpha}\frac{\eta-1}{\eta-3} \end{bmatrix},$$

and

$$\lambda I - A = \begin{bmatrix} \lambda + \frac{1+\alpha}{1-\alpha}\frac{\eta-1}{\eta-3} & \frac{1+\alpha}{1-\alpha}\frac{2}{\eta-3} & \frac{1+\alpha}{1-\alpha}\frac{2}{\eta-3} \\ \frac{1+\alpha}{1-\alpha}\frac{2}{\eta-3} & \lambda + \frac{1+\alpha}{1-\alpha}\frac{\eta-1}{\eta-3} & \frac{1+\alpha}{1-\alpha}\frac{2}{\eta-3} \\ \frac{1+\alpha}{1-\alpha}\frac{2}{\eta-3} & \frac{1+\alpha}{1-\alpha}\frac{2}{\eta-3} & \lambda + \frac{1+\alpha}{1-\alpha}\frac{\eta-1}{\eta-3} \end{bmatrix}. \tag{12.32}$$

Hence, we have

$$\det(\lambda I - A) = \left(\lambda + \frac{1+\alpha\,\eta+3}{1-\alpha\,\eta-3}\right)\left(\lambda + \frac{1+\alpha}{1-\alpha}\right)^2. \tag{12.33}$$

Then when $\det(\lambda I - A) = 0$, we have

$$\lambda_1 = -\frac{1+\alpha\,\eta+3}{1-\alpha\,\eta-3}, \text{ and } \lambda_2 = \lambda_3 = -\frac{1+\alpha}{1-\alpha}.$$

Due to $0 < \alpha < 1$, $\beta > 0$, and $\eta > 0$, $\eta \neq 3$, we obtain $|\lambda_1| > 1$, $|\lambda_2| = |\lambda_3| > 1$. □

12.3 Snapback Repellers and Chaos

In this section, we consider the snapback repellers of $R_0 \circ R_1$ which will cause the chaos in the sense of Li-Yorke for $R_0 \circ R_1$.

Let $f : I \to I$ be a continuously differentiable function in $I \subset \mathbb{R}$. We recall from [7] the definitions of homoclinic points and orbits for f. Let p be a repelling fixed point of f, that is,

$$f(p) = p, \quad |f'(p)| > 1.$$

Then there is an open interval about p on which f is one-to-one and satisfies the expansion property $|f(x) - p| > |x - p|$. We define the local unstable set at p, denoted by $W^u_{loc}(p)$, to be the maximal such open interval about p. A point $q \in I$ is said to be homoclinic to p if $q \in W^u_{loc}(p)$ and $f^n(q) = p$ for some $n \in \{1, 2, 3, ...\}$. For a homoclinic point q, the set $\{f^j(q) | j = 1, 2, ..., n\}$ is said the homoclinic orbit of q. The homoclinic orbit of q is said to be nondegenerate if $f'(x) \neq 0$ for all points x on the orbit. Otherwise, the homoclinic orbit is said to be degenerate (Figs. 12.4 and 12.5).

Theorem 12.3 *Let* $0 < \alpha < 1$, $\beta > 0$, *and* $\eta > 0$, $\eta \neq 3$.

(1) If $1 \le \eta < 3$ *and* α, β, η *satisfy the following inequality*

$$\sqrt{\frac{1+\alpha}{\beta}} < -\frac{(1+\alpha)(3+\eta)}{3(\eta-3)}\sqrt{\frac{1+\alpha}{3\beta}} \le \frac{3+\eta}{2\eta}\sqrt{\frac{3+\alpha\eta}{\beta\eta}}, \tag{12.34}$$

then the repelling fixed point 0 *of* y *has nondegenerate homoclinic orbits.*

(2) If $\eta > 3$ *and* α, β, η *satisfy the following inequality*

$$\sqrt{\frac{1+\alpha}{\beta}} < \frac{(1+\alpha)(\eta+3)}{3(\eta-3)}\sqrt{\frac{1+\alpha}{3\beta}} \le \frac{\eta+3}{6}\sqrt{\frac{3\alpha+\eta}{3\beta}}, \tag{12.35}$$

then the repelling fixed point 0 *of* y *has nondegenerate homoclinic orbits.*

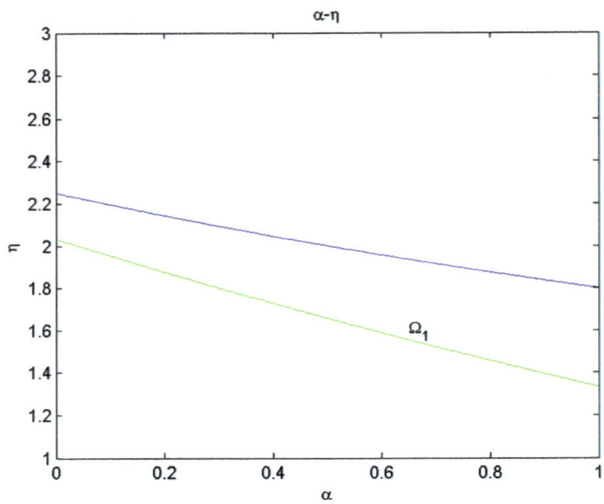

Fig. 12.4 The region Ω_1 represents the inequality (12.34)

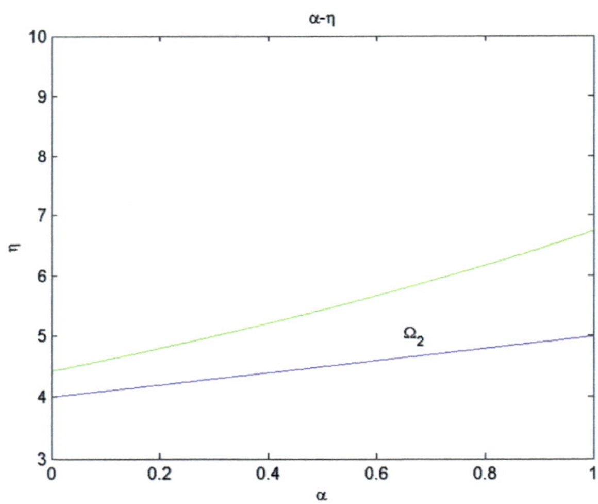

Fig. 12.5 The region Ω_2 represents the inequality (12.35)

Proof First, we consider $1 \leq \eta < 3$, according to Theorem 12.1, we know D_1 is invariant of y. Next, it follow from the (v) of Lemma 12.1 that

$$y'(0) = (G \circ F)'(0) = \frac{\eta + 3}{\eta - 3} F'(0) = -\frac{\eta + 3}{\eta - 3} \frac{1 + \alpha}{1 - \alpha} > 1,$$

where we have used $0 < \alpha < 1$ and $1 \leq \eta < 3$. Therefore 0 is a repelling point of y. It is easy to prove the existence of homoclinic points near 0. Due to the fact

$$\sqrt{\frac{1 + \alpha}{\beta}} < \frac{(1 + \alpha)(3 + \eta)}{3(3 - \eta)} \sqrt{\frac{1 + \alpha}{3\beta}} \leq \frac{3 + \eta}{2\eta} \sqrt{\frac{3 + \alpha\eta}{\beta\eta}},$$

it is easy to see that the homoclinic orbit of 0 is nondegenerate. Similarly, for $\eta > 3$, we prove the repelling fixed point 0 of y has nondegenerate homoclinic orbits. The proof is complete. □

Here is a definition of a snapback repeller of a differentiable function f from [21].

Definition 12.2 Suppose z is a fixed point of f with all eigenvalues of $Df(z)$ exceeding 1 in magnitude, and suppose there exists a point $x_0 \neq z$ in a repelling neighborhood of z, such that $x_M = z$ and $\det(Df(x_k)) \neq 0$ for $1 \le k \le M$, where $x_k = f^k(x_0)$. Then z is called a snapback repeller of f.

Theorem 12.4 *Let* $0 < \alpha < 1$, $\beta > 0$, *and* $\eta > 0$, $\eta \neq 3$.

(1) If $1 \le \eta < 3$ *and* α, β, η *satisfy (12.34),* $(0,0,0)^T$ *is a snap-back repeller of* $R_0 \circ R_1$.
(2) If $\eta > 3$ *and* α, β, η *satisfy (12.35),* $(0,0,0)^T$ *is a snap-back repeller of* $R_0 \circ R_1$.

Proof First, we consider $1 \le \eta < 3$. It is noted from Theorem 12.2 that $(0,0,0)^T$ is the fixed point of $R_0 \circ R_1$ with all eigenvalues of $D(R_0 \circ R_1)(0,0,0)^T$ exceeding 1 in magnitude. Let $B_r(X)$ denote the closed ball in $\mathbb{R}^n$ of radius r centered at the point X, and $B_r^0(X)$ its interior.

It follows from $F'(0) = -\dfrac{1+\alpha}{1-\alpha}$ given in (12.17) and the properties of continuous functions that there exists a positive constant $r > 0$, such that for each $v \in B_r^0(0)$, we have $|F'(v)| > 1$. According to Theorem 12.3, we can choose $0 \neq v_1 \in B_r^0(0)$ that satisfies

$$(G \circ F)^K(v_1) = \left(\frac{\eta+3}{\eta-3}F\right)^K (v_1) = 0,$$

where K is some positive constant. Let $v_1 = v_2 = v_3$. Due to (12.31), we have

$$Q \equiv D(R_0 \circ R_1)\left(\begin{bmatrix} v_1 \\ v_2 \\ v_3 \end{bmatrix}\right) = \begin{bmatrix} \frac{\eta-1}{\eta-3}F'(v_1) & \frac{2}{\eta-3}F'(v_1) & \frac{2}{\eta-3}F'(v_1) \\ \frac{2}{\eta-3}F'(v_1) & \frac{\eta-1}{\eta-3}F'(v_1) & \frac{2}{\eta-3}F'(v_1) \\ \frac{2}{\eta-3}F'(v_1) & \frac{2}{\eta-3}F'(v_1) & \frac{\eta-1}{\eta-3}F'(v_1) \end{bmatrix}$$

and

$$\det(\lambda I - Q) = \left(\lambda - \frac{\eta+3}{\eta-3}F'(v_1)\right)\left(\lambda - F'(v_1)\right)^2.$$

When $\det(\lambda I - Q) = 0$, we have $\lambda_4 = \dfrac{\eta+3}{\eta-3}F'(v_1)$, $\lambda_5 = \lambda_6 = F'(v_1)$. Due to $|F'(v_1)| > 1$, then we can obtain $|\lambda_4| > 1$, $|\lambda_5| = |\lambda_6| > 1$.

By (12.12) and the mathematical deduction, we have

$$(R_0 \circ R_1)^n \left(\begin{bmatrix} v_1 \\ v_2 \\ v_3 \end{bmatrix} \right) = \left[\left(\frac{\eta+3}{\eta-3} F \right)^n (v_1) \;\; \left(\frac{\eta+3}{\eta-3} F \right)^n (v_1) \;\; \left(\frac{\eta+3}{\eta-3} F \right)^n (v_1) \right]^T$$

Let $n = K$ and

$$\left(\frac{\eta+3}{\eta-3} F \right)^K (v_1) = 0.$$

Then we have $(R_0 \circ R_1)^K (v_1, v_2, v_3)^T = (0, 0, 0)^T$. Since the homoclinic orbit of v_1 is nondegenerate, that is,

$$\left(\left(\frac{\eta+3}{\eta-3} F \right)^j \right)' (v_1) \neq 0, \quad j = 1, 2, ..., K,$$

we have $\det \left(D(R_0 \circ R_1)^s (v_1, v_2, v_3)^T \right) \neq 0$, $s = 1, 2, ..., K$. Hence, $(0, 0, 0)^T$ is a snap-back repeller of $R_0 \circ R_1$. Similarly, for $\eta > 3$, $(0, 0, 0)^T$ is a snap-back repeller of $R_0 \circ R_1$. The proof is complete. □

We cite the following theorem from [22], [21], or [17].

Theorem 12.5 *Let D be an open set in R^N and $f : D \to D$ be a continuous map. If f possesses a snap-back repeller, then f is chaotic in the sense of Li-Yorke. That is, there exists*

(i) a positive integer n such that for each integer $p \geq n$, f has a point of period p;
(ii) a "scrambled set" of f, i.e., an uncountable set S containing no periodic points of f such that: $f(S) \subset S$, and

(a) for every $x, y \in S$ with $x \neq y$,

$$\limsup_{k \to \infty} ||f^k(x) - f^k(y)|| > 0,$$

(b) for every $x \in S$ and any periodic point y of f,

$$\limsup_{k \to \infty} ||f^k(x) - f^k(y)|| > 0;$$

(iii) an uncountable subset S_0 of S such that for every $x, y \in S_0$,

$$\liminf_{k \to \infty} \left\| f^k(x) - f^k(y) \right\| = 0.$$

Theorem 12.6 *Let $0 < \alpha < 1$, $\beta > 0$, and $\eta > 0$, $\eta \neq 3$.*

(1) If $1 \leq \eta < 3$ and α, β, η satisfy (12.34) then $R_0 \circ R_1$ is chaotic in the sense of Li-Yorke.

(2) If $\eta > 3$ and α, β, η satisfy (12.35) then $R_0 \circ R_1$ is chaotic in the sense of Li-Yorke.

Proof It follows from Theorem 12.4 that $(0, 0, 0)^T$ is the snap-back repeller of $R_0 \circ R_1$. Hence, by Theorem 12.5, $R_0 \circ R_1$ is chaotic. The proof is complete. □

12.4 A Numerical Example

In this section, we present the simulations to visualize the dynamics behavior of the solutions of system (12.1). According to Theorem 12.6, we choose $\alpha = 0.65$, $\beta = 1$, $\eta = 1.6$, which satisfy

$$\begin{cases} \sqrt{\dfrac{1+\alpha}{\beta}} \approx 1.2845, \quad \dfrac{(1+\alpha)(3+\eta)}{3(3-\eta)}\sqrt{\dfrac{1+\alpha}{3\beta}} \approx 1.3402, \\ \dfrac{3+\eta}{2\eta}\sqrt{\dfrac{3+\alpha\eta}{\beta\eta}} \approx 2.2842. \end{cases} \tag{12.36}$$

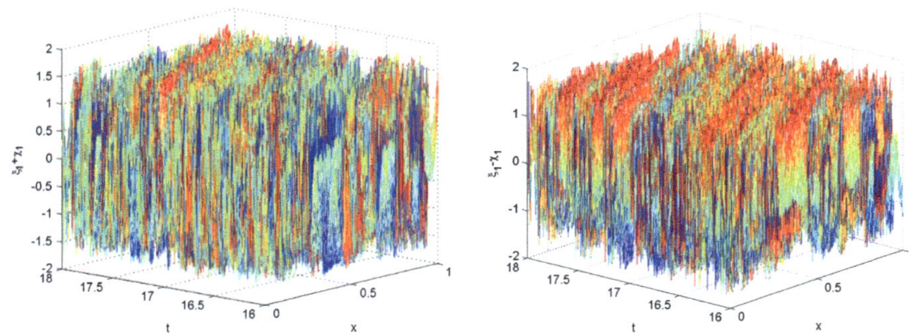

Fig. 12.6 Spatiotemporal profiles of $w_x^{(1)}(x, t)$ and $w_t^{(1)}(x, t)$

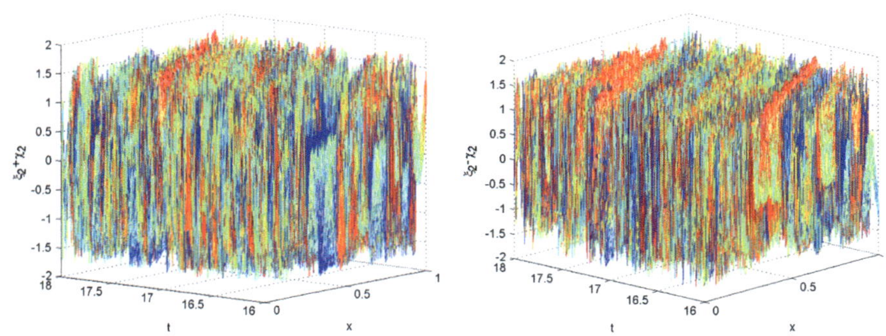

Fig. 12.7 Spatiotemporal profiles of $w_x^{(2)}(x, t) =$ and $w_t^{(2)}(x, t)$

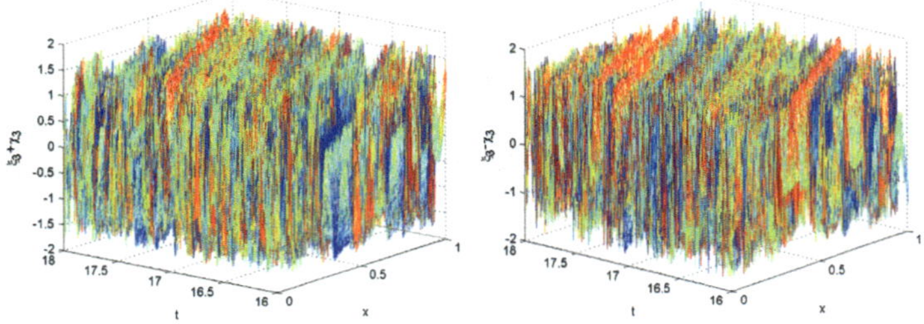

Fig. 12.8 Spatiotemporal profiles of $w_x^{(3)}(x,t)$ and $w_t^{(3)}(x,t)$

For $i = 1, 2, 3$, we choose

$$\begin{cases} w^{(i)}(x,0) = \frac{1}{8}(x - \frac{1}{2})^4 - \cos x, \\ w_t^{(i)}(x,0) = \frac{1}{2}(x - \frac{1}{2})^3 - \sin x, \end{cases} \quad \begin{cases} \xi_i = \frac{1}{2}(x - \frac{1}{2})^3, \\ \chi_i = \sin x. \end{cases}$$

We present the graphics in some detail, for $w_x^{(i)}, w_t^{(i)}, i = 1, 2, 3$ for time durations $t \in [16, 18]$. From Figs. 12.6, 12.7 and 12.8, it is found that $w_x^{(i)}, w_t^{(i)}, i = 1, 2, 3$, are extremely oscillatory in every direction of space and time, respectively.

References

1. Schiiff, S.J., Jerger, K., Duong, D.H., Chang, T., Spano, M.L., Ditto, W.L.: Controlling chaos in the brain. Nature **370**, 615–620 (1994)
2. Wang, X.Y., Liu, C.M., Xu, D.H., Liu, C.X.: Image encryption scheme using chaos and simulated annealing algorithm. Nonlinear Dyn. **84**, 1417–1429 (2016)
3. Wang, X.Y., Liu, C.M., Zhang, H.L.: An effective and fast image encryption algorithm based on chaos and interweaving of ranks. Nonlinear Dyn. **84**, 1595–1607 (2016)
4. Vaseghi, B., Pourmina, M.A., Mobayen, S.: Secure communication in wireless sensor networks based on chaos synchronization using adaptive sliding mode control. Nonlinear Dyn. **89**, 1689–1704 (2017)
5. Volos, C., Akgul, A., Pham, V.T., Stouboulos, I., Kyprianidis, I.: A simple chaotic circuit with a hyperbolic sine function and its use in a sound encryption scheme. Nonlinear Dyn. **89**, 1047–1061 (2017)
6. Li, T., Yorke, J.: Period three implies chaos. Am. Math. Mon. **82**, 985–992 (1975)
7. Devaney, R.L.: An Introduction to Chaotic Dynamical Systems, 2nd edn. Addison-Wesley, New York (1989)
8. Robinson, C.: Dynamical Systems, Stability, Symbolic Dynamics and Chaos, 2nd edn. CRC Press, Boca Raton (1999)
9. Chen, G., Huang, T.W., Juang, J., Ma, D.W.: Unbounded growth of total variations of snapshots of the 1D linear wave equation due to that chaotic behavior of iterates of composite nonlinear boundary reflection relation, Control of nonlinear distributed parameter systems. In: Lecture Notes in Pure and Applied Mathematics, vol. 218, pp. 15–43. Dekker, New York (2001)

10. Huang, Y.: A new characterization of nonisotropic chaotic vibrations of the one-dimnesional linear wave equaiton with a van der Pol boundary condition. J. Math. Anal. Appl. **288**, 78–96 (2003)
11. Huang, Y.: Growth rates of total variations of snapshots of the 1D linear wave equation with composite nonlinear boundary reflection relations, Internat. J. Bifur. Chaos Appl. Sci. Eng. **13**, 1183–1195 (2003)
12. Li, L.L., Huang, Y.: Growth rates of total variations of snapshots of 1D linear wave equations with nonlinear right-end boundary conditions. J. Math. Anal. Appl. **361**, 69–85 (2010)
13. Li, L.L.: Analyzing displacement term's memory effect in a nonlinear boundary value problem to prove chaotic vibration of the wave equation. J. Math. Anal. Appl. **429**, 758–773 (2015)
14. Li, L.L., Huang, T.W., Huang, X.Y.: Chaotic oscillations of the 1D wave equation due to extreme imbalance of self-regulations. J. Math. Anal. Appl. **450**, 1388–1400 (2017)
15. Hu, C.C.: Chaotic vibrations of the one-dimensional mixed wave system, Internat. J. Bifur. Chaos Appl. Sci. Eng. **19**, 579–590 (2009)
16. Zhang, L.J., Shi, Y.M., Zhang, X.: Chaotic dynamical behaviors of a one-dimensional wave equation. J. Math. Anal. Appl. **369**, 623–634 (2010)
17. Chen, G., Hsu, S.B., Zhou, J.X.: Snapback repellers as a cause of chaotic vibration of the wave equation with a van der Pol boundary condition and energy injecton at the middle of the span. J. Math. Phys. **39**, 6459–6489 (1998)
18. Chen, G., Hsu, S.B., Zhou, J.X.: Chaotic vibrations of the one-dimensional wave equation due to a self-excitation boundary condition Part I: Controlled hysteresis. Trans. Am. Math. Soc. **350**, 4265–4311 (1998)
19. Chen, Z.J., Huang, Y.: Functional envelopes relative to the point-open topology on a subset. Discret. Contin. Dyn. Syst. **37**, 99–118 (2017)
20. Chen, Z.J., Huang, T.W., Huang, Y., Xin, L.: Chaotic behaviors of one dimensional wave equations with van der Pol nonlinear boundary conditions. J. Math. Phys. **59**, 022704 (2018)
21. Marotto, F.R.: On redefining a snap-back repeller. Chaos Solitons Fractals **25**, 25–28 (2005)
22. Marotto, F.R.: Snap-back repellers imply chaos in $\mathbb{R}^n$. J. Math. Anal. Appl. **63**, 199–223 (1978)

Chapter 13
A New Spatial and Temporal Incremental Harmonic Balance Method for Obtaining Steady-State Responses of a One-Dimensional Continuous System

Xuefeng Wang and Weidong Zhu

Abstract A new spatial and temporal incremental harmonic balanced (STIHB) method is developed for obtaining steady-state responses of a one-dimensional continuous system. In the STIHB method, Galerkin procedure in the spatial coordinate and the harmonic balance procedure in the temporal coordinate are combined simultaneously to obtain the spatial and temporal harmonic balanced residual, and integrations in Galerkin procedures are replaced by the fast discrete sine transform (DST) or fast discrete cosine transform (DCT) in the spatial coordinate and the fast Fouriour transform (FFT) in the temporal coordinate. The harmonic balanced residual for an arbitrary second-order PDE can be automatically and efficiently obtained by a computer program when the expression of the PDE is given. The exact Jacobian matrix for the arbitrary PDE can be automatically and efficiently obtained by following a calculation routine when the linearized expression of the PDE is given, and it can be easily implemented by a computer program. The exact Jacobian matrix can also be used to study stability of steady-state responses, where no more extra derivations are needed.

13.1 Introduction

A governing partial differential equation (PDE) of a one-dimensional continuous vibratory system can be generally obtained by the extended Hamilton's principle. There are two classes of methods to deal with the PDE. The first class of methods are analytical methods such as perturbation methods [4–6]. This class of methods are usually not suitable for problems with strong nonlinearities, and significant derivation work is required. The second class of methods are local and global spatial

---

X. Wang
University of Alabama, Tuscaloosa, AL 35487, USA
e-mail: xwang201@eng.ua.edu

W. Zhu (✉)
University of Maryland, Baltimore County, 1000 Hilltop Circle, Baltimore, MD 21250, USA
e-mail: wzhu@umbc.edu

A. K. Abramian et al. (eds.), *Nonlinear Dynamics of Discrete and Continuous Systems*,
Advanced Structured Materials 139,
https://doi.org/10.1007/978-3-030-53006-8_13

discretization methods. The finite element method (FEM) is a local spatial discretization method; it is a good choice if the geometric shape of the system is complex or there are discontinuities in it, since dense meshs around critical regions can ensure convergence of solutions. Compared with the FEM, a global spatial discretization method is more convenient and faster if the geometric shape is simple and smooth. If eigenfunctions of the PDE are available, its exact solutions can be obtained. However, eigenfunctions of the PDE do not usually exist. Galerkin method employs a series of trial functions that satisfy all boundary conditions of the system as basis functions to determine approximated solutions [1]. A drawback of a global spatial discretization method is its derivation complexity. Derivation work on Galerkin procedure of nonlinear terms of the PDE significantly increases with the number of trial functions used. Chen et al. [7] proposed a technique that aims to reduce the calculation complexity. While amount of calculation is reduced, the calculation procedure is still complex and its derivation can vary with problems.

When a set of spatially discretized ordinary differential equations (ODEs) in terms of generalized coordinates are obtained by Galerkin procedure, it can be used to calculate transient and steady-state responses. Steady-state periodic responses can be obtained by the harmonic balance method [8], where generalized coordinates are expanded by a truncated Fourier series and the harmonic balance procedure or Galerkin procedure in the temporal coordinate is conducted to calculate coefficients of the truncated Fourier series, the harmonic balance Newton-Raphson (HBNR) method [9], where Galerkin procedure in the temporal coordinate is followed by Newton-Raphson method, and the incremental harmonic balance (IHB) method [10–12], where Newton-Raphson method is conducted before Galerkin procedure, which is amenable for computer implementation [13]. Ferri [14] showed the equivalance of the HBNR and IHB methods. Ling and Wu [15] incorporated the fast Fourier transform (FFT) into the IHB method to reduce computation time. Wang and Zhu [16] introduced a modified IHB method where the FFT and Broyden's method that is a quasi-Newton method are combined to avoid tedious derivations of the Jacobian matrix. There is a Jacobian-free Newton-Krylov (JFNK) method that can also be combined with the harmonic balance method to solve for steady-state responses of nonlinear ODEs [17]. A proper preconditioner is usually required for the JFNK method to make it efficient [18].

Stability of steady-state responses is always a matter of concern in dynamic analysis. A common method to study stability of steady-state responses of a set of ODEs is based on Floquet theory. A set of linearized ODEs for the set of original ODEs and its state-space form should be derived, which yield state-space equations of a linear-time-periodic (LTP) system. There are two ways to calculate eigenvalues of the LTP system. The first way is to calculate eigenvalues of its transformation matrix that can be obtained by a numerical method [19]. The second way is to calculate eigenvalues of Toeplitz form of its system matrix [2]. Both ways are dependent on derivations of the set of linearized ODEs, which can be complex and tedious when the degree of freedoms (DOF) of the set of original ODEs is large.

In this work, a new spatial and temporal incremental harmonic balance (STIHB) method is developed to solve for steady-state responses of the governing PDE of a general second-order one-dimensional continuous system, and it can be extended to fourth-order and multi-dimensional continuous systems. There are two parts in the STIHB method. In the first part, the spatial and temporal harmonic balanced residual of the PDE is obtained by simultaneously conducting Galerkin procedures on the spatial and temporal coordinates. The main advantage of the STIHB method is that derivations of the set of ODEs by Galerkin procedure in the spatial coordinate and the hamonic balanced residual of the resulting ODEs are replaced by the fast discrete sine transform (DST) or fast discrete cosine transform (DCT) in the spatial coordinate depending on boundary conditions of the system and the FFT in the temporal coordinate, which is referred to as a DST-FFT or DCT-FFT procedure, and it can be automatically and efficiently conducted by a computer program when the expression of the PDE is given. In the second part, a type of Newton methods is used to find solutions of the PDE to make the harmonic balanced residual vanish. The DST-FFT or DCT-FFT procedure combined with Broyden's method gives a simple version of the STIHB method. However, Newton-Raphson method using the exact Jacobian matrix can lead to faster convergence. It is shown here that when the linearized expression of the PDE is given, the exact Jacobian matrix can be automatically obtained via a calculation routine that can be easily implemented by a computer program, which gives a complex version of the STIHB method. It is also shown that the exact Jacobian matrix is directly related to Toeplitz form of the system matrix of the LTP system, which implies that stability of steady-state responses can be studied from the exact Jacobian matrix. Hence, stability analysis is free from derivations of the set of ODEs by Galerkin procedure.

The remaining part of this chapter is organized as follows: general expressions of the harmonic balanced residual of a general second-order governing PDE of a one-dimensional continuous system are presented in Sect. 13.2.1, and the simple version of the STIHB method, where the harmonic balanced residual is derived using the DST-FFT procedure, is described there. The complex version of the STIHB method, where the exact Jacobian matrix is derived, is described in Sect. 13.2.2. The method to study stability of steady-state responses when the exact Jacobian matrix is available is introduced in Sect. 13.2.3. The STIHB method is applied to the transverse vibration problem of a string with geometric nonlinearity in Sect. 13.3. In Sect. 13.4, frequency-response curves of the string with different parameters are calculated and stability of solutions on the curves are shown. Finally, some concluding remarks are presented in Sect. 13.5.

13.2 Description of the STIHB Method

13.2.1 Basic Equations for the STIHB Method

The STIHB method is developed here for a general one-dimensional second-order continuous system. With the normalized spatial coordinate $x \in [0, 1]$ and temporal coordinate $t \in [0, 2\pi]$, a general second-order PDE can be written as

$$F(x, t, w, w_t, w_{tt}, w_x, w_{xx}, w_{xt}, f) = 0$$
$$x \in [0, 1),\ t \in [0, 2\pi) \tag{13.1}$$

where w is the dependent variable, f is an external periodic excitation with the normalized fundamental frequency of one, and subscripts x and t denote spatial and temporal partial derivatives, respectively. Note that w_{xt} is included Eq. (13.1) so that one can deal with a translating string [20]; parametrically-excited continuous systems can also be studied [11]. Galerkin procedure for Eq. (13.1) in the spatial coordinate yields a set of ODEs with respect to t:

$$\int_0^2 \mathrm{d}x \mathbf{H}^T F(x, t, w, w_t, w_{tt}, w_x, w_{xx}, w_{xt}, f) = \mathbf{0}_{N\times 1} \tag{13.2}$$

where $\mathbf{H} = [\eta_1(x) \ \cdots \ \eta_N(x)]$ is a vector of orthonormal trial functions $\eta_n(x)$, in which N is the number of trial functions used, and the superscript T denotes transpose of a vector or matrix. To solve for steady-state responses of Eq. (13.1) by simultaneously conducting Galerkin procedures in the spatial and temporal coordinates, the dependent variable can be expressed by

$$w(x, t) = \sum_{n=1}^{N} \sum_{m=-M}^{M} a_{n,m} \eta_n(x) \phi_m(t) \tag{13.3}$$

where M is the truncated number of Fourier series in the temporal coordinate, $\phi_m(t) = \exp(\mathrm{j}mt)$ are basis functions of Fourier series, j denotes the imaginary unit, $a_{n,m}$ are coefficents for combined spatial and temporal bases $\eta_n(x)\phi_m(t)$, and $a_{n,-m}$ are complex conjugates of $a_{n,m}$. A vector form of $a_{n,m}$ is

$$\mathbf{q} = [\mathbf{a}_{-M}^T \ \cdots \ \mathbf{a}_M^T]^T \tag{13.4}$$

where $\mathbf{a}_m = [a_{1,m} \ \cdots \ a_{N,m}]^T$. Note that there are other forms of $a_{n,m}$; the form in Eq. (13.4) is consistent with the structure of Toeplitz matrix and it is amenable for stability analysis in Sect. 13.2.4. By collecting $\phi_m(t)$ in a vector form, $\Phi = [\phi_{-M}(t) \ \cdots \ \phi_M(t)]^T$, a compact form of $w(x, t)$ is

$$\begin{aligned} w(x, t) &= \sum_{m=-M}^{M} \phi_m(t) \mathbf{H}\mathbf{a}_m \\ &= \Phi^T [\mathbf{H}\mathbf{a}_{-M} \ \cdots \ \mathbf{H}\mathbf{a}_M]^T \\ &= \Phi^T (\mathbf{E}_{2M+1} \otimes \mathbf{H})\mathbf{q} \end{aligned} \tag{13.5}$$

where $\mathbf{E}_{2M+1}$ is the $2M + 1$ by $2M + 1$ identity matrix and $\otimes$ denotes Kronecker product. The form of $w(x, t)$ in Eq. (13.5) can be converted to

$$\begin{aligned} w(x,t) &= \Phi^T(\mathbf{E}_{2M+1}\otimes\mathbf{H})\mathbf{q} \\ &= (\Phi^T\otimes[1])(\mathbf{E}_{2M+1}\otimes\mathbf{H})\mathbf{q} \\ &= (\Phi^T\otimes\mathbf{H})\mathbf{q} \\ &= \big(([1]\,\Phi^T)\otimes(\mathbf{H}\mathbf{E}_N)\big)\mathbf{q} \\ &= \mathbf{H}(\Phi^T\otimes\mathbf{E}_N)\mathbf{q} \end{aligned} \tag{13.6}$$

where [1] is the matrix with only one element that is one, the mix-product property of Kronecker product is used in the third and fifth steps, and $\mathbf{E}_N$ is the N by N identity matrix. For convenience, $\Phi^T\otimes\mathbf{E}_N$ is denoted by $\Phi^T_\otimes$, which yields $w(x,t)=\mathbf{H}\Phi^T_\otimes\mathbf{q}$. The first- and second-order spatial derivatives of $w(x,t)$ are

$$w_x(x,t)=\mathbf{H}\mathbf{G}\Phi^T_\otimes\mathbf{q} \tag{13.7}$$

$$w_{xx}(x,t)=\mathbf{H}\mathbf{G}^2\Phi^T_\otimes\mathbf{q} \tag{13.8}$$

respectively, where $\mathbf{G}$ is a constant matrix to transform $\mathbf{H}$ to $\mathrm{d}\mathbf{H}/\mathrm{d}x$, which depends on the choice of $\mathbf{H}$. The first- and second-order temporal derivatives of $w(x,t)$ are

$$w_t(x,t)=\mathbf{H}\Phi^T_\otimes\mathbf{D}_\otimes\mathbf{q} \tag{13.9}$$

$$w_{tt}(x,t)=\mathbf{H}\Phi^T_\otimes\mathbf{D}^2_\otimes\mathbf{q} \tag{13.10}$$

respectively, where $\mathbf{D}$ is the constant diagonal matrix of $[-\mathrm{j}M\ \cdots\ \mathrm{j}M]$ to transform Φ^T to $\mathrm{d}\Phi^T/\mathrm{d}t$, and $(\Phi^T\mathbf{D})\otimes\mathbf{E}_N=\Phi^T_\otimes\mathbf{D}_\otimes$ by using the mix-product property of Kronecker product with the notation $\mathbf{D}_\otimes=\mathbf{D}\otimes\mathbf{E}_N$. The second-order mixed derivative of $w(x,t)$ is

$$w_{xt}(x,t)=\mathbf{H}\mathbf{G}\Phi^T_\otimes\mathbf{D}_\otimes\mathbf{q} \tag{13.11}$$

Substituting Eqs. (13.6)–(13.11) with a guess solutioin of $\mathbf{q}$ into Eq. (13.2), where the set of ODEs may not hold and its left-hand side is its residual, and conducting Galerkin procedure for the residual in the temporal coordinate yield

$$\mathbf{r}(\mathbf{q})=\int_0^{2\pi}\frac{\mathrm{d}t}{\pi}\int_0^2 dx\hat{\Phi}_\otimes\mathbf{H}^T F \tag{13.12}$$

where $\hat{\Phi}=[\phi_M(t)\ \cdots\ \phi_{-M}(t)]^T$ is orthonormal to Φ, i.e., $\int_0^{2\pi}\frac{\mathrm{d}t}{\pi}\Phi\hat{\Phi}^T=\mathbf{E}_{2M+1}$, and $\hat{\Phi}_\otimes=\hat{\Phi}\otimes\mathbf{E}_N$. Equation (13.12) is the basic form of the harmonic balanced residual of $F(x,t,w,w_t,w_{tt},w_x,w_{xx},w_{xt},f)$ at $\mathbf{q}$ and it is amenable for derivations of its Jacobian matrix. An increment of $\mathbf{r}(\mathbf{q})$ is $\Delta\mathbf{r}(\mathbf{q})$. Linearizing $F(x,t,w,w_t,w_{tt},w_x,w_{xx},w_{xt},f)$ with respect to w, w_x, w_{xx}, w_{xt}, w_t, and w_{tt} in $\Delta\mathbf{r}(\mathbf{q})$ yields

$$\begin{aligned}\Delta\mathbf{r}(\mathbf{q}) &= \int_0^{2\pi}\frac{\mathrm{d}t}{\pi}\int_0^2 \mathrm{d}x\hat{\Phi}_{\otimes}\mathbf{H}^T\Delta F\\ &= \int_0^{2\pi}\frac{\mathrm{d}t}{\pi}\int_0^2 \mathrm{d}x\hat{\Phi}_{\otimes}\mathbf{H}^T\left(\frac{\partial F}{\partial w}\frac{\partial w}{\partial \mathbf{q}}+\frac{\partial F}{\partial w_x}\frac{\partial w_x}{\partial \mathbf{q}}+\frac{\partial F}{\partial w_{xx}}\frac{\partial w_{xx}}{\partial \mathbf{q}}\right.\\ &\quad\left.+\frac{\partial F}{\partial w_{xt}}\frac{\partial w_{xt}}{\partial \mathbf{q}}+\frac{\partial F}{\partial w_t}\frac{\partial w_t}{\partial \mathbf{q}}+\frac{\partial F}{\partial w_{tt}}\frac{\partial w_{tt}}{\partial \mathbf{q}}\right)\Delta\mathbf{q}\\ &= \int_0^{2\pi}\frac{\mathrm{d}t}{\pi}\int_0^2 \mathrm{d}x\hat{\Phi}_{\otimes}\mathbf{H}^T\left(\frac{\partial F}{\partial w}\mathbf{H}\Phi_{\otimes}^T+\frac{\partial F}{\partial w_x}\mathbf{H}\mathbf{G}\Phi_{\otimes}^T+\frac{\partial F}{\partial w_{xx}}\mathbf{H}\mathbf{G}^2\Phi_{\otimes}^T\right.\\ &\quad\left.+\frac{\partial F}{\partial w_{xt}}\mathbf{H}\mathbf{G}\Phi_{\otimes}^T\mathbf{D}_{\otimes}+\frac{\partial F}{\partial w_t}\mathbf{H}\Phi_{\otimes}^T\mathbf{D}_{\otimes}+\frac{\partial F}{\partial w_{tt}}\mathbf{H}\Phi_{\otimes}^T\mathbf{D}_{\otimes}^2\right)\Delta\mathbf{q}\end{aligned} \tag{13.13}$$

where $\Delta\mathbf{q}$ is an increment of $\mathbf{q}$. The Jacobian matrix of $\mathbf{r}(\mathbf{q})$ with respect to $\mathbf{q}$ is

$$\begin{aligned}\mathbf{J}(\mathbf{q}) &= \int_0^{2\pi}\frac{\mathrm{d}t}{\pi}\int_0^2 \mathrm{d}x\hat{\Phi}_{\otimes}\mathbf{H}^T\left(\frac{\partial F}{\partial w}\mathbf{H}\Phi_{\otimes}^T+\frac{\partial F}{\partial w_x}\mathbf{H}\mathbf{G}\Phi_{\otimes}^T+\frac{\partial F}{\partial w_{xx}}\mathbf{H}\mathbf{G}^2\Phi_{\otimes}^T\right.\\ &\quad\left.+\frac{\partial F}{\partial w_{xt}}\mathbf{H}\mathbf{G}\Phi_{\otimes}^T\mathbf{D}_{\otimes}+\frac{\partial F}{\partial w_t}\mathbf{H}\Phi_{\otimes}^T\mathbf{D}_{\otimes}+\frac{\partial F}{\partial w_{tt}}\mathbf{H}\Phi_{\otimes}^T\mathbf{D}_{\otimes}^2\right)\end{aligned} \tag{13.14}$$

where $\Delta\mathbf{r}(\mathbf{q}) = \mathbf{J}(\mathbf{q})\Delta\mathbf{q}$ is used.

13.2.1.1 Simple Version of the STIHB Method Using an Approximated Jacobian Matrix

Derivations in Galerkin procedures in Eqs. (13.12) and (13.14) are complex and tedious, or even undoable when there is some complex nonlinear function. Approximations of the Jacobian matrix $\mathbf{J}(\mathbf{q})$ and harmonic balanced residual $\mathbf{r}(\mathbf{q})$ can be used in the STIHB method. The harmonic balanced residual $\mathbf{r}(\mathbf{q})$ can be approximated by the DST-FFT or DCT-FFT procedure. To illustrate this procedure, the vector form of $a_{n,m}$ in Eq. (13.4) is converted to a matrix form

$$\mathbf{Q} = [\mathbf{a}_{-M} \ \cdots \ \mathbf{a}_M] \tag{13.15}$$

and the form of the dependent variable in Eq. (13.5) is converted to

$$w(x,t) = \sum_{m=-M}^{M}\phi_m(t)\mathbf{H}\mathbf{a}_m = \mathbf{H}\mathbf{Q}\mathbf{\Phi} \tag{13.16}$$

Sampling $w(x,t)$ at discretized spatial points $\left\{x_k := \frac{k}{N_s}\right\}_{k=1,\ \ldots,\ N_s-1}$ with the integer $N_s > N$ and temporal points $\left\{t_i := \frac{2\pi i}{M_s}\right\}_{i=0,\ \ldots,\ M_s-1}$ with the integer $M_s > 2M$ yields

$$\mathbf{w} = \bar{\mathbf{H}}\mathbf{Q}\bar{\mathbf{\Phi}} \tag{13.17}$$

where elements of $\mathbf{w}$ at the kth row and ith column are $w(x_k, t_i)$,

$$\bar{\mathbf{H}} = \begin{bmatrix} \eta_1\left(\frac{1}{N_s}\right) & \cdots & \eta_N\left(\frac{1}{N_s}\right) \\ \vdots & \ddots & \vdots \\ \eta_1\left(\frac{N_s-1}{N_s}\right) & \cdots & \eta_N\left(\frac{N_s-1}{N_s}\right) \end{bmatrix} \tag{13.18}$$

$$\bar{\mathbf{\Phi}} = \begin{bmatrix} \exp\left(\mathrm{j}(-M)\frac{2\pi 0}{M_s}\right) & \cdots & \exp\left(\mathrm{j}(-M)\frac{2\pi(M_s-1)}{M_s}\right) \\ \vdots & \ddots & \vdots \\ \exp\left(\mathrm{j}M\frac{2\pi 0}{M_s}\right) & \cdots & \exp\left(\mathrm{j}M\frac{2\pi(M_s-1)}{M_s}\right) \end{bmatrix} \tag{13.19}$$

The first- and second-order spatial derivatives and first- and second-order temporal derivatives of $w(x, t)$ that are sampled at $\{x_k\}_{k=1,\ ...,\ N_s-1}$ and $\{t_i\}_{i=0,\ ...,\ M_s-1}$ are

$$\mathbf{w}_x = \bar{\mathbf{H}}\mathbf{G}\mathbf{Q}\bar{\mathbf{\Phi}} \tag{13.20}$$

$$\mathbf{w}_{xx} = \bar{\mathbf{H}}\mathbf{G}^2\mathbf{Q}\bar{\mathbf{\Phi}} \tag{13.21}$$

$$\mathbf{w}_{xt} = \bar{\mathbf{H}}\mathbf{G}\mathbf{Q}\mathbf{D}\bar{\mathbf{\Phi}} \tag{13.22}$$

$$\mathbf{w}_t = \bar{\mathbf{H}}\mathbf{Q}\mathbf{D}\bar{\mathbf{\Phi}} \tag{13.23}$$

$$\mathbf{w}_{tt} = \bar{\mathbf{H}}\mathbf{Q}\mathbf{D}^{\Theta}\bar{\mathbf{\Phi}} \tag{13.24}$$

respectively. The matrix form of the harmonic balanced residual of $F(x, t, w, w_t, w_{tt}, w_x, w_{xx}, w_{xt}, f)$ is

$$\mathbf{R}(\mathbf{Q}) = \int_0^{2\pi} \frac{\mathrm{d}t}{\pi} \int_0^2 \mathrm{d}x \mathbf{H}^T F \hat{\mathbf{\Phi}}^T \tag{13.25}$$

where $\mathbf{R}(\mathbf{Q})$ is a N by $2M + 1$ matrix, which is a variation of $\mathbf{r}(\mathbf{q})$ in Eq. (13.12). Elements of $\mathbf{R}(\mathbf{Q})$ at the nth row and mth column are coefficients $r_{n,m}$ of bases $\eta_n(x)\phi_m(t)$ in F with w, w_x, w_{xx}, w_{xt}, w_t, and w_{tt} evaluated at $\mathbf{Q}$. The relationship between $\mathbf{r}(\mathbf{q})$ and $\mathbf{R}(\mathbf{Q})$ is

$$\mathbf{r}(\mathbf{q}) = [\mathbf{R}_{-M}^T \ \cdots \ \mathbf{R}_M^T]^T \tag{13.26}$$

where $\mathbf{R}_m = [r_{1,m} \ \cdots \ r_{N,m}]^T$. The harmonic balanced residual can be used to approximate $F(x, t, w, w_t, w_{tt}, w_x, w_{xx}, f)$:

$$\tilde{F}(x,t,w,w_t,w_{tt},w_x,w_{xx},w_{xt},f)=\sum_{n=1}^{N}\sum_{m=-M}^{M} r_{n,m}\eta_n(x)\phi_m(t)=\mathbf{HR(Q)}\Phi \tag{13.27}$$

Replacing $w, w_t, w_{tt}, w_x, w_{xt}$, and w_{xx} by $\mathbf{w}, \mathbf{w}_t, \mathbf{w}_{tt}, \mathbf{w}_x, \mathbf{w}_{xt}$, and $\mathbf{w}_{xx}$, respectively, in $F(x,t,w,w_t,w_{tt},w_x,w_{xx},w_{xt},f)$, substituting Eqs. (13.17) and (13.20–13.24) into the resulting equation, and evaluating x and t at $\{x_k\}_{k=1,\ldots,N_s-1}$ and $\{t_i\}_{i=0,\ldots,M_s-1}$, respectively, yield

$$\mathbf{F}=F\left(x_k,t_i,\mathbf{w},\mathbf{w}_t,\mathbf{w}_{tt},\mathbf{w}_x,\mathbf{w}_{xx},\mathbf{w}_{xt},\mathbf{f}\right) \tag{13.28}$$

where $\mathbf{f}$ and $\mathbf{F}$ are N_s-1 by M_s matrices whose elements at the kth row and ith column are values of f and $F(x,t,w,w_t,w_{tt},w_x,w_{xx},w_{xt},f)$ evaluated at x_k and t_i, respectively. On the other hand, if $\tilde{F}(x,t,w,w_t,w_{tt},w_x,w_{xx},w_{xt},f)$ is a good approximation, $\mathbf{F}$ can be obtained by sampling $\tilde{F}(x,t,w,w_t,w_{tt},w_x,w_{xx},w_{xt},f)$ at $\{x_k\}_{k=1,\ldots,N_s-1}$ and $\{t_i\}_{i=0,\ldots,M_s-1}$ in Eq. (13.27), which yields

$$\mathbf{F}=\bar{\mathbf{H}}\mathbf{R(Q)}\bar{\mathbf{\Phi}} \tag{13.29}$$

With use of orthonormal relations $\int_0^2 \mathrm{d}x\mathbf{H}^T\mathbf{H}=\bar{\mathbf{H}}^T\bar{\mathbf{H}}=\mathbf{E}_N$ and $\int_0^{2\pi}\frac{\mathrm{d}t}{\pi}\Phi\hat{\Phi}^T=\bar{\Phi}\hat{\bar{\Phi}}^T=\mathbf{E}_{2M+1}$, where

$$\hat{\bar{\Phi}}=\begin{bmatrix}\exp\left(\mathrm{j}M\frac{2\pi 0}{M_s}\right) & \cdots & \exp\left(\mathrm{j}M\frac{2\pi(M_s-1)}{M_s}\right)\\ \vdots & \ddots & \vdots\\ \exp\left(\mathrm{j}(-M)\frac{2\pi 0}{M_s}\right) & \cdots & \exp\left(\mathrm{j}(-M)\frac{2\pi(M_s-1)}{M_s}\right)\end{bmatrix} \tag{13.30}$$

in the spatial and temporal coordinates, respectively, in Eq. (13.29), the harmonic balanced residual in Eq. (13.25) can be obtained by

$$\mathbf{R(Q)}=\bar{\mathbf{H}}^T\mathbf{F}\hat{\bar{\Phi}}^T \tag{13.31}$$

If trial functions in $\mathbf{H}$ are $\eta_n(x)=\sin(n\pi x)$, the left multiplication of $\mathbf{F}$ by $\bar{\mathbf{H}}$ is the DST of every column in $\mathbf{F}$, which can be implemented by a fast algorithm [3], and the right multiplication of $\mathbf{F}$ by $\hat{\bar{\Phi}}$ is the discrete Fourier transform of every row of $\mathbf{F}$, which can be efficiently implemented by the FFT. Combination of the two transforms is the DST-FFT procedure, which can be used to obtain the harmonic balanced residual $\mathbf{R(Q)}$ without integrations, and the basic form of the harmonic balanced residual $\mathbf{r(q)}$ can be obtained from Eq. (13.26). If trial functions in $\mathbf{H}$ are $\eta_n(x)=\cos(n\pi x)$, the fast DST can be replaced by the fast DCT, and the DST-FFT procedure becomes the DCT-FFT procedure.

A quasi-Newton method, called Broyden's method, can be used to find $\mathbf{q}$ using Eq. (13.14) to make $\mathbf{r(q)}$ vanish without deriving the exact Jacobian matrix $\mathbf{J(q)}$.

The nonlinear function $F(x, t, w, w_t, w_{tt}, w_x, w_{xx}, w_{xt}, f)$ can be divided into two parts:

$$F(x, t, w, w_t, w_{tt}, w_x, w_{xx}, w_{xt}, f) = F_{\text{lin}} + F_{\text{nonl}} \tag{13.32}$$

where F_{lin} and F_{nonl} are linear and nonlinear parts of $F(x, t, w, w_t, w_{tt}, w_x, w_{xx}, w_{xt}, f)$, respectively. First, an initial guess of $\mathbf{q}$ can be chosen as $\mathbf{q}_0 = \mathbf{0}_{(2MN+N)\times 1}$, and an initial guess of $\mathbf{J}(\mathbf{q})$ can be chosen as the Jacobian matrix of the harmonic balanced residual for F_{lin}, which is denoted by $\mathbf{J}_0$. The harmonic balanced residual in the kth iteration is $\mathbf{r}_k = \mathbf{r}(\mathbf{q}_k)$, where $\mathbf{q}_k$ is a trial solution of $\mathbf{q}$ in this iteration. If Euclidean norm of $\mathbf{r}_k$ is larger than a preset tolerance, a trial solution of $\mathbf{q}$ can be updated by

$$\mathbf{q}_{k+1} = \mathbf{q}_k - \mathbf{J}_k^{-1}\mathbf{r}_k \tag{13.33}$$

where $\mathbf{J}_k$ is an approximation of $\mathbf{J}(\mathbf{q})$ in the kth iteration and the superscript -1 denotes inverse of a matrix. If the norm of $\mathbf{r}_{k+1} = \mathbf{r}(\mathbf{q}_{k+1})$ is less than the preset tolerance, $\mathbf{q}_{k+1}$ is an acceptable solution of $\mathbf{q}$. Otherwise, approximation of $\mathbf{J}(\mathbf{q})$ is updated by

$$\mathbf{J}_{k+1} = \mathbf{J}_k + \frac{(\mathbf{y}_k - \mathbf{J}_k\mathbf{s}_k)\mathbf{s}_k^T}{\mathbf{s}_k^T\mathbf{s}_k} \tag{13.34}$$

where $\mathbf{s}_k = \mathbf{q}_{k+1} - \mathbf{q}_k$ and $\mathbf{y}_k = \mathbf{r}_{k+1} - \mathbf{r}_k$. A new trial solution of $\mathbf{q}$ can be obtained from Eq. (13.33). The above procedure is repeated until the norm of the harmonic balanced residual is less than the preset tolerance, which means that the solution of $\mathbf{q}$ is found, or the number of iterations is very large, which means that the procedure to find the solution of $\mathbf{q}$ is not convergent. This is the simple version of the STIHB method.

13.2.2 Complex Version of the STIHB Method Using the Exact Jacobian Matrix

The harmonic balanced residual $\mathbf{r}(\mathbf{q})$ in the complex version of the STIHB method is calculated in the same way shown in Sect. 13.2.2, where $\mathbf{q}$ is a guess of a steady-state solution, but the exact Jacobian matrix $\mathbf{J}(\mathbf{q})$ is derived when expressions of $\frac{\partial F}{\partial w}$, $\frac{\partial F}{\partial w_x}$, $\frac{\partial F}{\partial w_{xx}}$, $\frac{\partial F}{\partial w_{xt}}$, $\frac{\partial F}{\partial w_t}$, and $\frac{\partial F}{\partial w_{tt}}$ are given. Six terms in $\mathbf{J}(\mathbf{q})$ in Eq. (13.14) are

$$\mathbf{J}_w = \int_0^{2\pi} \frac{dt}{\pi} \int_0^2 dx \hat{\mathbf{\Phi}}_\otimes \mathbf{H}^T \frac{\partial F}{\partial w} \mathbf{H} \mathbf{\Phi}_\otimes^T \tag{13.35}$$

$$\mathbf{J}_{w_x} = \int_0^{2\pi} \frac{dt}{\pi} \int_0^2 dx \hat{\mathbf{\Phi}}_\otimes \mathbf{H}^T \frac{\partial F}{\partial w_x} \mathbf{H}\mathbf{G} \mathbf{\Phi}_\otimes^T \tag{13.36}$$

$$\mathbf{J}_{w_{xx}} = \int_0^{2\pi} \frac{dt}{\pi} \int_0^2 dx \hat{\Phi}_\otimes \mathbf{H}^T \frac{\partial F}{\partial w_{xx}} \mathbf{H}\mathbf{G}^\ominus \Phi_\otimes^T \tag{13.37}$$

$$\mathbf{J}_{w_{xt}} = \int_0^{2\pi} \frac{dt}{\pi} \int_0^2 dx \hat{\Phi}_\otimes \mathbf{H}^T \frac{\partial F}{\partial w_x} \mathbf{H}\mathbf{G}\Phi_\otimes^T \mathbf{D}_\otimes \tag{13.38}$$

$$\mathbf{J}_{w_t} = \int_0^{2\pi} \frac{dt}{\pi} \int_0^2 dx \hat{\Phi}_\otimes \mathbf{H}^T \frac{\partial F}{\partial w_t} \mathbf{H}\Phi_\otimes^T \mathbf{D}_\otimes \tag{13.39}$$

$$\mathbf{J}_{w_{tt}} = \int_0^{2\pi} \frac{dt}{\pi} \int_0^2 dx \hat{\Phi}_\otimes \mathbf{H}^T \frac{\partial F}{\partial w_{tt}} \mathbf{H}\Phi_\otimes^T \mathbf{D}_\otimes^2 \tag{13.40}$$

Using Eq. (13.28) with $F(x, t, w, w_t, w_{tt}, w_x, w_{xx}, w_{xt}, f)$ replaced by $\frac{\partial F}{\partial w}$, $\frac{\partial F}{\partial w_x}$, $\frac{\partial F}{\partial w_{xx}}$, $\frac{\partial F}{\partial w_{xt}}$, $\frac{\partial F}{\partial w_t}$, and $\frac{\partial F}{\partial w_{tt}}$ yields $\mathbf{F}_w$, $\mathbf{F}_{w_x}$, $\mathbf{F}_{w_{xx}}$, $\mathbf{F}_{w_{xt}}$, $\mathbf{F}_{w_t}$, and $\mathbf{F}_{w_{tt}}$, respectively. Using the DST-FFT procedure in Eq. (13.31) with $\mathbf{F}$ replaced by $\mathbf{F}_w$, $\mathbf{F}_{w_x}$, $\mathbf{F}_{w_{xx}}$, $\mathbf{F}_{w_{xt}}$, $\mathbf{F}_{w_t}$, and $\mathbf{F}_{w_{tt}}$, and rearranging the resulting matrices using Eq. (13.26) with $\mathbf{R}$ replaced by the resulting matrices yield $\hat{\mathbf{r}}_w$, $\hat{\mathbf{r}}_{w_x}$, $\hat{\mathbf{r}}_{w_{xx}}$, $\hat{\mathbf{r}}_{w_{xt}}$, $\hat{\mathbf{r}}_{w_t}$, and $\hat{\mathbf{r}}_{w_{tt}}$, respectively. Six terms of partial derivatives of F in $\mathbf{J}(\mathbf{q})$ are then expressed by

$$\frac{\partial F}{\partial w} = \mathbf{H}\mathbf{G}\Phi_\otimes^T \mathbf{r}_w \tag{13.41}$$

$$\frac{\partial F}{\partial w_x} = \mathbf{H}\Phi_\otimes^T \mathbf{r}_{w_x} \tag{13.42}$$

$$\frac{\partial F}{\partial w_{xx}} = \mathbf{H}\mathbf{G}\Phi_\otimes^T \mathbf{r}_{w_{xx}} \tag{13.43}$$

$$\frac{\partial F}{\partial w_{xt}} = \mathbf{H}\Phi_\otimes^T \mathbf{r}_{w_{xt}} \tag{13.44}$$

$$\frac{\partial F}{\partial w_t} = \mathbf{H}\mathbf{G}\Phi_\otimes^T \mathbf{r}_{w_t} \tag{13.45}$$

$$\frac{\partial F}{\partial w_{tt}} = \mathbf{H}\mathbf{G}\Phi_\otimes^T \mathbf{r}_{w_{tt}} \tag{13.46}$$

where $\mathbf{r}_w = (\mathbf{E}_{2M+1} \otimes \mathbf{G})^{-1}\hat{\mathbf{r}}_w$, $\hat{\mathbf{r}}_{w_x} = \mathbf{r}_{w_x}$, $\mathbf{r}_{w_{xx}} = (\mathbf{E}_{2M+1} \otimes \mathbf{G})^{-1}\hat{\mathbf{r}}_{w_{xx}}$, $\hat{\mathbf{r}}_{w_{xt}} = \mathbf{r}_{w_{xt}}$, $\mathbf{r}_{w_t} = (\mathbf{E}_{2M+1} \otimes \mathbf{G})^{-1}\hat{\mathbf{r}}_{w_t}$, and $\mathbf{r}_{w_{tt}} = (\mathbf{E}_{2M+1} \otimes \mathbf{G})^{-1}\hat{\mathbf{r}}_{w_{tt}}$. Since

$$
\begin{aligned}
\mathbf{H}^T\mathbf{H}\mathbf{G}\Phi_\otimes^T &= \mathbf{H}^T\mathbf{H}\mathbf{G}(\Phi^T \otimes \mathbf{E}_N) \\
&= \big([1] \otimes (\mathbf{H}^T\mathbf{H}\mathbf{G})\big)(\Phi^T \otimes \mathbf{E}_N) \\
&= (\Phi^T\mathbf{E}_{2M+1}) \otimes (\mathbf{E}_N\mathbf{H}^T\mathbf{H}\mathbf{G}) \\
&= (\Phi^T \otimes \mathbf{E}_N)\big(\mathbf{E}_{2M+1} \otimes (\mathbf{H}^T\mathbf{H}\mathbf{G})\big) \\
&= \Phi_\otimes^T\big(\mathbf{E}_{2M+1} \otimes (\mathbf{H}^T\mathbf{H}\mathbf{G})\big)
\end{aligned} \tag{13.47}
$$

where the mix-product property of Kronecker product is used in the third and fourth steps, one has

$$
\mathbf{H}^T\mathbf{H}\mathbf{G}\Phi_\otimes^T = \Phi_\otimes^T\big(\mathbf{E}_{2M+1} \otimes (\mathbf{H}^T\mathbf{H}\mathbf{G})\big) \tag{13.48}
$$

Similarly, one has

$$
\mathbf{H}^T\mathbf{H}\Phi_\otimes^T = \Phi_\otimes^T\big(\mathbf{E}_{2M+1} \otimes (\mathbf{H}^T\mathbf{H})\big) \tag{13.49}
$$

Substituting Eqs. (13.41)–(13.46) into Eqs. (13.35)–(13.40) and using Eqs. (13.48) and (13.49) in the resulting equations yield

$$
\begin{aligned}
\mathbf{J}_w &= \int_0^{2\pi}\frac{\mathrm{d}t}{\pi}\int_0^2 \mathrm{d}x\,\hat{\Phi}_\otimes\mathbf{H}^T\mathbf{H}\mathbf{G}\Phi_\otimes^T\mathbf{r}_w\mathbf{H}\Phi_\otimes^T \\
&= \int_0^{2\pi}\frac{\mathrm{d}t}{\pi}\int_0^2 \mathrm{d}x\,\hat{\Phi}_\otimes\Phi_\otimes^T\big(\mathbf{E}_{2M+1} \otimes (\mathbf{H}^T\mathbf{H}\mathbf{G})\big)\mathbf{r}_w\mathbf{H}\Phi_\otimes^T \\
&= \int_0^{2\pi}\frac{\mathrm{d}t}{\pi}\hat{\Phi}_\otimes\Phi_\otimes^T\mathbf{S}_w\Phi_\otimes^T
\end{aligned} \tag{13.50}
$$

$$
\begin{aligned}
\mathbf{J}_{w_x} &= \int_0^{2\pi}\frac{\mathrm{d}t}{\pi}\int_0^2 \mathrm{d}x\,\hat{\Phi}_\otimes\mathbf{H}^T\mathbf{H}\Phi_\otimes^T\mathbf{r}_{w_x}\mathbf{H}\mathbf{G}\Phi_\otimes^T \\
&= \int_0^{2\pi}\frac{\mathrm{d}t}{\pi}\int_0^2 \mathrm{d}x\,\hat{\Phi}_\otimes\Phi_\otimes^T\big(\mathbf{E}_{2M+1} \otimes (\mathbf{H}^T\mathbf{H})\big)\mathbf{r}_{w_x}\mathbf{H}\mathbf{G}\Phi_\otimes^T \\
&= \int_0^{2\pi}\frac{\mathrm{d}t}{\pi}\hat{\Phi}_\otimes\Phi_\otimes^T\mathbf{S}_{w_x}\Phi_\otimes^T
\end{aligned} \tag{13.51}
$$

$$
\begin{aligned}
\mathbf{J}_{w_{xx}} &= \int_0^{2\pi}\frac{\mathrm{d}t}{\pi}\int_0^2 \mathrm{d}x\,\hat{\Phi}_\otimes\mathbf{H}^T\mathbf{H}\mathbf{G}\Phi_\otimes^T\mathbf{r}_{w_{xx}}\mathbf{H}\mathbf{G}^\Theta\Phi_\otimes^T \\
&= \int_0^{2\pi}\frac{\mathrm{d}t}{\pi}\int_0^2 \mathrm{d}x\,\hat{\Phi}_\otimes\Phi_\otimes^T\big(\mathbf{E}_{2M+1} \otimes (\mathbf{H}^T\mathbf{H}\mathbf{G})\big)\mathbf{r}_{w_{xx}}\mathbf{H}\mathbf{G}^\Theta\Phi_\otimes^T \\
&= \int_0^{2\pi}\frac{\mathrm{d}t}{\pi}\hat{\Phi}_\otimes\Phi_\otimes^T\mathbf{S}_{w_{xx}}\Phi_\otimes^T
\end{aligned} \tag{13.52}
$$

$$\begin{aligned}
\mathbf{J}_{w_{xt}} &= \int_0^{2\pi} \frac{\mathrm{d}t}{\pi} \int_0^2 \mathrm{d}x\, \hat{\Phi}_{\otimes} \mathbf{H}^T \mathbf{H} \Phi_{\otimes}^T \mathbf{r}_{w_{xt}} \mathbf{H} \mathbf{G} \Phi_{\otimes}^T \mathbf{D}_{\otimes} \\
&= \int_0^{2\pi} \frac{\mathrm{d}t}{\pi} \int_0^2 \mathrm{d}x\, \hat{\Phi}_{\otimes} \Phi_{\otimes}^T \big(\mathbf{E}_{2M+1} \otimes (\mathbf{H}^T \mathbf{H})\big) \mathbf{r}_{w_{xt}} \mathbf{H} \mathbf{G} \Phi_{\otimes}^T \mathbf{D}_{\otimes} \\
&= \int_0^{2\pi} \frac{\mathrm{d}t}{\pi} \hat{\Phi}_{\otimes} \Phi_{\otimes}^T \mathbf{S}_{w_{xt}} \Phi_{\otimes}^T \mathbf{D}_{\otimes}
\end{aligned} \tag{13.53}$$

$$\begin{aligned}
\mathbf{J}_{w_t} &= \int_0^{2\pi} \frac{\mathrm{d}t}{\pi} \int_0^2 \mathrm{d}x\, \hat{\Phi}_{\otimes} \mathbf{H}^T \mathbf{H} \mathbf{G} \Phi_{\otimes}^T \mathbf{r}_{w_t} \mathbf{H} \Phi_{\otimes}^T \mathbf{D}_{\otimes} \\
&= \int_0^{2\pi} \frac{\mathrm{d}t}{\pi} \int_0^2 \mathrm{d}x\, \hat{\Phi}_{\otimes} \Phi_{\otimes}^T \big(\mathbf{E}_{2M+1} \otimes (\mathbf{H}^T \mathbf{H} \mathbf{G})\big) \mathbf{r}_{w_t} \mathbf{H} \Phi_{\otimes}^T \mathbf{D}_{\otimes} \\
&= \int_0^{2\pi} \frac{\mathrm{d}t}{\pi} \hat{\Phi}_{\otimes} \Phi_{\otimes}^T \mathbf{S}_{w_t} \Phi_{\otimes}^T \mathbf{D}_{\otimes}
\end{aligned} \tag{13.54}$$

$$\begin{aligned}
\mathbf{J}_{w_{tt}} &= \int_0^{2\pi} \frac{\mathrm{d}t}{\pi} \int_0^2 \mathrm{d}x\, \hat{\Phi}_{\otimes} \mathbf{H}^T \mathbf{H} \mathbf{G} \Phi_{\otimes}^T \mathbf{r}_{w_{tt}} \mathbf{H} \Phi_{\otimes}^T \mathbf{D}_{\otimes}^2 \\
&= \int_0^{2\pi} \frac{\mathrm{d}t}{\pi} \int_0^2 \mathrm{d}x\, \hat{\Phi}_{\otimes} \Phi_{\otimes}^T \big(\mathbf{E}_{2M+1} \otimes (\mathbf{H}^T \mathbf{H} \mathbf{G})\big) \mathbf{r}_{w_{tt}} \mathbf{H} \Phi_{\otimes}^T \mathbf{D}_{\otimes}^2 \\
&= \int_0^{2\pi} \frac{\mathrm{d}t}{\pi} \hat{\Phi}_{\otimes} \Phi_{\otimes}^T \mathbf{S}_{w_{tt}} \Phi_{\otimes}^T \mathbf{D}_{\otimes}^2
\end{aligned} \tag{13.55}$$

respectively, where

$$\mathbf{S}_w = \int_0^2 \mathrm{d}x \big(\mathbf{E}_{2M+1} \otimes (\mathbf{H}^T \mathbf{H} \mathbf{G})\big) \mathbf{r}_w \mathbf{H} \tag{13.56}$$

$$\mathbf{S}_{w_x} = \int_0^2 \mathrm{d}x \big(\mathbf{E}_{2M+1} \otimes (\mathbf{H}^T \mathbf{H})\big) \mathbf{r}_{w_x} \mathbf{H} \mathbf{G} \tag{13.57}$$

$$\mathbf{S}_{w_{xx}} = \int_0^2 \mathrm{d}x \big(\mathbf{E}_{2M+1} \otimes (\mathbf{H}^T \mathbf{H} \mathbf{G})\big) \mathbf{r}_{w_{xx}} \mathbf{H} \mathbf{G}^2 \tag{13.58}$$

$$\mathbf{S}_{w_{xt}} = \int_0^2 \mathrm{d}x \big(\mathbf{E}_{2M+1} \otimes (\mathbf{H}^T \mathbf{H})\big) \mathbf{r}_{w_{xt}} \mathbf{H} \mathbf{G} \tag{13.59}$$

$$\mathbf{S}_{w_t} = \int_0^2 \mathrm{d}x \big(\mathbf{E}_{2M+1} \otimes (\mathbf{H}^T \mathbf{H} \mathbf{G})\big) \mathbf{r}_{w_t} \mathbf{H} \tag{13.60}$$

$$\mathbf{S}_{w_{tt}} = \int_0^2 \mathrm{d}x \big(\mathbf{E}_{2M+1} \otimes (\mathbf{H}^T \mathbf{H} \mathbf{G})\big) \mathbf{r}_{w_{tt}} \mathbf{H} \tag{13.61}$$

Since $\big(\mathbf{E}_{2M+1} \otimes (\mathbf{H}^T\mathbf{HG})\big)\mathbf{r}_w$ is a column vector and $\mathbf{H}$ is a row vector, one has

$$\big(\mathbf{E}_{2M+1} \otimes (\mathbf{H}^T\mathbf{HG})\big)\mathbf{r}_w\mathbf{H} = \big(\mathbf{E}_{2M+1} \otimes (\mathbf{H}^T\mathbf{HG})\big)\mathbf{r}_w \otimes \mathbf{H} \tag{13.62}$$

Using $\mathbf{H} = \mathbf{HE}_N$ and the mix-product property of Kronecker product in Eq. (13.62) yields

$$\big(\mathbf{E}_{2M+1} \otimes (\mathbf{H}^T\mathbf{HG})\big)\mathbf{r}_w\mathbf{H} = \big(\mathbf{E}_{2M+1} \otimes (\mathbf{H}^T\mathbf{HG}) \otimes \mathbf{H}\big)(\mathbf{r}_w \otimes \mathbf{E}_N) \tag{13.63}$$

Similarly, one has

$$\big(\mathbf{E}_{2M+1} \otimes (\mathbf{H}^T\mathbf{H})\big)\mathbf{r}_{w_x}\mathbf{HG} = \big(\mathbf{E}_{2M+1} \otimes (\mathbf{H}^T\mathbf{H}) \otimes (\mathbf{HG})\big)(\mathbf{r}_{w_x} \otimes \mathbf{E}_N) \tag{13.64}$$

$$\big(\mathbf{E}_{2M+1} \otimes (\mathbf{H}^T\mathbf{H})\big)\mathbf{r}_{w_{xx}}\mathbf{HG}^2 = \big(\mathbf{E}_{2M+1} \otimes (\mathbf{H}^T\mathbf{H}) \otimes (\mathbf{HG}^2)\big)(\mathbf{r}_{w_{xx}} \otimes \mathbf{E}_N) \tag{13.65}$$

Substituting Eq. (13.63) into Eqs. (13.56), (13.60), and (13.61) yields

$$\mathbf{S}_w = (\mathbf{E}_{2M+1} \otimes \Pi_1)\mathbf{r}_{w,\otimes} \tag{13.66}$$

$$\mathbf{S}_{w_t} = (\mathbf{E}_{2M+1} \otimes \Pi_1)\mathbf{r}_{w_t,\otimes} \tag{13.67}$$

$$\mathbf{S}_{w_{tt}} = (\mathbf{E}_{2M+1} \otimes \Pi_1)\mathbf{r}_{w_{tt},\otimes} \tag{13.68}$$

respectively, where $\Pi_1 = \int_0^2 \mathrm{d}x\big((\mathbf{H}^T\mathbf{HG}) \otimes \mathbf{H}\big)$, $\mathbf{r}_{w,\otimes} = \mathbf{r}_w \otimes \mathbf{E}_N$, $\mathbf{r}_{w_t,\otimes} = \mathbf{r}_{w_t} \otimes \mathbf{E}_N$, and $\mathbf{r}_{w_{tt},\otimes} = \mathbf{r}_{w_{tt}} \otimes \mathbf{E}_N$. Substituing Eqs. (13.64) and (13.65) into Eqs. (13.57)–(13.59) yields

$$\mathbf{S}_{w_x} = (\mathbf{E}_{2M+1} \otimes \Pi_2)\mathbf{r}_{w_x,\otimes} \tag{13.69}$$

$$\mathbf{S}_{w_{xx}} = (\mathbf{E}_{2M+1} \otimes \Pi_3)\mathbf{r}_{w_{xx},\otimes} \tag{13.70}$$

$$\mathbf{S}_{w_{xt}} = (\mathbf{E}_{2M+1} \otimes \Pi_2)\mathbf{r}_{w_{xt},\otimes} \tag{13.71}$$

respectively, where $\Pi_2 = \int_0^2 \mathrm{d}x\big((\mathbf{H}^T\mathbf{H}) \otimes (\mathbf{HG})\big)$, $\Pi_3 = \int_0^2 \mathrm{d}x\big((\mathbf{H}^T\mathbf{H}) \otimes (\mathbf{HG}^2)\big)$, $\mathbf{r}_{w_x,\otimes} = \mathbf{r}_{w_x} \otimes \mathbf{E}_N$, $\mathbf{r}_{w_{xx},\otimes} = \mathbf{r}_{w_{xx}} \otimes \mathbf{E}_N$, and $\mathbf{r}_{w_{xt},\otimes} = \mathbf{r}_{w_{xt}} \otimes \mathbf{E}_N$. Hence, $\mathbf{S}_w$, $\mathbf{S}_{w_x}$, $\mathbf{S}_{w_{xx}}$, $\mathbf{S}_{w_{xt}}$, $\mathbf{S}_{w_t}$, and $\mathbf{S}_{w_{tt}}$ can be obtained from Eqs. (13.66), (13.69), (13.70), (13.71), (13.67), and (13.68), respectively, with $\mathbf{r}_w$, $\mathbf{r}_{w_x}$, $\mathbf{r}_{w_{xx}}$, $\mathbf{r}_{w_{xt}}$, $\mathbf{r}_{w_t}$, and $\mathbf{r}_{w_{tt}}$ given in Eqs. (13.41)–(13.46), respectively. Note that $\Phi_\otimes^T\mathbf{S}_w$, $\Phi_\otimes^T\mathbf{S}_{w_x}$, $\Phi_\otimes^T\mathbf{S}_{w_{xx}}$, $\Phi_\otimes^T\mathbf{S}_{w_{xt}}$, $\Phi_\otimes^T\mathbf{S}_{w_t}$, and $\Phi_\otimes^T\mathbf{S}_{w_{tt}}$ are N by N periodic matrices.

Generally, a N by N periodic matrix $\mathbf{S}(t)$ with the normalized fundamental frequency can be expressed by

$$\mathbf{S}(t) = \sum_{m=-M}^{M} \exp(\mathrm{j}mt)\mathbf{S}_m = \Phi_\otimes^T\mathbf{S}_T \tag{13.72}$$

where M harmonic functions are included to describe $\mathbf{S}(t)$ and $\mathbf{S}_T = [\mathbf{S}_{-M}^T \ \cdots \ \mathbf{S}_M^T]^T$ is a coefficient matrix. Its truncated Toeplitz form is

$$\mathbf{S}_{\mathcal{T}} = \begin{bmatrix} \mathbf{S}_0 & \cdots & \mathbf{S}_{-M} & \cdots & \mathbf{0} \\ \vdots & \ddots & \vdots & \ddots & \vdots \\ \mathbf{S}_M & \cdots & \mathbf{S}_0 & \cdots & \mathbf{S}_{-M} \\ \vdots & \ddots & \vdots & \ddots & \vdots \\ \mathbf{0} & \cdots & \mathbf{S}_M & \cdots & \mathbf{S}_0 \end{bmatrix} \tag{13.73}$$

A N-dimensional vector $\xi(t)$ with the normalized fundamental frequency can be expressed by

$$\xi(t) = \sum_{m=-M}^{M} \exp(\mathrm{j}mt)\xi_m = \Phi_{\otimes}^T \xi_T \tag{13.74}$$

where $\xi_T = [\xi_{-M}^T \ \cdots \ \xi_M^T]^T$ is a coefficient vector. The multiplication of $\mathbf{S}(t)$ and $\xi(t)$ including M harmonic functions is

$$\mathbf{S}(t)\xi(t) = \sum_{m=-M}^{M} \exp(\mathrm{j}mt)\zeta_m = \Phi_{\otimes}^T \zeta_T \tag{13.75}$$

where $\zeta_T = [\zeta_{-M}^T \ \cdots \ \zeta_M^T]^T$. Conducting Galerkin procedure on two sides of Eq. (13.75) yields

$$\int_0^{2\pi} \frac{\mathrm{d}t}{\pi} \hat{\Phi}_{\otimes} \mathbf{S}(t)\xi(t) = \int_0^{2\pi} \frac{\mathrm{d}t}{\pi} \hat{\Phi}_{\otimes} \Phi_{\otimes}^T \zeta_T = \zeta_T \tag{13.76}$$

Substituting Eq. (13.74) into Eq. (13.76) yields

$$\int_0^{2\pi} \frac{\mathrm{d}t}{\pi} \hat{\Phi}_{\otimes} \mathbf{S}(t) \Phi_{\otimes}^T \xi_T = \mathbf{S}_{\mathcal{T}} \xi_T \tag{13.77}$$

where $\mathbf{S}_{\mathcal{T}} \xi_T = \zeta_T$ due to the property of Toeplitz transform [2] is used. Since Eq. (13.77) is satisfied for an arbitrary ξ_T, one has

$$\int_0^{2\pi} \frac{\mathrm{d}t}{\pi} \hat{\Phi}_{\otimes} \Phi_{\otimes}^T \mathbf{S}_T \Phi_{\otimes}^T = \mathbf{S}_{\mathcal{T}} \tag{13.78}$$

Using Eq. (13.78) in Eqs. (13.50)–(13.55) with $\mathbf{S}_T$ replaced by $\mathbf{S}_w$, $\mathbf{S}_{w_x}$, $\mathbf{S}_{w_{xx}}$, $\mathbf{S}_{w_{xt}}$, $\mathbf{S}_{w_t}$, and $\mathbf{S}_{w_{tt}}$ yields $\mathbf{J}_w = \mathbf{S}_{w,\mathcal{T}}$, $\mathbf{J}_{w_x} = \mathbf{S}_{w_x,\mathcal{T}}$, $\mathbf{J}_{w_{xx}} = \mathbf{S}_{w_{xx},\mathcal{T}}$, $\mathbf{J}_{w_{xt}} = \mathbf{S}_{w_{xt},\mathcal{T}}$, $\mathbf{J}_{w_t} = \mathbf{S}_{w_t,\mathcal{T}} \mathbf{D}_{\otimes}$, and $\mathbf{J}_{w_{tt}} = \mathbf{S}_{w_{tt},\mathcal{T}} \mathbf{D}_{\otimes}^2$, respectively, where truncated Toeplitz forms $\mathbf{S}_{w,\mathcal{T}}$, $\mathbf{S}_{w_x,\mathcal{T}}$, $\mathbf{S}_{w_{xx},\mathcal{T}}$, $\mathbf{S}_{w_{xt},\mathcal{T}}$, $\mathbf{S}_{w_t,\mathcal{T}}$, and $\mathbf{S}_{w_{tt},\mathcal{T}}$ of $\Phi_{\otimes}^T \mathbf{S}_w$, $\Phi_{\otimes}^T \mathbf{S}_{w_x}$, $\Phi_{\otimes}^T \mathbf{S}_{w_{xx}}$, $\Phi_{\otimes}^T \mathbf{S}_{w_{xt}}$, $\Phi_{\otimes}^T \mathbf{S}_{w_t}$, and $\Phi_{\otimes}^T \mathbf{S}_{w_{tt}}$ can be directly constructed by $\mathbf{S}_w$, $\mathbf{S}_{w_x}$, $\mathbf{S}_{w_{xx}}$, $\mathbf{S}_{w_{xt}}$, $\mathbf{S}_{w_t}$, and $\mathbf{S}_{w_{tt}}$, respectively. The final expression of $\mathbf{J}(\mathbf{q})$ is

$$\mathbf{J}(\mathbf{q}) = \mathbf{S}_{w,\mathcal{T}} + \mathbf{S}_{w_x,\mathcal{T}} + \mathbf{S}_{w_{xx},\mathcal{T}} + \mathbf{S}_{w_{xt},\mathcal{T}}\mathbf{D}_{\otimes} + \mathbf{S}_{w_t,\mathcal{T}}\mathbf{D}_{\otimes} + \mathbf{S}_{w_{tt},\mathcal{T}}\mathbf{D}_{\otimes}^2 \tag{13.79}$$

In conclusion, when expressions of $\frac{\partial F}{\partial w}$, $\frac{\partial F}{\partial w_x}$, $\frac{\partial F}{\partial w_{xx}}$, $\frac{\partial F}{\partial w_{xt}}$, $\frac{\partial F}{\partial w_t}$, and $\frac{\partial F}{\partial w_{tt}}$ are given and $\eta_n(x)$ are sine functions, the calculation routine to obtain $\mathbf{J}(\mathbf{q})$ is: 1) sampling $\frac{\partial F}{\partial w}$, $\frac{\partial F}{\partial w_x}$, $\frac{\partial F}{\partial w_{xx}}$, $\frac{\partial F}{\partial w_{xt}}$, $\frac{\partial F}{\partial w_t}$, and $\frac{\partial F}{\partial w_{tt}}$ at $\{x_k\}_{k=1,\ \ldots,\ N_s-1}$ and $\{t_i\}_{i=0,\ \ldots,\ M_s-1}$ to construct $\mathbf{F}_w$, $\mathbf{F}_{w_x}$, $\mathbf{F}_{w_{xx}}$, $\mathbf{F}_{w_{xt}}$, $\mathbf{F}_{w_t}$, and $\mathbf{F}_{w_{tt}}$, respectively; 2) using the DST-FFT procedure for $\mathbf{F}_w$, $\mathbf{F}_{w_x}$, $\mathbf{F}_{w_{xx}}$, $\mathbf{F}_{w_{xt}}$, $\mathbf{F}_{w_t}$, and $\mathbf{F}_{w_{tt}}$ to calculate $\mathbf{r}_w$, $\mathbf{r}_{w_x}$, $\mathbf{r}_{w_{xx}}$, $\mathbf{r}_{w_{xt}}$, $\mathbf{r}_{w_t}$ and $\mathbf{r}_{w_{tt}}$, respectively; 3) using Eqs. (13.66)–(13.71) to calculate $\mathbf{S}_w$, $\mathbf{S}_{w_t}$, $\mathbf{S}_{w_{tt}}$, $\mathbf{S}_{w_x}$, $\mathbf{S}_{w_{xx}}$, and $\mathbf{S}_{w_{xt}}$, respectively; 4) constructing truncated Toeplitz forms $\mathbf{S}_{w,\mathcal{T}}$, $\mathbf{S}_{w_t,\mathcal{T}}$, $\mathbf{S}_{w_{tt},\mathcal{T}}$, $\mathbf{S}_{w_x,\mathcal{T}}$, $\mathbf{S}_{w_{xx},\mathcal{T}}$, and $\mathbf{S}_{w_{xt},\mathcal{T}}$ for $\Phi_{\otimes}^T\mathbf{S}_w$, $\Phi_{\otimes}^T\mathbf{S}_{w_t}$, $\Phi_{\otimes}^T\mathbf{S}_{w_{tt}}$, $\Phi_{\otimes}^T\mathbf{S}_{w_x}$, $\Phi_{\otimes}^T\mathbf{S}_{w_{xx}}$, and $\Phi_{\otimes}^T\mathbf{S}_{w_{xt}}$, respectively; and 5) using Eq. (13.79) to calculate $\mathbf{J}(\mathbf{q})$. When $\eta_n(x)$ are cosine functions, the same calculation routine to obtain $\mathbf{J}(\mathbf{q})$ can be used except that the DST-FFT procedure is replaced by the DCT-FFT procedure. This is the complex version of the STIHB method. The exact Jacobian matrix in Eq. (13.79) is used to study stability of steady-state responses in Sect. 13.2.4, where derivations of the set of ODEs in Eq. (13.2) are no longer needed.

13.2.3 Stability of Steady-State Responses

A steady-state solution of Eq. (13.1), $w_{ss} = \mathbf{H}\mathbf{q}_{ss}$, is given, where $\mathbf{q}_{ss} = \{q_{ss,1}(t), \cdots, q_{ss,N}(t)\}^T$ is a vector of generalized coordinates, in which $q_{ss,n}(t)$ are periodic functions with the normalized fundamental frequency. Substituting a perturbed solution, $w_{ss} + \delta w = \mathbf{H}\mathbf{q}_{ss} + \mathbf{H}\delta\mathbf{q}$, into Eq. (13.2) and linearizing the set of the resulting ODEs about $\mathbf{H}\mathbf{q}_{ss}$ yield

$$\begin{aligned}\int_0^2 \mathrm{d}x\mathbf{H}^T\Bigg(\left(\frac{\partial F}{\partial w}\mathbf{H} + \frac{\partial F}{\partial w_x}\mathbf{H}\mathbf{G} + \frac{\partial F}{\partial w_{xx}}\mathbf{H}\mathbf{G}^2\right)\delta\mathbf{q}\\ + \left(\frac{\partial F}{\partial w_{xt}}\mathbf{H}\mathbf{G} + \frac{\partial F}{\partial w_t}\mathbf{H}\right)\delta\dot{\mathbf{q}} + \frac{\partial F}{\partial w_{tt}}\mathbf{H}\delta\ddot{\mathbf{q}}\Bigg) = \mathbf{0}_{N\times 1}\end{aligned} \tag{13.80}$$

A state-space form of Eq. (13.80) with $\gamma = [\delta\mathbf{q}^T, \delta\dot{\mathbf{q}}^T]^T$ as a state vector is

$$\dot{\gamma} = \mathbf{A}(t)\gamma \tag{13.81}$$

where

$$\mathbf{A}(t) = \begin{bmatrix} \mathbf{0}_{N\times N} & \mathbf{E}_N \\ -\tilde{\mathbf{M}}(t)^{-1}\tilde{\mathbf{S}}(t) & -\tilde{\mathbf{M}}(t)^{-1}\tilde{\mathbf{B}}(t) \end{bmatrix} \tag{13.82}$$

$$\tilde{\mathbf{S}}(t) = \int_0^2 \mathrm{d}x\mathbf{H}^T\left(\frac{\partial F}{\partial w}\mathbf{H} + \frac{\partial F}{\partial w_x}\mathbf{H}\mathbf{G} + \frac{\partial F}{\partial w_{xx}}\mathbf{H}\mathbf{G}^2\right) \tag{13.83}$$

$\tilde{\mathbf{B}}(t) = \int_0^2 \mathrm{d}x \mathbf{H}^T (\frac{\partial F}{\partial w_{xt}} \mathbf{H}\mathbf{G} + \frac{\partial F}{\partial w_t} \mathbf{H})$, and $\tilde{\mathbf{M}}(t) = \int_0^2 \mathrm{d}x \mathbf{H}^T \frac{\partial F}{\partial w_{tt}} \mathbf{H}$. Since $\mathbf{A}(t)$ is a periodic matrix with the normalized fundamental frequency, stability of Eq. (13.81) can be studied by examining eigenvalues of its transformation matrix, $\lambda = [\lambda_1 \cdots \lambda_N]$, but its calculation can be time-consuming and integrations in derivations of $\tilde{\mathbf{S}}(t)$, $\tilde{\mathbf{B}}(t)$, and $\tilde{\mathbf{M}}(t)$ cannot be avoided. An alternative method to study stability is to first calculate eigenvalues of Toeplitz form of $\mathbf{A}(t)$. The complex plane can be divided into inifinte strips $\big(-\mathrm{j}(2k-1)\pi,\ \mathrm{j}(2k+1)\pi\big]$ with the integer $k \in (-\infty, +\infty)$, and locations of eigenvalues of Toeplitz form of $\mathbf{A}(t)$ are repeated in every strip. Eigenvalues in the fundamental strip $(-\mathrm{j}\pi,\ \mathrm{j}\pi]$ are reflections of λ from other strips, and their real parts are equal to those of λ. Hence, stability of Eq. (13.81) can be studied by examining eigenvalues of Toeplitz form of $\mathbf{A}(t)$ in the fundamental strip. To avoid derivations of $\tilde{\mathbf{S}}(t)$, $\tilde{\mathbf{B}}(t)$, and $\tilde{\mathbf{M}}(t)$ in this method, Eq. (13.82) is written as $\mathbf{A}(t) = \mathbf{M}(t)^{-1}\tilde{\mathbf{A}}(t)$, where

$$\mathbf{M}(t) = \begin{bmatrix} \tilde{\mathbf{M}}(t) & \mathbf{0}_{N\times N} \\ \mathbf{0}_{N\times N} & \tilde{\mathbf{M}}(t) \end{bmatrix} \tag{13.84}$$

$$\tilde{\mathbf{A}}(t) = \begin{bmatrix} \mathbf{0}_{N\times N} & \tilde{\mathbf{M}}(t) \\ -\tilde{\mathbf{S}}(t) & -\tilde{\mathbf{B}}(t) \end{bmatrix} \tag{13.85}$$

The property of Toeplitz transform shows that Toeplitz form of $\mathbf{A}(t)$ is the multiplication of Toeplitz forms of $\mathbf{M}(t)^{-1}$ and $\tilde{\mathbf{A}}(t)$. If the truncated Toeplitz form of $\mathbf{M}(t) = \sum_{m=-M}^{M} \exp(\mathrm{j}mt)\mathbf{M}_m$

$$\mathbf{M}_{\mathcal{T}} = \begin{bmatrix} \mathbf{M}_0 & \cdots & \mathbf{M}_{-M} & \cdots & \mathbf{0} \\ \vdots & \ddots & \vdots & \ddots & \vdots \\ \mathbf{M}_M & \cdots & \mathbf{M}_0 & \cdots & \mathbf{M}_{-M} \\ \vdots & \ddots & \vdots & \ddots & \vdots \\ \mathbf{0} & \cdots & \mathbf{M}_M & \cdots & \mathbf{M}_0 \end{bmatrix} \tag{13.86}$$

is invertible, the truncated Toeplitz form of $\mathbf{A}(t) = \sum_{m=-M}^{M} \exp(\mathrm{j}mt)\mathbf{A}_m$ can be calculated by $\mathbf{A}_{\mathcal{T}} = \mathbf{M}_{\mathcal{T}}^{-\Delta}\tilde{\mathbf{A}}_{\mathcal{T}}$, where

$$\tilde{\mathbf{A}}_{\mathcal{T}} = \begin{bmatrix} \tilde{\mathbf{A}}_0 & \cdots & \tilde{\mathbf{A}}_{-M} & \cdots & \mathbf{0} \\ \vdots & \ddots & \vdots & \ddots & \vdots \\ \tilde{\mathbf{A}}_M & \cdots & \tilde{\mathbf{A}}_0 & \cdots & \tilde{\mathbf{A}}_{-M} \\ \vdots & \ddots & \vdots & \ddots & \vdots \\ \mathbf{0} & \cdots & \tilde{\mathbf{A}}_M & \cdots & \tilde{\mathbf{A}}_0 \end{bmatrix} \tag{13.87}$$

is the truncated Toeplitz form of $\tilde{\mathbf{A}}(t) = \sum_{m=-M}^{M} \exp(\mathrm{j}mt)\tilde{\mathbf{A}}_m$. Using Eqs. (13.72) and (13.78) with $\mathbf{S}(t)$ replaced by $\tilde{\mathbf{S}}(t)$, $\tilde{\mathbf{B}}(t)$, and $\tilde{\mathbf{M}}(t)$ yields their truncated Toeplitz forms

$$\tilde{\mathbf{S}}_{\mathcal{T}} = \int_0^{2\pi} \frac{\mathrm{d}t}{\pi} \hat{\Phi}_{\otimes} \int_0^2 \mathrm{d}x \mathbf{H}^T \left(\frac{\partial F}{\partial w} \mathbf{H} + \frac{\partial F}{\partial w_x} \mathbf{H}\mathbf{G} + \frac{\partial F}{\partial w_{xx}} \mathbf{H}\mathbf{G}^2 \right) \Phi_{\otimes}^T \tag{13.88}$$

$$\tilde{\mathbf{B}}_{\mathcal{T}} = \int_0^{2\pi} \frac{\mathrm{d}t}{\pi} \hat{\Phi}_{\otimes} \int_0^2 \mathrm{d}x \mathbf{H}^T \left(\frac{\partial F}{\partial w_{xt}} \mathbf{H}\mathbf{G} + \frac{\partial F}{\partial w_t} \mathbf{H} \right) \Phi_{\otimes}^T \tag{13.89}$$

$$\tilde{\mathbf{M}}_{\mathcal{T}} = \int_0^{2\pi} \frac{\mathrm{d}t}{\pi} \hat{\Phi}_{\otimes} \int_0^2 \mathrm{d}x \mathbf{H}^T \frac{\partial F}{\partial w_{tt}} \mathbf{H} \Phi_{\otimes}^T \tag{13.90}$$

respectively. Comparing $\tilde{\mathbf{S}}_{\mathcal{T}}$, $\tilde{\mathbf{B}}_{\mathcal{T}}$, and $\tilde{\mathbf{M}}_{\mathcal{T}}$ with Eqs. (13.35)–(13.40) and using $\mathbf{J}_w = \mathbf{S}_{w,\mathcal{T}}$, $\mathbf{J}_{w_x} = \mathbf{S}_{w_x,\mathcal{T}}$, $\mathbf{J}_{w_{xx}} = \mathbf{S}_{w_{xx},\mathcal{T}}$, $\mathbf{J}_{w_{xt}} = \mathbf{S}_{w_{xt},\mathcal{T}}$, $\mathbf{J}_{w_t} = \mathbf{S}_{w_t,\mathcal{T}}\mathbf{D}_{\otimes}$, and $\mathbf{J}_{w_{tt}} = \mathbf{S}_{w_{tt},\mathcal{T}}\mathbf{D}_{\otimes}^2$ yield

$$\tilde{\mathbf{S}}_{\mathcal{T}} = \mathbf{S}_{w,\mathcal{T}} + \mathbf{S}_{w_x,\mathcal{T}} + \mathbf{S}_{w_{xx},\mathcal{T}} \tag{13.91}$$

$$\tilde{\mathbf{B}}_{\mathcal{T}} = \mathbf{S}_{w_{xt},\mathcal{T}} + \mathbf{S}_{w_t,\mathcal{T}} \tag{13.92}$$

$$\tilde{\mathbf{M}}_{\mathcal{T}} = \mathbf{S}_{w_{tt},\mathcal{T}} \tag{13.93}$$

respectively. When $\mathbf{S}_w$, $\mathbf{S}_{w_t}$, $\mathbf{S}_{w_{tt}}$, $\mathbf{S}_{w_x}$, $\mathbf{S}_{w_{xx}}$, and $\mathbf{S}_{w_{xt}}$ are calculated from Eqs. (13.66)–(13.71), $\tilde{\mathbf{A}}_{\mathcal{T}}$ can be obtained with

$$\tilde{\mathbf{A}}_m = \begin{bmatrix} \mathbf{0}_{N\times N} & \tilde{\mathbf{M}}_m \\ -\tilde{\mathbf{S}}_m & -\tilde{\mathbf{B}}_m \end{bmatrix} \tag{13.94}$$

where $\mathbf{S}_w + \mathbf{S}_{w_x} + \mathbf{S}_{w_{xx}} = [\tilde{\mathbf{S}}_{-M}^T \ \cdots \ \tilde{\mathbf{S}}_M^T]^T$, $\mathbf{S}_{w_{xt}} + \mathbf{S}_{w_t} = [\tilde{\mathbf{B}}_{-M}^T \ \cdots \ \tilde{\mathbf{B}}_M^T]^T$, and $\mathbf{S}_{w_{tt}} = [\tilde{\mathbf{M}}_{-M}^T \ \cdots \ \tilde{\mathbf{M}}_M^T{}^T$, and $\mathbf{M}_{\mathcal{T}}^{-\Delta}$ can be obtained with

$$\mathbf{M}_m = \begin{bmatrix} \tilde{\mathbf{M}}_m & \mathbf{0}_{N\times N} \\ \mathbf{0}_{N\times N} & \tilde{\mathbf{M}}_m \end{bmatrix} \tag{13.95}$$

The truncated Toeplitz form of $\mathbf{A}(t)$ is calculated by $\mathbf{A}_{\mathcal{T}} = \mathbf{M}_{\mathcal{T}}^{-\Delta}\tilde{\mathbf{A}}_{\mathcal{T}}$, and eigenvalues of $\mathbf{A}_{\mathcal{T}}$ in the fundamental strip are used to study stability of $w_{ss} = \mathbf{H}\mathbf{q}_{ss}$. In a special case, if $\frac{\partial F}{\partial w_{tt}} = m_{w_{tt}}$ is a constant, calculation of $\mathbf{A}_{\mathcal{T}}$ can be simplified using $\tilde{\mathbf{M}}(t) = \int_0^2 \mathrm{d}x \mathbf{H}^T \frac{\partial F}{\partial w_{tt}} \mathbf{H} = m_{w_{tt}} \mathbf{E}_N$ and consequently $\mathbf{M}_{\mathcal{T}} = m_{w_{tt}} \mathbf{E}_{2N(2M+1)}$, where $\mathbf{E}_{2N(2M+1)}$ is the $2N(2M+1)$ by $2N(2M+1)$ identity matrix. The simplified expression of the truncated Toeplitz form of $\mathbf{A}(t)$ is $\mathbf{A}_{\mathcal{T}} = m_{w_{tt}}^{-1}\tilde{\mathbf{A}}_{\mathcal{T}}$.

13.3 Steady-State Responses of a Fixed-Fixed String with Geometric Nonlinearity and Their Stability Analysis

The STIHB method is demonstrated by studying the transverse vibration of a fixed-fixed string with geometric nonlinearity. Its governing equation with the normalized spatial and temporal coordinates is [20, 21],

$$F(x, t, w_t, w_{tt}, w_x, w_{xx}, w_{xt}, f) = \omega^2 w_{tt} + \omega c_d w_t - w_{xx} - k_d w_x^2 w_{xx} \\ - y_0 \sin(\pi x)\cos t = 0 \tag{13.96}$$

where $w(x,t)$ is the transverse displacement of the string, $f = y_0 \sin(\pi x)\cos t$ is the external excitation, ω is the angular excitation frequency before normalization, c_d is the damping coefficient, and k_d is the nonlinear stiffness coefficient. Boundary conditions of the string are

$$w(0,t) = w(1,t) = 0 \tag{13.97}$$

and orthonormal trial functions in $\mathbf{H}$ can be $\eta_n(x) = \sin(n\pi x)$. There are two tasks in the STIHB method before using Newton-Raphson method to find steady-state responses of Eq. (13.96). The first task is to construct the harmonic balanced residual, and the second task is to construct the Jacobian matrix. A steady-state response of the string can be $w(x,t) = \mathbf{HQ\Phi}$, where $\mathbf{Q}$ is a guess solution of coefficients of combined spatial and temporal bases $\sin(n\pi x)\exp(\mathrm{j}mt)$ of $w(x,t)$ in the matrix form, and procedures to construct the harmonic balanced residual of $F(x, t, w_t, w_{tt}, w_x, w_{xx}, w_{xt}, f)$ with the given $\mathbf{Q}$ are:

1. Selecting the number of trial functions in the spatial coordinate and truncated number of Fourier series in the temporal coordinate to be N and M, respectively;
2. Discretizing spatial and temporal coordinates to be $\left\{x_k := \frac{k}{N_s}\right\}_{k=1,\ \dots,\ N_s-1}$ with $N_s > N$ and $\left\{t_i := \frac{2\pi i}{M_s}\right\}_{i=0,\ \dots,\ M_s-1}$ with $M_s > 2M$, respectively;
3. Replacing $w, w_t, w_{tt}, w_x, w_{xx}$, and w_{xt} by $\mathbf{w}, \mathbf{w}_t, \mathbf{w}_{tt}, \mathbf{w}_x, \mathbf{w}_{xx}$, and $\mathbf{w}_{xt}$, respectively, on the left-hand side of Eq. (13.96) and substituting Eq. (13.17) and Eqs. (13.20)–(13.24) into the resulting equation yield

$$\mathbf{F} = \omega^2 \bar{\mathbf{H}}\mathbf{Q}\mathbf{D}^{\Theta}\bar{\mathbf{\Phi}} + \omega c_d \bar{\mathbf{H}}\mathbf{Q}\mathbf{D}\bar{\mathbf{\Phi}} - \bar{\mathbf{H}}\mathbf{G}^2\mathbf{Q}\bar{\mathbf{\Phi}} \\ - k_d(\bar{\mathbf{H}}\mathbf{G}\mathbf{Q}\bar{\mathbf{\Phi}}) \circ (\bar{\mathbf{H}}\mathbf{G}\mathbf{Q}\bar{\mathbf{\Phi}}) \circ (\bar{\mathbf{H}}\mathbf{G}^2\mathbf{Q}\bar{\mathbf{\Phi}}) - \mathbf{f} \tag{13.98}$$

where $\circ$ denotes Hadamard product,

$$\bar{\mathbf{H}}\mathbf{G} = \begin{bmatrix} \pi\cos\left(\pi\frac{1}{N_s}\right) & \cdots & N\pi\cos\left(N\pi\frac{1}{N_s}\right) \\ \vdots & \ddots & \vdots \\ \pi\cos\left(\pi\frac{N_s-1}{N_s}\right) & \cdots & N\pi\cos\left(N\pi\frac{N_s-1}{N_s}\right) \end{bmatrix} \tag{13.99}$$

$$\bar{\mathbf{H}}\mathbf{G}^2 = \begin{bmatrix} -\pi^2 \sin\left(\pi \frac{1}{N_s}\right) & \cdots & -N^2\pi^2 \sin\left(N\pi \frac{1}{N_s}\right) \\ \vdots & \ddots & \vdots \\ -\pi^2 \sin\left(\pi \frac{N_s-1}{N_s}\right) & \cdots & -N^2\pi^2 \sin\left(N\pi \frac{N_s-1}{N_s}\right) \end{bmatrix} \tag{13.100}$$

$$\mathbf{f} = y_0 \begin{bmatrix} \sin(\pi \frac{1}{N_s}) \cos\left(\frac{2\pi 0}{M_s}\right) & \cdots & \sin(\pi \frac{1}{N_s}) \cos\left(\frac{2\pi (M_s-1)}{M_s}\right) \\ \vdots & \ddots & \vdots \\ \sin(\pi \frac{N_s-1}{N_s}) \cos\left(\frac{2\pi 0}{M_s}\right) & \cdots & \sin(\pi \frac{N_s-1}{N_s}) \cos\left(\frac{2\pi (M_s-1)}{M_s}\right) \end{bmatrix}_{(N_s-1)\times M_s} \tag{13.101}$$

4. Constructing the harmonic balanced residual $\mathbf{r}(\mathbf{q})$ using Eq. (13.31) to conduct the DST-FFT procedure for $\mathbf{F}$ and using Eq. (13.26) to convert the resulting matrix form of the harmonic balanced residual to its vector form.

There are two ways to obtain the Jacobian matrix. In the simple version of the STIHB method, an approximated Jacobian matrix can be updated by Eq. (13.34) with Broyden's method in each iteration. In the complex version of the STIHB method, the exact Jacobian matrix can be constructed with terms in the linearized equation of Eq. (13.96) given by $\frac{\partial F}{\partial w} = 0$, $\frac{\partial F}{\partial w_{xt}} = 0$, $\frac{\partial F}{\partial w_x} = -2k_d w_x w_{xx}$, $\frac{\partial F}{\partial w_{xx}} = -1 - k_d w_x^2$, $\frac{\partial F}{\partial w_t} = \omega c_d$, and $\frac{\partial F}{\partial w_{tt}} = \omega^2$. Since $\frac{\partial F}{\partial w}$, $\frac{\partial F}{\partial w_{xt}}$, $\frac{\partial F}{\partial w_t}$, and $\frac{\partial F}{\partial w_{tt}}$ are constants, $\mathbf{J}_w$, $\mathbf{J}_{w_{xt}}$, $\mathbf{J}_{w_t}$, and $\mathbf{J}_{w_{tt}}$ can be directly obtained. Procedures to construct the exact Jacobian matrix are:

1. Constructing $\mathbf{J}_w = \mathbf{0}_{N(2M+1)\times N(2M+1)}$, $\mathbf{J}_{w_{xt}} = \mathbf{0}_{N(2M+1)\times N(2M+1)}$

$$\begin{aligned} \mathbf{J}_{w_t} &= \int_0^{2\pi} \frac{\mathrm{d}t}{\pi} \int_0^2 \mathrm{d}x \hat{\Phi}_\otimes \mathbf{H}^T \frac{\partial F}{\partial w_t} \mathbf{H} \Phi_\otimes^T \mathbf{D}_\otimes \\ &= \omega c_d \int_0^{2\pi} \frac{\mathrm{d}t}{\pi} \hat{\Phi}_\otimes \int_0^2 \mathrm{d}x \mathbf{H}^T \mathbf{H} \Phi_\otimes^T \mathbf{D}_\otimes \\ &= \omega c_d \mathbf{D}_\otimes \end{aligned} \tag{13.102}$$

$$\begin{aligned} \mathbf{J}_{w_{tt}} &= \int_0^{2\pi} \frac{\mathrm{d}t}{\pi} \int_0^2 \mathrm{d}x \hat{\Phi}_\otimes \mathbf{H}^T \frac{\partial F}{\partial w_{tt}} \mathbf{H} \Phi_\otimes^T \mathbf{D}_\otimes^2 \\ &= \omega^2 \int_0^{2\pi} \frac{\mathrm{d}t}{\pi} \int_0^2 \mathrm{d}x \hat{\Phi}_\otimes \mathbf{H}^T \mathbf{H} \Phi_\otimes^T \mathbf{D}_\otimes^2 \\ &= \omega^2 \mathbf{D}_\otimes^2 \end{aligned} \tag{13.103}$$

2. Replacing w_x and w_{xx} in $\frac{\partial F}{\partial w_x} = -2k_d w_x w_{xx}$ by $\mathbf{w}_x$ and $\mathbf{w}_{xx}$, respectively, to obtain $\mathbf{F}_{w_x}$ and substituting Eqs. (13.20) and (13.21) into the resulting equation yield

$$\mathbf{F}_{w_x} = -2k_d(\bar{\mathbf{H}}\mathbf{G}\mathbf{Q}\bar{\mathbf{\Phi}}) \circ (\bar{\mathbf{H}}\mathbf{G}^2\mathbf{Q}\bar{\mathbf{\Phi}}) \tag{13.104}$$

3. Constructing the harmonic balanced residual $\mathbf{r}_{w_x}$ using Eq. (13.31) to conduct the DST-FFT procedure for $\mathbf{F}_{w_x}$ and using Eq. (13.26) to convert the resulting matrix form of the harmonic balanced residual to its vector form $\mathbf{r}_{w_x}$;
4. Using Eq. (13.69) to calculate $\mathbf{S}_{w_x}$ and Eq. (13.78) with $\mathbf{S}_{\mathcal{T}}$ replaced by $\mathbf{S}_{w_x}$ to calculate $\mathbf{J}_{w_x} = \mathbf{S}_{w_x,\mathcal{T}}$;
5. Replacing w_x by $\mathbf{w}_x$ in $\frac{\partial F}{\partial w_{xx}} = -1 - k_d w_x^2$ to obtain $\mathbf{F}_{w_{xx}}$ and substituting Eq. (13.20) into the resulting equation to yield

$$\mathbf{F}_{w_{xx}} = -\mathbf{O} - k_d(\bar{\mathbf{H}}\mathbf{G}\mathbf{Q}\bar{\mathbf{\Phi}}) \circ (\bar{\mathbf{H}}\mathbf{G}\mathbf{Q}\bar{\mathbf{\Phi}}) \tag{13.105}$$

   where $\mathbf{O}$ is a $N_s - 1$ by M_s matrix whose elements are all one;
6. Constructing the harmonic balanced residual $\mathbf{r}_{w_{xx}}$. Using Eq. (13.31) to conduct the DCT-FFT procedure for $\mathbf{F}_{w_{xx}}$ with $\mathbf{H}$ replaced by $\mathbf{H}_c = [\cos(\pi x) \cdots \cos(N\pi x)]$, premultiplying the resulting matrix by $\mathbf{I}^{-1}$, where $\mathbf{I}$ is the diagonal matrix of $[\pi \ \cdots \ N\pi]$, and using Eq. (13.26) to convert the resulting matrix form of the harmonic balanced residual to its vector form $\mathbf{r}_{w_{xx}}$;
7. Using Eq. (13.70) to calculate $\mathbf{S}_{w_{xx}}$ and Eq. (13.78) with $\mathbf{S}_{\mathcal{T}}$ replaced by $\mathbf{S}_{w_{xx}}$ to calculate $\mathbf{J}_{w_{xx}} = \mathbf{S}_{w_{xx},\mathcal{T}}$;
8. Constructing the exact Jacobian matrix

$$\mathbf{J}(\mathbf{q}) = \omega c_d \mathbf{D}_{\otimes} + \omega^2 \mathbf{D}_{\otimes}^2 + \mathbf{S}_{w_x,\mathcal{T}} + \mathbf{S}_{w_{xx},\mathcal{T}} \tag{13.106}$$

   where $\mathbf{q}$ is the vector form of $\mathbf{Q}$.

With the harmonic balanced residual and exact Jacobian matrix constructed by the above procedures, Newton-Raphson method can be used to find solutions of $w(x,t)$, which make the residual vanish.

When a steady-state solution $w_{ss}(x,t)$ is obtained, its stability can be studied by examining eigenvalues of Toeplitz form of $\mathbf{A}(t)$ in the fundamental strip. Since $\frac{\partial F}{\partial w_t} = \omega c_d$ and $\frac{\partial F}{\partial w_{tt}} = \omega^2$ are constants, substituting $\tilde{\mathbf{B}}(t) = \omega c_d \mathbf{E}_N$ and $\tilde{\mathbf{M}}(t) = \omega^2 \mathbf{E}_N$ into Eq. (13.82) yields

$$\mathbf{A}(t) = \begin{bmatrix} \mathbf{0}_{N\times N} & \mathbf{E}_N \\ -\frac{1}{\omega^2}\tilde{\mathbf{S}}(t) & -\frac{c_d}{\omega}\mathbf{E}_N \end{bmatrix} \tag{13.107}$$

Toeplitz form of $\mathbf{A}(t)$ is

$$\mathbf{A}_{\mathcal{T}} = \begin{bmatrix} \mathbf{A}_0 & \cdots & \mathbf{A}_{-M} & \cdots & \mathbf{0} \\ \vdots & \ddots & \vdots & \ddots & \vdots \\ \mathbf{A}_M & \cdots & \mathbf{A}_0 & \cdots & \mathbf{A}_{-M} \\ \vdots & \ddots & \vdots & \ddots & \vdots \\ \mathbf{0} & \cdots & \mathbf{A}_M & \cdots & \mathbf{A}_0 \end{bmatrix} \tag{13.108}$$

where

$$\mathbf{A}_0 = \begin{bmatrix} \mathbf{0}_{N\times N} & \mathbf{E}_N \\ -\tilde{\mathbf{S}}_0 & -\frac{c_d}{\omega}\mathbf{E}_N \end{bmatrix} \tag{13.109}$$

$$\mathbf{A}_m = \begin{bmatrix} \mathbf{0}_{N\times N} & \mathbf{0}_{N\times N} \\ -\tilde{\mathbf{S}}_m & -\mathbf{0}_{N\times N} \end{bmatrix}, \quad m \neq 0 \tag{13.110}$$

$$\tilde{\mathbf{S}}_{\mathcal{T}} = \begin{bmatrix} \tilde{\mathbf{S}}_0 & \cdots & \tilde{\mathbf{S}}_{-M} & \cdots & \mathbf{0} \\ \vdots & \ddots & \vdots & \ddots & \vdots \\ \tilde{\mathbf{S}}_M & \cdots & \tilde{\mathbf{S}}_0 & \cdots & \tilde{\mathbf{S}}_{-M} \\ \vdots & \ddots & \vdots & \ddots & \vdots \\ \mathbf{0} & \cdots & \tilde{\mathbf{S}}_M & \cdots & \tilde{\mathbf{S}}_0 \end{bmatrix} = \mathbf{S}_{w_x,\mathcal{T}} + \mathbf{S}_{w_{xx},\mathcal{T}} \tag{13.111}$$

If there exist eigenvalues of $\mathbf{A}_{\mathcal{T}}$ in the fundamental strip $(-\mathrm{j}\pi,\ \mathrm{j}\pi]$, whose real parts are larger than zero, the steady-state solution $w_{ss}(x, t)$ is unstable; if their imaginary parts are $\mathrm{j}\pi$, which means that exponents of these eigenvalues are less then -1, there exists a period-doubling bifurcation; and if their imaginary parts are neither zero nor $\mathrm{j}\pi$, which means that exponents of these eigenvalues that escape the unit circle are complex, there exists Hopf bifurcation.

13.4 Results and Discussions

Some parameters used in the following calculation are $c_d = 0.8$, $y_0 = 0.2$, $M = 19$, $N_s = 64$, and $M_s = 128$. When $k_d = 10$, the string has weak geometric nonlinearity. Comparison of convergence between the simple and complex versions of the STIHB method is shown in Table 13.1, where numbers of iterations to obtain steady-state solutions with $N = 5$ at $\omega = 1.5, 2.5, 3.5$, and 4.5 by the two versions of the STIHB method are listed. It clearly shows that convergence of the complex version is much faster than that of the simple version, especially in a frequency region around a resonant frequency of the string with a large value of $||\mathbf{q}||_2$, where $||\cdot||_2$ is Euclidean norm of a vector. Frequency-response curves of the string for the given parameters are shown in Fig. 13.1, where $N = 5$ and $N = 10$. It shows that the curve with $N = 5$ overlaps that with $N = 10$, which means that use of five trial functions is good enough to solve for Eq. (13.96) with the weak nonlinearity. Stability of solutions on the curves is checked and all solutions are stable. In the case with strong nonlinearity with $k_d = 110$, parts of frequency-response curves with $N = 5$ and $N = 10$ slightly deviate from those with higher N, as shown in Fig. 13.2. Frequency-response curves with $N = 20$ and $N = 30$ almost overlap each other, which means that solutions obtained with $N = 30$ can be very close to real solutions, although those obtained with $N = 5$ and $N = 10$ can also be acceptable. Stable and unstable solutions with $N = 30$ are indicated in Fig. 13.3, and there is no period-doubling or Hopf bifurcation.

Table 13.1 Numbers of iterations to obtain steady-state solutions with the simple and complex versions of the STIHB method

ω (rad/s)	$\|\|\mathbf{q}\|\|_2$	Simple	Complex
1.5	0.0255	7	3
2.5	0.0449	10	4
3.5	0.0609	16	5
4.5	0.0183	5	3

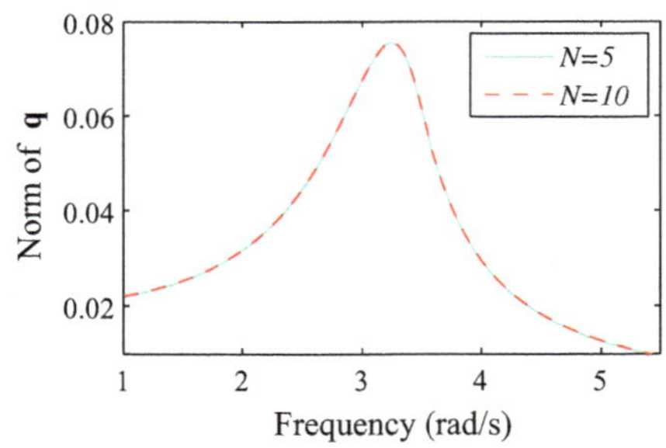

Fig. 13.1 Frequency-response curves for the case of weak nonlinearity with $k_d = 10$; $N = 5$ and 10

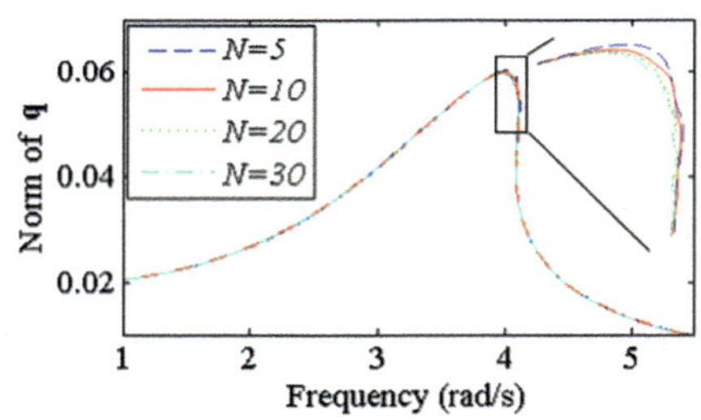

Fig. 13.2 Frequency-response curves for the case of strong nonlinearity with $k_d = 110$; $N = 5$, 10, 20, and 30

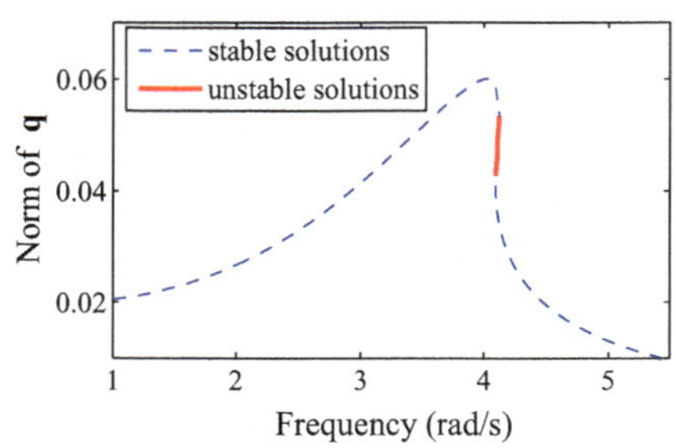

Fig. 13.3 Stability of solutions on the frequency-response curve with $N = 30$ for the case of strong nonlinearity with $k_d = 110$

A solution $w(x, t) = \mathbf{HQ\Phi}$ obtained with $N = 5$ in $\mathbf{H}$ makes the norm of the harmonic balanced residual calculated from Eq. (13.31) with the same N less than a preset tolerance, which is 10^{-10} here. However, using this solution to evaluate the norm of the harmonic balanced residual with $N = 10$ yields a relatively large value, whose magnitude can be about y_0 in a frequency region around a resonant frequency. This means that a solution obtained by Galerkin procedure in the spatial coordinate with five trial functions, which makes residuals vanish at the first five spatial frequencies of trial functions, may not make them vanish at higher spatial frequencies. When $N = 30$, residuals at higher spatial frequencies are tens of times smaller than those when $N = 5$. This shows that solutions obtained with $N = 30$ can be very close to real solutions. However, even if residuals at higher spatial frequencies are large when N is relatively small, solutions with this N may not deviate much from real solutions, which is illustrated below.

If $w_0 = \mathbf{HQ}_\Gamma \mathbf{\Phi}$ is obtained with $N = N_c$ and $\mathbf{R}(\mathbf{Q}_0)$ includes N_t trial functions with $N_t > N_c$, residuals in $\mathbf{R}(\mathbf{Q}_0)$ at the first N_c spatial frequencies are less than the preset tolerance. Substituting w_0 and $\mathbf{R}(\mathbf{Q}_0)$ into Eq. (13.96) yields

$$F_0(w_0) = F(x, t, w_{0,t}, w_{0,tt}, w_{0,x}, w_{0,xx}, w_{0,xt}, f) = \mathbf{HR}(\mathbf{Q})\Phi \tag{13.112}$$

The real solution is assumed to be $w_r = \delta w + w_0$. Substituting it into Eq. (13.96) yields $F_0(\delta w + w_0) = 0$. Linearizing $F_0(\delta w + w_0) = 0$ about w_0 and substituting Eq. (13.112) into the resulting equation yield

$$\frac{\partial F_0}{\partial w_t}\delta w_t + \frac{\partial F_0}{\partial w_{tt}}\delta w_{tt} + \frac{\partial F_0}{\partial w_x}\delta w_x + \frac{\partial F_0}{\partial w_{xx}}\delta w_{xx} + \frac{\partial F_0}{\partial w_{xt}}\delta w_{xt} + \mathbf{HR}(\mathbf{Q})\Phi = 0 \tag{13.113}$$

where $\mathbf{HR}(\mathbf{Q})\Phi$ can be considered as a higher-frequency excitation. When there is a cut-off frequency N_{cut} of Eq. (13.113) in the spatial coordinate, which means that its solutions are insensitive to excitations with spatial frequencies higher than N_{cut} in the spatial coordinate, and N_c is larger than N_{cut}, which means that significant components in $\mathbf{HR}(\mathbf{Q})\Phi$ are at spatial frequencies higher than N_{cut}, the magnitude of δw can be very small even if that of $\mathbf{HR}(\mathbf{Q})\Phi$ is large, and w_0 can be a good approximation of w_r.

13.5 Conclusion

The STIHB method is developed to automatically and efficiently calculate steady-state responses of a general one-dimensional second-order continuous system. Galerkin procedure in the spatial coordinate to obtain a set of ODEs and the harmonic balance procedure for the set of ODEs in the temporal coordinate to obtain the harmonic balanced residual are combined and implemented by the DST-FFT or DCT-FFT procedure, which can be automatically and efficiently obtained by a computer program, where numbers of basis functions in the spatial and temporal coordinates can be arbitrarily selected. The only part of the program that needs to be specified is the expression of the governing PDE of the system. The simple version of the STIHB method can be used to calculate steady-state responses by combining the DST-FFT or DCT-FFT procedure with Broyden's method. The complex version of the STIHB method uses the exact Jacobian matrix in Newton-Raphson method, which can also be automatically and efficiently obtained by following a calculation routine, and the only part of the routine that needs to be specified is the linearized expression of the PDE. The complex version of the STIHB method yields faster convergence than the simple version, and stability of steady-state responses can be analyzed by constructing the exact Jacobian matrix in Toeplitz form of the system matrix of the set of linearized ODEs in the state-space form. Hence, derivations of the set of ODEs for stability analysis can be avoided. The STIHB method is demonstrated by studying the transverse vibration of a string with geometric nonlinearity. In the case with weak nonlinearity, solutions obtained with $N = 5$ are good enough and all solutions are stable. In the case with strong nonlinearity, parts of frequency-response curves with $N = 5$ and $N = 10$ slightly deviate from those with higher N and there are unstablle solutions. Solutions obtained with a larger N can be more accurate, but it is sufficient to use a relatively small N when N is larger than N_{cut}.

Acknowledgements The authors would like to thank the support from the National Science Foundation through Award CMMI-1335397.

References

1. Meirovitch, L.: Principles and Techniques of Vibrations. Prentice-Hall, Upper Saddle River, NJ (1997)
2. Wereley, N.M.: Analysis and Control of Linear Periodically Time Varying Systems. Ph.D. Thesis, Massachusetts Institute of Technology (1990)
3. Chivukula, R.K., Reznik, Y.A.: Fast computing of discrete cosine and sine transforms of types VI and VII. In: Proceedings, of SPIE 8135, Applications of Digital Image Processing XXXIV (2011)
4. Ozhan, B.B., Pakdemirli, M.: A general solution procedure for the forced vibrations of a continuous system with cubic nonlinearites: primary resonance case. J. Sound Vib. **325**(4–5), 894–906 (2009)
5. Wickert, J.A.: Non-linear vibration of a traveling tensioned beam. Int. J. Non-Linear Mech. **27**(3), 503–517 (1992)
6. Nayfeh, A.H.: Reduced-order models of weakly nonlinear spatially continuous systems. Nonlinear Dyn. **16**(2), 105–125 (1998)
7. Chen, L., Zhao, W., Ding, H.: On Galerkin discretization of axially moving nonlinear strings. Acta, Mech. Solida Sin. **22**(4), 369–376 (2009)
8. Szemplinska-Stupnicka, W.: The generalized harmonic balance method for determining the combination resonance in the parametric dynamic systems. J. Sound Vib. **58**(3), 347–361 (1978)
9. Tamura, H., Tsuda, Y., Sueoka, A.: Higher approximation of steady oscillations in nonlinear systems with single degree of freedom : suggested multi-harmonic balance method. Bull. JSME **24**(195), 1616–1625 (1981)
10. Lau, S.L., Cheung, Y.K., Wu, S.Y.: A variable parameter incrementation method for dynamic instability of linear and nonlinear elastic systems. J. Appl. Mech. **49**(4), 849–853 (1982)

11. Xu, G.Y., Zhu, W.D.: Nonlinear and time-varying dynamics of high-dimensional models of a translating beam with a stationary load subsystem. J. Vib. Acoust. **132**(6), 061012–17 (2010)
12. Huang, J.L., Zhu, W.D.: Nonlinear dynamics of a high-dimensional model of a rotating Euler–Bernouli beam under the gravity load. J. Appl. Mech. **81**(10), 101007–20 (2014)
13. Cheung, Y.K., Chen, S.H., Lau, S.L.: Application of the incremental harmonic balance method to cubic non-linearity systems. J. Sound Vib. **140**(2), 273–286 (1990)
14. Ferri, A.A.: On the equivalence of the incremental harmonic balance method and the harmonic Balance-Newton-Raphson method. J. Appl. Mech. **53**(2), 455–457 (1986)
15. Ling, F.H., Wu, X.X.: Fast Galerkin method and its application to determine periodic solutions of non-linear oscillators. Int. J. Non-Linear Mech. **22**(2), 89–98 (1987)
16. Wang, X.F., Zhu, W.D.: A modified incremental harmonic balance method based on the fast Fourier transform and Broyden's method. Nonlinear Dyn. **81**(1/2), 981–989 (2015)
17. Rizzoli, V., Mastri, F., Cecchetti, C., Sgallari, F.: Fast and robust inexact newton approach to the harmonic-balance analysis of nonlinear microwave circuits. Microw. Guid. Wave Lett. **7**(10), 359–361 (1997)
18. Knoll, D.A., Keyes, D.E.: Jacobian-free Newton-Krylov methods: a survey of approaches and applications. J. Comput. Phys. **193**(2), 357–397 (2004)
19. Hsu, C.S., Cheng, W.H.: Applications of the theory of impulsive parametric excitation and new treatments of general parametric excitation problems. J. Appl. Mech. **40**(1), 78–86 (1973)
20. Mote Jr., C.D.: On the nonlinear oscillation of an axially moving string. J. Appl. Mech. **33**(2), 463–464 (1966)
21. Zhu, W.D., Wu, K.: Dynamic stability of a class of second-order distributed structural systems with sinusoidally varying velocities. J. Appl. Mech. **80**(6), 061008–15 (2013)

Chapter 14
The Stability of Non-linear Power Systems

Kaihua Xi, Johan L. A. Dubbeldam, Feng Gao, Hai Xiang Lin, and Jan H. van Schuppen

Abstract The power system is one of the most complicated man-made non-linear systems which plays an important role for human being since it was first made in the 19th century. In the past decade, the integration of renewable power sources such as wind energy and solar energy has increased rapidly due to their sustainability. However, these energy sources are weather dependent which cannot be controlled or even predicted precisely. A challenge brought by this transition to renewable power generation is the uncertain fluctuations that negatively affects the stability of the power system, which leads to the important problem: how to improve by control the stability of the system such that it remains stable when subjected to considerable fluctuations in the energy supply? Hence, research is needed into the stability metrics of the non-linear power system and control strategies for the stability improvement. In this chapter, we describe the linear and non-linear stability analysis of power systems and summarize the corresponding control strategies for stability improvement.

K. Xi (✉)
School of Mathematics, Shandong University, Jinan 250100, Shandong, China
e-mail: kxi@sdu.edu.cn

F. Gao
School of Electrical Engineering, Shandong University, Jinan 250061, Shandong, China
e-mail: fgao@sdu.edu.cn

J. L. A. Dubbeldam · H. X. Lin · J. H. van Schuppen
Delft Institute of Applied Mathematics, Delft University of Technology, Van Mourik Broekmanweg 6, 2628 XE Delft, The Netherlands
e-mail: J.L.A.Dubbeldam@tudelft.nl

H. X. Lin
e-mail: H.X.Lin@tudelft.nl

J. H. van Schuppen
e-mail: J.H.vanschuppen@tudelft.nl

A. K. Abramian et al. (eds.), *Nonlinear Dynamics of Discrete and Continuous Systems*, Advanced Structured Materials 139,
https://doi.org/10.1007/978-3-030-53006-8_14

14.1 Introduction

In order to decrease the CO_2 emissions from the traditional fossil fuel power plants, there are more and more wind farms and Photo Voltaic (PV) farms established on the generation side and rooftop solar PV panels installed at houses of consumers on the distribution side almost all over the world in the past decade. The rapid increase of the weather dependent power energy, which is also called variable renewable energy, brings several challenges to the power system. It is well known that these renewable power generation depends on the weather which cannot be controlled or even accurately predicted. In this case, unlike the traditional power system where the uncertainties usually come from the consumer side only, the uncertainties now come from both the generation and the load side and thus will be harder to manage. These fluctuations do not only deteriorate the quality of power supply, but also decrease the power system stability [28].

Since power systems rely on the synchronous machines (e.g., rotor-generators driven by steam or gas turbines) for power generation, a requirement for normal system operation is that all the synchronous machines remain in synchronization. The ability of a power system to maintain the synchronization when subjected to severe transient disturbance such as short-circuit of transmission lines, loss of generation, is called *transient stability* [1, 10, 18, 22, 51], which we will also refer to as synchronization stability in this chapter. The synchronous state is actually an *equilibrium point* of the system, which has been widely studied in the field of complex network [25, 41, 47]. Synchronization stability has been studied mainly by linearizing about a stable equilibrium [2, 12, 32, 34, 39]. The framework developed by Pecora et al. has greatly facilitated these computations. However, as fluctuations induced by changing weather can have enormous impact, a linearization approach will often not be sufficient. From the perspective of non-linear systems, this stability measures the ability that the state stays in the basin of attraction after disturbances. This stability is influenced by the nonlinearity of the power system. The basin of attraction (also called the *stability region*) of a nonlinear system is defined as the set of the initial states of the trajectories which converge to the equilibrium as the time goes to infinity [19, 37]. For a nonlinear system with a small basin of attraction, the trajectory usually has a small escape time from the region when subjected to disturbances. *Stability margin* is another definition corresponding to the non-linear stability, which measures the distance from a stable state to the state of losing synchronization [10, 17, 51]. The larger the stability margin, the more stable is the power system against disturbances.

For a power system, the non-linear stability depends on the severity of the disturbance. Renewable energy such as wind power and solar power is often strongly affected by the weather and consequently causes power fluctuations and frequency fluctuations of a large-scale power system. These continuous fluctuations of the frequency may further lead the system to lose the synchronization. In order to further increase the integration of the renewable energy, the problem of *increasing the synchronization stability to avoid losing frequency synchronization caused by various disturbances* is receiving more and more attention. It is obvious that the decrease of the strength of the disturbance can effectively increase the stability. The strategies on how to suppress these disturbances is not the focus of the chapter. We pay attention to the possible strategies for improving the stability by changing the power system itself, which consists of synchronous power generators, power transmission lines and loads.

Control of power systems can enhance the system stability. The control objectives for control of a power system include to deliver electric power to customers of the network operator, to maintain the stability of the power system, preferably in the domain of attraction of the current steady state, and to minimize the cost of the operation of the power system. In practice, the power transmission in the first control objective depends on the location of the power generation and loads, the network topology and the transmission line capacity. The latter two control objectives are separated for frequency control into primary, secondary and tertiary frequency control, respectively, see [22, 53]. The primary control which also called droop control keeps the synchronization of the frequency at a value which may deviate from the nominal value. The secondary control the synchronized frequency to the nominal frequency and the tertiary control determine the set point stabilized by the primary and secondary control. The secondary control and the tertiary control jointly determine the set point of the power system. In addition, the secondary control affects the dynamics.

The control objective of maintaining the state of the system within a domain of attraction of a steady state motivates research to explore ways to characterize the stability region, the boundary of the domain of attraction and the stability margin [11, 58]. For a power system with control, the stability depends on factors such as

(i) the topology of the network, which can be changed by adding new lines and nodes to the network or configuring the capacity of the lines,
(ii) the inertia of the synchronous machines, which may be changed by placing or removing virtual inertia to the nodes in the network,
(iii) the damping coefficients of the synchronous machine, which includes the droop control gain parameter that can be configured in droop control [22],
(iv) and power generation and load, which can be controlled by changing the mechanical power generation or switching on or off the power consumption.

To accomplish the control objective of keeping the system stability, these four factors can be changed based on characteristics of the stability region. The first step for the stability improvement is to find a metric for the stability, which can point to those factors that are best changed. The need for improving the stability of a power systems motivates research into the mathematics of stability analysis of power systems.

In this chapter, we focus on the improvement of the stability of power systems. We give a survey on the recent development of the stability analysis and summarize the potential stability metrics for the stability and corresponding strategies for the stability improvement. The chapter is organized as follows. Section 14.2 introduces the model of the power systems. Section 14.3 discusses the necessary condition for the existence of synchronization state. The linear stability and non-linear stability of the synchronization state are described in Sects. 14.4 and 14.5 respectively. We conclude the chapter in Sect. 14.6.

14.2 The Model of Power Systems

There are three main components in power systems, namely power generators, transmission network, and loads. We consider the power system described by a graph $\mathcal{G} = (\mathcal{V}, \mathcal{E})$ with nodes $\mathcal{V}$ and edges $\mathcal{E} \subseteq \mathcal{V} \times \mathcal{V}$ where a node represents a bus and edge (i, j) represents the direct transmission line connection between node i and node j. Each bus is locally connected to either energy sources, or energy loads, or to both. We denote the number of nodes in the network by n.

We focus on the power system with lossless transmission lines, of which the dynamics can be equivalently described by the following swing equations [4, 9, 31],

$$\dot{\delta}_i = \omega_i, \ i \in \mathcal{V}, \tag{14.1a}$$

$$M_i \dot{\omega}_i = P_i - D_i \omega_i - \sum_{j \in \mathcal{V}} B_{ij} \sin(\delta_i - \delta_j), \ i \in \mathcal{V}, \tag{14.1b}$$

where δ_i is the phase angle at node i, ω_i is the frequency deviation from the nominal frequency, e.g., 50 or 60 Hz, $M_i > 0$ is the moment inertia of the machine, P_i is the power supplied by synchronous machines or by renewable energy sources if $P_i > 0$, and is power load if $P_i < 0$, $D_i > 0$ is the damping coefficient including droop control gain parameter, $B_{ij} = \hat{B}_{ij} V_i V_j$ which can be viewed as the weight of the edges in the graph $\mathcal{G}$. Since the control of the voltage and of the frequency can be decoupled when the transmission lines are lossless [45], we do not model the dynamics of the voltages and assume the voltage of each bus is a constant which can be derived from power flow calculation [33, 42].

Throughout the discussion of this chapter, we assume the network is undirected and connected. The Laplacian matrix of the network is in the form

$$L_n = \begin{pmatrix} \sum_{j=1}^n B_{1j} & -B_{12} & \cdots & -B_{1n} \\ -B_{21} & \sum_{j=1}^n B_{2j} & \cdots & -B_{2n} \\ \vdots & \vdots & \ddots & \vdots \\ -B_{n1} & -B_{n2} & \cdots & \sum_{j=1}^n B_{nj} \end{pmatrix} \tag{14.2}$$

Because the line weight $B_i j$ for all $(i, j) \in \mathcal{E}$ are positive, L_n is non-negative definite. The eigenvalues of $L_n \in \mathbb{R}^{n \times n}$ are denoted by $0 < \sigma_2 \leq \cdots \leq \sigma_n$. Herein, the second smallest eigenvalue σ_2 measures the connectivity of the network [50].

It has been demonstrated in [43, 44] that frequency droop controlled Micro-Grids which have some sort of energy storage and lossless transmission lines can also be modeled by second-order swing equations (14.1). Some other models are also applied for the synchronization stability analysis of power networks, see [36] for details of the comparison of these models.

By selecting one node as infinite bus with constant phase and the other one as a synchronous generator in a two-node network, the following *Single Machine Infinite Bus* (SMIB) model can be obtained,

$$\dot{\delta} = \omega, \tag{14.3a}$$

$$M\dot{\omega} = P - D\omega - B\sin\delta, \tag{14.3b}$$

which can be directly derived from (14.1). Here P and B are the transmitted power and the line capacity respectively, the voltages are also assumed constant, δ is the angle difference between the synchronous machine and the infinite bus, which should be kept in a small range in order to stay in the synchronization state. This requires $\omega = 0$ and $P = B\sin\delta$ at the synchronized state. The diagram of this SMIB model is shown in Fig. 14.1.

The modeling of fluctuations affecting a power system are discussed next, because such fluctuations strongly motivate current research in stability analysis and control of power systems.

In practice, continuous fluctuations act 24 h a day, though their intensity varies during the day and depend on the weather and on human behavior. The energy loads fluctuate all the time due to consumers switching on electricity devices. The power generation of the synchronous generators are operated to balance these fluctuations and keep the stability of the power system.

Weather dependent power sources as wind turbines, wind parks, and the sun via photo-voltaic panels, generate power with large fluctuations. The wind power produced varies with the intensities of the wind force, the solar power produced varies with the sun intensities received on earth and with the cloud covers between the sun and the PV panel. The strength of these fluctuations are much stronger than those in the traditional power system, which bring great challenges to the operation of the power systems. Modeling of the fluctuations and procedures of system identification may help to obtain realistic models for control of power systems to suppress the fluctuations. Effective suppression of the fluctuations are important for the synchronous stability of the power systems.

Fig. 14.1 The SMIB model

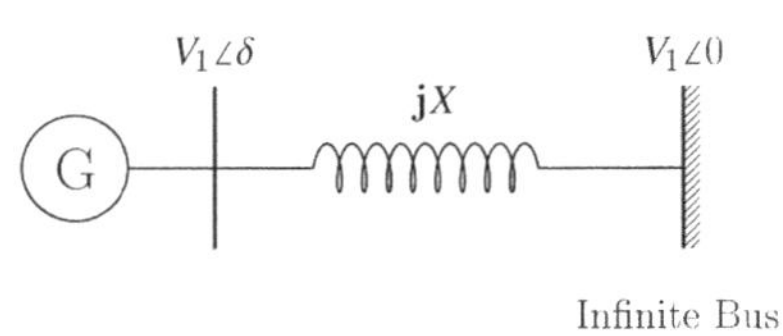

Abrupt changes in the transmission network may occur, which could cause serious blackouts. These events have happened more frequently in the past decade than before. Examples of such abrupt changes are the breakage of power lines, for example due to freezing rain on the lines, break down of part of a power plant with synchronous machines, or instability of the power network due to a network node experiencing a relatively large power disturbance. For such abrupt changes there are special control procedures, like islanding of the power network, control of each power network island, and later return to the normal state by joining the power network islands. These procedures are not further discussed in this chapter. The focus of this chapter is on analysis of power system stability and on control for the improvement of power system stability.

14.3 The Synchronous State

The stability of a nonlinear system usually refers to the ability of the system to stay in the basin of attraction of a synchronous state. In this section, the focus is on the existence of the synchronous state of the power system.

The equilibrium point of the system (14.1) is referred to as the *synchronous state* defined as follows.

Definition 14.1 Assuming that power generation and loads are constant, the synchronous state of the system satisfies for all $i \in \mathcal{V}$

$$\omega_i = \omega_{syn}, \tag{14.4a}$$

$$\dot{\omega}_i = 0, \tag{14.4b}$$

$$\delta_i = \omega_{syn} t + \delta_i^*, \tag{14.4c}$$

$$\dot{\delta}_i = \omega_{syn}, \tag{14.4d}$$

where $\omega_{syn} \in \mathbb{R}$ is the synchronized frequency deviation, δ_i^* is the phase angle of node i at the steady state. In the synchronous state, all the phase distances $|\delta_i - \delta_j| = |\delta_i^* - \delta_j^*|$ are constant.

The terminology *phase locking* and *phase cohesiveness* are also used to describe this synchronization state of frequency [15]. In particular, the phase locking state with $|\delta_i^* - \delta_j^*| = 0$ for $i, j = 1, \ldots, n$, is called *phase synchronized* state. In practice, the synchronous state does not exist for the power system due to the continuously fluctuating power loads. However, it is practical to assume the power loads are constant on small time-scales which lead to a synchronous state.

By summing all the equations for $i = 1, \ldots, n$, the explicit formula of the synchronized frequency ω_{syn} can be obtained as follows

$$\omega_{syn} = \frac{\sum_{i=1}^n P_i}{\sum_{i=1}^n D_i}, \tag{14.5}$$

where $\sum_{i=1}^{n} P_i$ is referred to as the *power imbalance*. It can be obtained that if the power imbalance is zero, the synchronized frequency deviation is zero. The restoration of the frequency deviation to zero is the task of secondary frequency control, see [16, 20, 53, 54, 56].

14.3.1 Existence of the Synchronous State

For the SMIB model, the equilibrium point satisfies

$$\omega = 0, \ \ \sin \delta^* = \frac{P}{B}. \tag{14.6}$$

at which the phase angle difference between the machine and infinite bus is δ^*. It is obvious that if $B < P$, this equilibrium point does not exist and the system converges to a *non-synchronous limit cycle* which can be characterized by

$$\omega_{ns} \approx \frac{P}{D} + \frac{DB}{P} \cos\left(\frac{P}{D}t\right) \tag{14.7}$$

when $|P|/D^2 \gg 1$ and $P^2/D^2 \gg B$, see [31]. For this SMIB model, the *critical line capacity* is $K_c = P$, which is the power that has to be transmitted to the load. This critical line capacity is also called *critical coupling* [14], which is defined as the smallest line capacity for the existence of an equilibrium point. If $B > K_c$, it is obvious that in a period of sin function there are two equilibrium points which satisfy (14.6), However, it is far more complex to obtain an explicit formula of the critical capacity for the power system (14.1) than for the SMIB model. It is obvious that the existence of the synchronous state depends on the power injection (load), the topology of the network and the line capacities. Hence, the critical line capacity depends on the power injection (load) and the network topology.

Due to the importance of the synchronization in complex network, the Kuramoto model is widely studied for the condition of the synchronization. The *first-order non-uniform Kuramoto model* is as follows

$$\dot{\delta}_i = \omega_i - K \sum_{j=1}^{n} a_{ij} \sin(\delta_i - \delta_j), \ i = 1, \dots, n. \tag{14.8}$$

where $a_{ij} = 1$ if node i and j is connected, otherwise $a_{ij} = 0$. Note that the model (14.1) is also referred to as *non-uniform second-order Kuramoto model*. The corresponding Laplacian matrix is denoted by $\boldsymbol{L}_a$. Here ω_i has a different meaning from the one in the power system (14.1), which denotes a force to the oscillator i. In literature, the critical line capacity K_c has been widely applied to the study of the impact of the parameter of the system on the existence of the synchronous state. If the critical coupling strength $K > K_c$, it satisfies $\dot{\theta}_i = \omega_s$ for all the nodes in the network where $\omega_s = \sum_i^n \omega_i/n$, which can be obtained by summing all the equation in (14.8). For completed network with $a_{ij} = 1$ for $i, j = 1, \ldots, n$, the upper bound and lower bound of K_c can be obtained explicitly [15]. For a general network, the lower bound of K_c can be obtained from the necessary condition or the sufficient condition for the existence of the synchronous state of (14.8), see [15]. The critical coupling strength depends on the distribution of ω_i and the network topology, the synchronization can improved via decreasing K_c by changing the topology and the distribution of the frequency ω_i. In order to connect these conditions to the power systems, Dörfler and Bullo [14] have proven the equivalence of the synchronization of the power network (14.1) and the Kuramoto network (14.8). With this equivalence, the existence condition of the synchronous state for the power system (14.1) can be deduced from those of the Kuramoto model (14.8).

Remark 14.1 With the lower bound of K_c as a stability metric, an optimization framework can be formed with the controllable factors, i.e., the network topology and the power generation and loads, as decision variables. Because the inertia and the damping coefficient have no influence on the synchronous state when $\omega_{syn} = 0$, this metric cannot be applied for the improvement of the stability by controlling the virtual inertia and damping coefficients.

The stability of the synchronous state can be determined by the Lyapunov method, which will be further described in Sect. 14.4. In principle there may be more than one stable synchronous state [13, 27, 29, 38, 40, 55] due to cycles in the network.

For a cyclic power network with alternating nodes of loads and generators as shown in Fig. 14.2. There are even number of nodes in the network and the power injection $P_i = -2P$ for even nodes and $P_i = 2P$ for odd nodes. This alternating distribution of power leads to $\sum_{i=1}^n P_i = 0$. The model of this network can be deduced from (14.1) as

$$\dot{\delta}_i = \omega_i, \tag{14.9a}$$

$$\dot{\omega}_i = P_i - D\omega_i - B[\sin(\delta_i - \delta_{i+1}) + \sin(\delta_i - \delta_{i-1})]. \tag{14.9b}$$

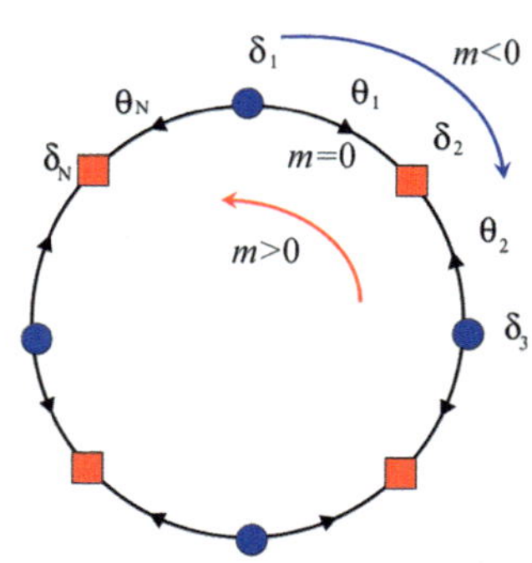

Fig. 14.2 A cyclic network with alternating consumer and generator nodes. Circle nodes are generators and square nodes are consumers. There may be stable equilibria with the power transported around the cycle clockwise with $m < 0$ and counterclockwise with $m > 0$

Denote the phase differences between neighbors by $\theta_1 = \delta_1 - \delta_n \pmod{2\pi}$ and $\theta_{i+1} = \delta_{i+1} - \delta_i \pmod{2\pi}$. The equilibria of this ring network are given by $\theta_i = \theta_1$ for odd i, and $\theta_i = \theta_2$ for even i, where

$$\theta_1 = \arcsin\left[\frac{P}{B\cos\frac{2m\pi}{n}}\right] + \frac{2\pi m}{n} \tag{14.10a}$$

$$\theta_2 = -\arcsin\left[\frac{P}{B\cos\frac{2m\pi}{n}}\right] + \frac{2\pi m}{n}, \tag{14.10b}$$

and m is an integer such that

$$|m| \leq \lfloor \frac{n}{2\pi} \arccos\left(\sqrt{\frac{P}{B}}\right) \rfloor.$$

The total number of stable equilibria is given by

$$N_s = 1 + 2\lfloor \frac{n}{2\pi} \arccos\left(\sqrt{\frac{P}{B}}\right) \rfloor. \tag{14.11}$$

where $\lfloor x \rfloor$ denotes the floor value of x, that is, the largest integer value which is smaller than or equal to x. When $P = 0$, N_s reaches the upper bound derived in [13].

Remark 14.2 It can be seen from formula (14.11) that the number of the stable synchronous state increase linearly as the size n of the cycle increases. For the synchronous state with $m \neq 0$, power loop occurs in the cycle. For practical purposes the case $m = 0$ is desirable for transport of electricity, as in this case direct transport of power from the generator to the consumer is realized. Direct transport from generator to consumer minimizes energy losses that always accompany the transport of electrical power. Possible ways to avoid the clockwise-counterclockwise power loops is to control the power generation or loads, such that the phase angles are in the security range (14.19) which will be described in Sect. 14.4.

Besides these stable synchronous states, there may be more than 2^n synchronous states for the power network, which depends on the distribution of the power generation and loads and the topology, see [3, 8, 26, 30] for details. Because the unstable equilibrium points are on the potential energy boundary, it is important to find these equilibrium points for analyzing the nonlinear stability, detail will be further described in Sect. 14.5.

14.3.2 Braess' Paradox in Power Grids

A surprising finding in the synchronization of power grids is that adding more connections does not always improve the synchronization in the grid, but could also destroy an existing stable synchronized state.

A similar phenomenon was reported in the 1968 by Braess [5] in the context of traffic flow. It turned out that adding a new road to an existing traffic plan may sometimes lead to increased congestion of the traffic flow, in contrast to the expectation.

To illustrate how adding a new connection to a power grid can destroy the synchronization, we consider a configuration of a network consisting of 2 clusters of 4 nodes, which are coupled at the top and bottom nodes of each cluster; see Fig. 14.3. The same configuration was also considered in [52]. We assume each line to have the same capacity $B_{ij} = K$ for all $(i, j) \in \mathcal{E}$. The flow between nodes i and j is given by $F_{ij} = K\sin(\delta_i - \delta_j)$. For clarity, the consumer nodes with consumption P are depicted green and the generation nodes, with generation P, are blue. A straightforward study of this simple network, shows that an equilibrium configuration can be obtained when all blue nodes except the upper left one have phase $-\pi/2$ and all green nodes except the bottom right one have phase $\pi/2$. The upper left on lower right nodes both have phase 0. By taking the capacity of each line K to be equal to P, this configuration allows each line between a generator and a consumer to carry P units of power. Additionally, power P is transferred from the top right to the top left node as wel as from the bottom right to the bottom left node. We remark that this configuration is critical in the sense that a small increase in power generation or demand cannot be accommodated by the network.

When a connection between the upper left and lower right nodes is added, an overflow occurs, that is, the power flowing from the upper left node to the consumer nodes below is larger than the critical capacity K_c and therefore synchronization is lost. It has recently been shown that Braess' paradox can be prevented by using secondary control [48]. It turns out that all nodes need to be controlled, that is, both the generator and the consumer nodes, in order to prevent the Braess' paradox from happening. This demonstrates that the network topology is extremely important in order to guarantee reliable operation of the power grid, and that not only generator

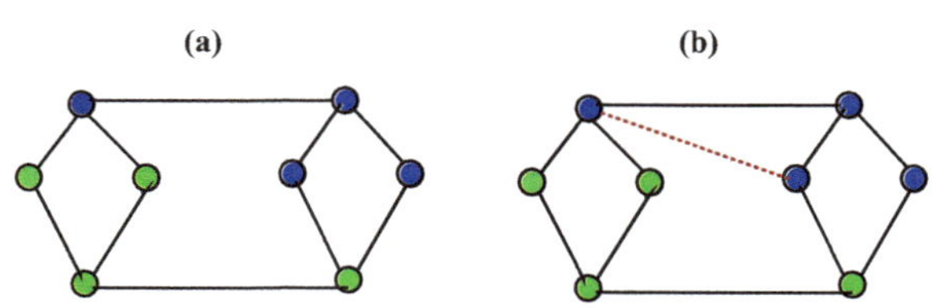

Fig. 14.3 A schematic of a simple network operating at critical capacity $K = K_c = P$. The system synchronizes in **a**, but adding a new connection shown in red between two generators induces an overload and destroys the synchronization of the system

nodes, but also consumer nodes should receive sufficient attention when introducing additional renewable power generators and consumers in the network.

Besides the work done on Braess' paradox in the group of Marc Timme [48, 52], a linear stability analysis has been carried out by Coletta and Jacquod [12] for simple linear chain networks. The work corroborates the results of Refs. [48, 52], by proving that for one-dimensional chain networks power can flow from consumer to generator and thereby surmounting the line capacity. This effect is equivalent to the Braess' paradox in a two-dimensional situation. Moreover, it was shown numerically that Braess' paradox can actually occur in real power grids, such as the UK power grid and the European Grid.

So far these calculations have all been carried out for purely capacitive networks, that is, without dissipative losses. As distributed generation of power is surging, dissipative effects will probably be more prevalent, which requires a more elaborate analysis including dissipative effects.

Remark 14.3 In the investigation of the Braess' Paradox in power grids, the existence condition of the synchronous state plays an important role such as in the finding and curing of it [48, 52]. Beside the critical line capacity K_c, the linear stability which will be discussed in Sect. 14.4 is also used as a stability metric in the study of the impact of the network topology[12, 52]. Possible ways to avoid the Braess' Paradox is to avoid the decrease of these metrics when adding new lines to the network. Control of power generation and loads is another way to curing this paradox.

14.4 Stability of the Linearized System

A non-linear system is linearly stable at an equilibrium state if the linearized system at that equilibrium state, determined by the Jacobian, is exponentially stable. The linear stability considers the local convergence speed at the neighborhood of the stable equilibrium point. This linear stability can be qualified by the real part of the eigenvalues. In this section, we introduce the linearization of the system and the dependence of the eigenvalues on the parameters of the power system. How to increase the linear stability by changing the parameters of the system will also be described.

Assume that there exists a synchronous state for the power system, which is denoted by $(\boldsymbol{\delta}^*, \mathbf{0})$. After linearization at the synchronous state, we derive

$$\dot{\boldsymbol{\delta}} = \boldsymbol{\omega}, \tag{14.12a}$$

$$\boldsymbol{M}\dot{\boldsymbol{\omega}} = -\boldsymbol{L}_c\boldsymbol{\delta} - \boldsymbol{D}\boldsymbol{\omega}. \tag{14.12b}$$

where $\boldsymbol{\delta} = \text{col}(\delta_i) \in \mathbb{R}^n$, $\boldsymbol{\omega} = \text{col}(\omega_i) \in \mathbb{R}^n$, $M = \text{diag}(M_i) \in \mathbb{R}^{n\times n}$, $\boldsymbol{L}_c = \big(B_{ij}\cos(\delta_i - \delta_j)\big) \in \mathbb{R}^{n\times n}$, $\boldsymbol{D} = \text{diag}(D_i) \in \mathbb{R}^{n\times n}$. Here, col$(\cdot)$ denotes a column vector and diag$(\cdot)$ denotes a diagonal matrix.

In a power system, the *small signal stability* is the ability of the power system to maintain the synchronization when subjected to small disturbances. The behavior of the power system is best such that, after a small disturbance acting on the power system, the state of the system returns to the synchronous state. Preferably this return should be quickly. The small signal stability analysis is based on the linearization to provide valuable information about the characteristics of the system and help configure the corresponding parameters.

After linearizing the SMIB model (14.3) at an equilibrium point, we obtain

$$\dot{\delta} = \omega, \tag{14.13a}$$

$$M\dot{\omega} = -D\omega - \overline{B}\delta, \tag{14.13b}$$

where $\overline{B} = B\cos\delta_i^*$. The eigenvalues of the linear system can be calculated as

$$\lambda = \frac{-D \pm \sqrt{D^2 - 4\overline{B}}}{2} \tag{14.14}$$

from which it can be obtained that when $\overline{B} = B\cos\delta_0^* > 0$, all the eigenvalues have negative real part. Thus the system is stable at the equilibrium δ_0 according to the second Lyapunov method for determining the stability of a nonlinear system. However, with $\overline{B} = B\cos\delta_1^* < 0$, there is one eigenvalue which has positive real part. This means that the system is unstable at the equilibrium point δ_1^*. Hence, a security condition can be obtained for the stability of the equilibrium point as $\cos\delta^* > 0$, which can be further expressed as $-\frac{\pi}{2} < \delta^* < \frac{\pi}{2}$. For the SMIB model, it is obvious that there is only one stable equilibrium point in the security range of phase angle.

The linear stability analysis of the system (14.1) is much more complex than the SMIB mode because of the high dimension. The system (14.12) has $2n$ equations, and there are $2n$ eigenvalues which depend on $\boldsymbol{M}, \boldsymbol{L}_c, \boldsymbol{D}$. In practice, both $\boldsymbol{M}$ and $\boldsymbol{D}$ are positive definite for power systems. $\boldsymbol{L}_c$ involves the topology of the network and the line capacities. It has been proven in [57] that with positive definite $\boldsymbol{M}$ and $\boldsymbol{D}$ the sign of the real part of the eigenvalues depends on the eigenvalues of $\boldsymbol{L}_c$. This is explained as follows.

The system (14.12) can be written in the compact form

$$\begin{pmatrix} \dot{\boldsymbol{\delta}} \\ \dot{\boldsymbol{\omega}} \end{pmatrix} = \begin{pmatrix} \boldsymbol{0} & \boldsymbol{I} \\ -\boldsymbol{L}_m & \boldsymbol{\beta} \end{pmatrix} \begin{pmatrix} \boldsymbol{\delta} \\ \boldsymbol{\omega} \end{pmatrix} \tag{14.15}$$

where $\boldsymbol{L}_m = \boldsymbol{M}^{-1}\boldsymbol{L}_c, \boldsymbol{\beta} = \text{diag}(D_i/M_i) \in \mathbb{R}^n$. We assume that all the components of $\boldsymbol{\beta}$ are identical, i.e., $\beta_i = \beta$. Let $\boldsymbol{Q} \in \mathbb{R}^{n\times n}$ be the matrix formed by the eigenvectors of $\boldsymbol{L}_m$ such that

$$\boldsymbol{Q}^{-1}\boldsymbol{L}_m\boldsymbol{Q} = \boldsymbol{\Lambda}$$

where $\boldsymbol{\Lambda}$ is a diagonal matrix with the diagonal component being the eigenvalues $0 = \lambda_1 < \lambda_2 \le \lambda_2 \cdots \le \lambda_n$ of $\boldsymbol{L}_m$ as its columns. Here all the eigenvalues of $\boldsymbol{L}_m$ are real even though $\boldsymbol{L}_m$ is not symmetric [34]. Let $\boldsymbol{X}_1 = Q^{-1}\boldsymbol{\delta}$ and $\boldsymbol{X}_2 = Q^{-1}\boldsymbol{\omega}$. These formulas transform (14.15) to

$$\begin{pmatrix} \dot{\boldsymbol{X}}_1 \\ \dot{\boldsymbol{X}}_2 \end{pmatrix} = \begin{pmatrix} \boldsymbol{0} & \boldsymbol{I} \\ -\boldsymbol{\Lambda} & \boldsymbol{\beta} \end{pmatrix} \begin{pmatrix} \boldsymbol{X}_1 \\ \boldsymbol{X}_2 \end{pmatrix} \tag{14.16}$$

which consists of n decoupled sub-systems as follows

$$\begin{pmatrix} \dot{X}_{1i} \\ \dot{X}_{2i} \end{pmatrix} = \begin{pmatrix} 0 & 1 \\ -\lambda_i & \beta \end{pmatrix} \begin{pmatrix} X_{1i} \\ X_{2i} \end{pmatrix}, \ i = 1, \ldots, n. \tag{14.17}$$

Because the eigenvalue $\lambda_1 = 0$, we do not consider the subsystem $i = 1$ which in fact does not influence the synchronization due to the phase rotation. For the subsystems $i = 2, \cdots, n$, the eigenvalues of (14.12) can be calculated as follows

$$\alpha_{i\pm} = -\frac{\beta}{2} \pm \frac{1}{2}\sqrt{\beta^2 - 4\lambda_i}, \tag{14.18}$$

which has a similar form as the eigenvalues of the SMIB model in (14.14). It can be easily observed that with $\beta > 0$, the sign of the real part of α_i is determined by λ_i, and the synchronous state is Lyapunov stable if and only if λ_i is positive for $i = 2, \ldots, n$. Because $\boldsymbol{M}$ is a diagonal positive definite matrix which does not impact the non-negative definite of $\boldsymbol{L}_m$, the number of eigenvalues of L_m with a positive real part equals the number of eigenvalues of L_c with a positive real part. Thus, the synchronous state $(\boldsymbol{\delta}^*, \boldsymbol{0})$ is Lyapunov stable if and only if $\boldsymbol{L}_c$ is non-negative definite.

By Lyapunov stability theory, a synchronous state is unstable if there is an eigenvalue λ_i with strictly positive real part for the linearized system at this state. The unstable synchronous state is called of *type* j if the number of eigenvalues λ_i with strictly positive real-part is j. In other words, the dimension of the unstable manifold of the type j equilibrium point is j. It can be observed from (14.18) that if $\lambda_i < 0$, α_{i+} has a positive real part, then it will lead to an unstable manifold and that the number of the eigenvalues of the power network with positive real part equals that of the negative eigenvalues of $\boldsymbol{L}_m$. Because the number of the eigenvalues of $\boldsymbol{L}_m$ with positive real part equals to that of $\boldsymbol{L}_c$, the synchronous state $(\boldsymbol{\delta}^*, \boldsymbol{0})$ is of type j if $\boldsymbol{L}_c$ has j negative eigenvalues. This statement can be applied to the determination of type j equilibrium point of the power system with special topology, such as acyclic network and cyclic network, which will be further discussed in Sect. 14.5.

From the above discussions it is clear that the eigenvalues of the Laplacian matrix $\boldsymbol{L}_c$ play an important role in the stability analysis of power systems. It is well known that if the weights $B_{ij}\cos(\delta_i^* - \delta_j^*)$ for $(i, j) \in \mathcal{E}$ are positive, all the non-zero eigenvalues of $\boldsymbol{L}_c$ are positive. For the unstable synchronous state, there exist lines with negative weight. The characteristic of the eigenvalues of the Laplacian matrix of weighted network with *negative weight* has been investigated in [6], in which more details on the determination of the number of negative eigenvalues can be found.

Statement 14.1 For general configuration of $M_i > 0$ and $D_i > 0$ in the power system (14.1), it also holds that the synchronous state is stable if and only if all the non-zero eigenvalues of $\boldsymbol{L}_c$ are positive. Since $\boldsymbol{L}_c$ is the Laplacian matrix of the network with weight $B_{ij}\cos(\delta_i^* - \delta_j^*)$ for all the edge $(i, j) \in \mathcal{E}$, it can be derived that if all the weights are positive, $\boldsymbol{L}_c$ is non-negative definite. With $B_{ij}\cos(\delta_i^* - \delta_j^*) > 0$, the security condition for stability can be obtained

$$|\delta_i^* - \delta_j^*| < \frac{\pi}{2}, \ \forall (i, j) \in \mathcal{E}, \tag{14.19}$$

which is a well-known sufficient condition for the Lapunov stability of the synchronous state. □

For details of the proof of the above statement, we refer to [46, 57].

Statement 14.2 The synchronous state in this security range is unique and stable for the lossless power network. However, this is not true for lossy power networks, see [46] for details. □

The linear stability of the system is qualified by the absolute value of $\mathrm{Re}(\alpha_{i+})$ for $i = 2, \ldots, n$. Figure 14.4 illustrates how $\mathrm{Re}(\alpha_{i+})$ depends on β and α_i. For each subsystem described by (14.17), there is a minimum for $\mathrm{Re}(\alpha_{i+})$ with respect to β. This minimum value $\mathrm{Re}(\alpha_{i+}) = -\sqrt{\lambda_i}$ is obtained when setting $\beta = 2\sqrt{\lambda_i}$ in (14.18). If λ_i increases then the minimal value decreases, hence $\mathrm{Re}(\alpha_{i+})$ is limited by the second smallest eigenvalue λ_2 of $\boldsymbol{L}_m$. Thus, the optimal configuration of β can be obtained as

$$\beta_{\mathrm{opt}} = 2\sqrt{\lambda_2}. \tag{14.20}$$

From $\lambda_i > \lambda_2$ for $i = 3, \ldots, n$, it can be derived that if $\beta = \beta_{\mathrm{opt}}$, the real-part of all the eigenvalues of (14.12) are all identical, i.e.,

$$\mathrm{Re}(\alpha_{i+}) = \frac{\beta_{\mathrm{opt}}}{2}, \ i = 2, \ldots, n.$$

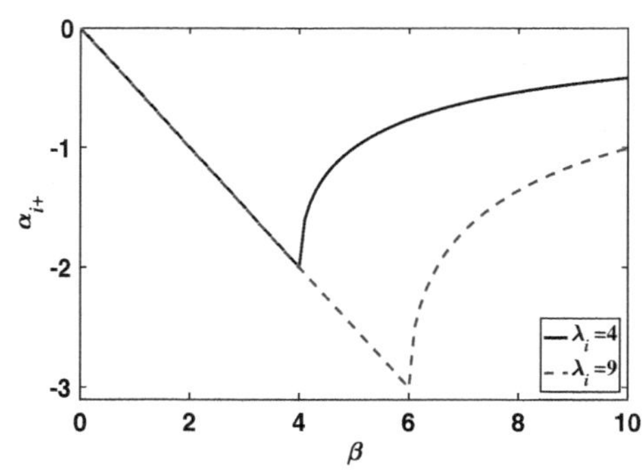

Fig. 14.4 The real part of α_{i+} with respect to β and λ_i

If β is smaller than β_{opt}, the real part $\text{Re}(\alpha_i+)$ can be increased by increasing the damping coefficient as $D_i = \beta M_i$ due to the independence of $\boldsymbol{L}_m$ on D_i, and the optimal configuration is $D_i = 2\sqrt{\lambda_2}/M_i$.

Statement 14.3 Since $\boldsymbol{L}_m = \boldsymbol{M}^{-1}\boldsymbol{L}_c$, then the eigenvalue λ_2 of $\boldsymbol{L}_m$ increases if the eigenvalues of $\boldsymbol{L}_c$ increase. Due to the fact that $\boldsymbol{L}_c$ is determined by the synchronous state, the topology and the capacity of transmission lines, the linear stability of the power system can be improved by controlling the power injection, well-designed topology and replacement of the transmission lines with low capacity by those with high capacity. Once these parameters are determined, the damping coefficient D_i can be determined by $D_i = 2\sqrt{\lambda_2}M_i$. □

Algorithms for maxmizing the second smallest eigenvalues by determining the state have been investigated in [21] which can be referred to for details.

Statement 14.4 The optimal configuration $\beta = \beta_{\text{opt}}$ is formulated with the assumption that all the components of $\boldsymbol{\beta}$ are identical. It has been shown that for non-identical β_i this setting is optimal along any given direction in the β_i-space for many power systems [34, 35]. Hence, this configuration and increasing the second smallest eigenvalue is appropriate for enhancing the stability. For details of the analysis, we refer to [34, 35]. □

The impact of the Braess' paradox on the linear stability has been investigated in [12], in which it is shown that adding a line to the network may decrease the linear stability of a power network. The linear stability of cyclic power network has been studied in [55]. An analytic formula of the eigenvalues of the linearized system is obtained, which demonstrated that the linear stability decreases as the size of the network increases. Simulations with various networks showed that the linear stability decreases as the heterogeneity of the power injection increases. In other words, the linear stability can be increased by reducing the heterogeneity of the power injection (loads).

14.5 The Nonlinear Stability

In this section, we introduce the synchronization stability of power systems. The stability region of power systems has been analyzed by Chiang et al. [11] and Zaborszky et al. [58]. However, because of the large-scale and complexity of the power network, the basin of attraction, related to transient stability, has a high computational complexity for numerical approximation. In this section, we introduce the *energy barrier* which is a conservative estimate of the stability margin of the power system.

Inspired by the *direct method* to estimate whether the power system is stable after a disturbance [10, 23, 24, 51], we explain how we can use the energy barrier method to determine the transient stability in the case of the SMIB model.

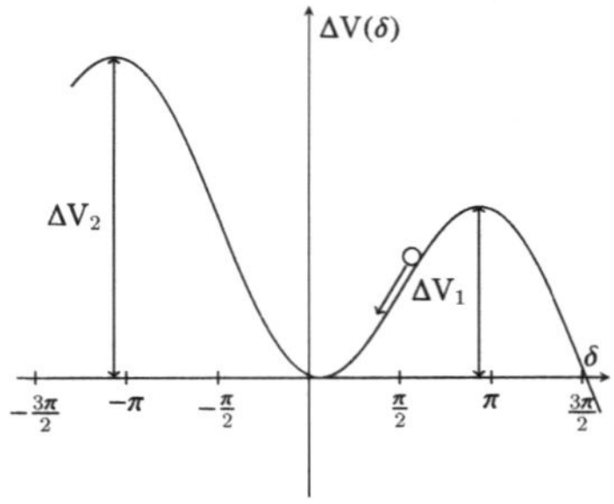

Fig. 14.5 The potential energy landscape of the SMIB system

The potential energy of this system is

$$\mathrm{V}(\delta) = -B\cos\delta - P\delta.$$

Figure 14.5b plots the potential energy difference $\mathrm{V}(\delta) - \mathrm{V}(\delta_0)$ where $\delta_0 = \arcsin P/B$ is the phase angle difference at the steady state. In the figure, the position and the speed of the ball displayed are δ and ω respectively. The potential energy possesses three extreme points in the range $(-3\pi/2, 3\pi/2)$, which include two unstable equilibria and one stable equilibrium. It can be observed that the trajectory will converge to the minimum of $\mathrm{V}(\delta)$ if its kinetic energy is smaller than the potential energy $\Delta\mathrm{V}_1$ and $\Delta\mathrm{V}_2$. If obtaining enough energy from a disturbance to overcome the potential energy, the trajectory will escape from the valley and thus the system desynchronizes. Hence, the energy barrier $\Delta\mathrm{V}_1$ and $\Delta\mathrm{V}_2$ which are the potential energy differences between the two unstable equilibria and the stable equilibrium, can be used to measure the synchronization stability, which have the following formula [10]

$$\Delta\mathrm{V}_1 = P(-\pi + 2\arcsin\frac{P}{B}) + 2\sqrt{B^2 - P^2}, \tag{14.21a}$$

$$\Delta\mathrm{V}_2 = P(\pi + 2\arcsin\frac{P}{B}) + 2\sqrt{B^2 - P^2}. \tag{14.21b}$$

From the above equations it is immediately clear that $\Delta\mathrm{V}_1$ decreases while $\Delta\mathrm{V}_2$ increases as the transmitted power P increases. As shown in Fig. 14.5b, it is much easier for the trajectory to overcome $\Delta\mathrm{V}_1$ to escape from the valley than $\Delta\mathrm{V}_2$. So $\Delta\mathrm{V}_1$ provides a conservative approximation of the basin of attraction and can be used to measure the transient stability.

For the power network (14.1), the calculation of the energy barrier is far more complex. The potential energy $V(\boldsymbol{\delta})$ is defined as

$$V(\boldsymbol{\delta}) = -B\sum_{(i,j)\in\mathcal{E}}\cos(\delta_i - \delta_j) - \sum_{i=1}^{N} P_i\delta_i. \tag{14.22}$$

The primary idea behind estimating the region of attraction of a stable equilibrium by the direct method, is that this region is bounded by a manifold $\mathcal{M}$ of the type-1 equilibria that reside on the potential energy boundary surface (PEBS) of the stable equilibrium. The PEBS can be viewed as the stability boundary of the associated gradient system [10, 51]

$$\frac{d\delta_i}{dt} = -\frac{\partial V(\boldsymbol{\delta})}{\partial \delta_i}. \tag{14.23}$$

The *closest equilibrium* is defined as the one with the lowest potential energy on the PEBS. By calculating the closest equilibrium with potential energy $V_{\min}$ and equating this to the total energy, it is guaranteed that points within the region bounded by the manifold $\mathcal{M} = \{(\delta, \omega)|E(\delta, \omega) = V_{\min}\}$, will always converge to the stable equilibrium point contained in $\mathcal{M}$. Various algorithms for the calculation of the closest equilibrium points are proposed, see [23, 24].

The idea of estimating the region of stability by type-1 equilibria is probably best illustrated by considering a simple example of a three-node network depicted in Fig. 14.6a. The 6 unstable equilibria are local minima on the potential energy boundary surface (PEBS) plotted by the black dash-dotted line. These minima are all type-1. The equilibrium 1 and 4, 2 and 5, 3 and 6 are caused by θ_1, θ_2 and θ_3 exceeding $\pi/2$ respectively. Because equilibrium point 1 has the smallest energy, it is the closest equilibrium point on the PEBS.

A small perturbation in the direction to saddle point 1, depicted by the red dashed curve leads to desynchronization, whereas a larger perturbation in a different direction (blue solid curve) eventually decays toward the stable equilibrium point and hence the system stays synchronized. This shows the conservativity of the direct method and the challenges in calculating the region of stability, as it depends on both the direction and size of the perturbation. One approach to this problem is to determine the so-

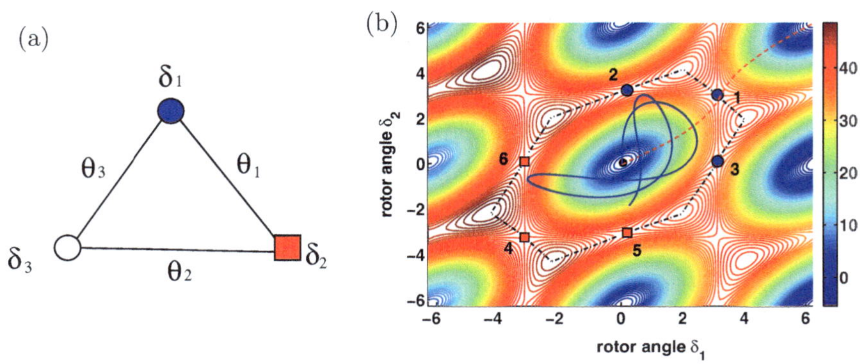

Fig. 14.6 **a** The 3-node power grid. **b** The potential energy of the three nodes power grid as a function of δ_i where $P_1/K = 0.125$, $P_2/K = -0.125$, and $P_3/K = 0$

called *controlling unstable equilibrium point*, which was developed. The method is not considered in this paper and we focus on the energy barrier, see [7, 11, 49]

It is obvious that if the potential energy of the type-1 equilibrium point is larger, the system can stand more serious disturbances. Hence the potential energy of all the type-1 equilibrium points can be used to measure the transient stability. However, to calculate the energy barrier, it is necessary to find all the type-1 equilibria which is actually a NP hard problem for a general network with many cycles. This idea is applied to a cyclic power network to investigate the impact of the cycles on the transient stability by the authors [55].

We focus on the stable equilibrium point with power flows in all the lines being P and the phase angle differences being $\arcsin P/B$, which is the same as in the SMIB model.

For this cyclic network, similar as in the SMIB model, the energy barrier can be calculated as

$$\Delta V_I^c = P\left(-\pi + 2\arcsin\frac{P}{B}\right) + 2B\sqrt{1-\frac{P^2}{B^2}} + \Delta U_I, \tag{14.24a}$$

$$\Delta U_I = \frac{2B}{n}\left(\frac{\pi}{2} - \arcsin\frac{P}{B}\right)^2\sqrt{1-\frac{P^2}{B^2}} + O\left(n^{-2}\right). \tag{14.24b}$$

This energy is a conservative approximation of the minimum energy from the disturbance that destroys the stability. When the system loses synchronization, there must be a line in which the phase angle difference is larger than $\pi/2$. When comparing this energy with that of the SMIB model in (14.21b), it can be found that ΔV_I^c is larger than ΔV_1. The minimum energy that leads to phase angle differences in branch lines exceeding $\pi/2$, are the same as ΔV_1 when the power transmission is P. Thus, the lines in a cycle are stronger than a branch line when they transfer the same amount of power. This explains in a micro-perceptive why dead-ends in a network undermine the stability.

Statement 14.5 With this finding, The stability of a nonlinear power system can be improved by either forming small cycles in the power network or by control so that the power-line branches transfer less power than the power lines in the cycles. □

From the comparison, it can also be deduced that the phase angle differences of the lines in cycles can be larger than those of lines which are not in cycles. In other words, with the same line capacity, the lines in cycles can transmit more power than those that are not in cycles.

Remark 14.4 However, this energy barrier focus on the potential energy landscape, in which the impact of the inertia and damping coefficients on the stability are not considered. This makes the energy barrier very conservative for the estimation of stability margin. In addition, for the large scale power networks with complex topology, finding all the type-1 equilibrium points is a challenging numerical problem due to the exponentially increase of the number of equilibrium points with the size of the network.

14.6 Conclusion

In this chapter, stability metrics and corresponding strategies to improve the stability of non-linear power systems have been introduced. The controllable factors that impact the stability include the inertia of synchronous machines, the damping coefficients, the topology of the network which involves the line capacity and the power generation and loads.

From the existence condition of the synchronous state, stability metric K_c can be extracted for the synchronization stability improvement by changing the topology of the network, the power generation and loads. Because this metric focuses on the synchronous state, the impact of the inertia of the synchronous machines and the damping coefficients are not reflected by this metric. Due to the equivalence between the power system and the Kuramoto model, the result obtained from the study of the existence of the Kuramoto model can be applied to the power system.

If a synchronous state exists for a power network, its local stability can be determined by the small signal stability based on the Lyapunov method. The stability of the linearized system is measured by the absolute value of the real part of the eigenvalues. With the optimal configuration method of $\beta_i = \beta_{\text{opt}}$, the linear stability can be enhanced by changing all the four factors. Note that it is demonstrated in [35] that the point with this setting in the $\boldsymbol{\beta}$-space is not a true local optimum for the linear stability. The method to find the optimum of the linear stability still needs further investigation. In addition, the linear stability formalism can only explore the local landscape of the stability region.

The energy barrier for the stability margin estimation is inspired by the direct method for the estimation of the system after a disturbance. It has been found that forming small cycles can increasing the stability margin. However, similar as the basin stability, this energy barrier is hard to be applied as a stability metric that can be used to form an optimization framework. In addition, the energy barrier focuses on the potential energy of the power network, it only reflects the impact of the topology and the power generation and loads on the stability.

The stability of the power network can be improved via various strategies. However, it is obvious that the optimal solution from these metrics are non-identical due to that none of them can includes all the influential factors of the stability. How these solution related to the stability region and what are the relationship between these solutions still need further study.

References

1. Anderson, P.M., Fouad, A.A.: Power System Control and Stability. Wiley-IEEE Press (2002)
2. Arenas, A., Díaz-Guilera, A., Kurths, J., Moreno, Y., Zhou, C.: Synchronization in complex networks. Phys. Rep. **469**(3), 93–153 (2008)
3. Baillieul, J., Byrnes, C.: Geometric critical point analysis of lossless power system models. IEEE Trans. Circuits Syst. **29**(11), 724–737 (1982)

4. Bergen, A.R., Hill, D.J.: A structure preserving model for power system stability analysis. IEEE Trans. Power App. Syst. **1**, 25–35 (1981)
5. Braess, D.: Uber ein paradoxon aus der verkehrsplanung. Unternehmensforschung Operations Research **12** (1968)
6. Bronski, J.C., DeVille, L.: Spectral theory for dynamics on graphs containing attractive and repulsive interactions. SIAM J. Appl. Math. **74**(1), 83–105 (2014)
7. Chang, H.D., Chu, C.C., Cauley, G.: Direct stability analysis of electric power systems using energy functions: theory, applications, and perspective. Proc. IEEE **83**(11), 1497–1529 (1995)
8. Chen, T., Davis, R., Mehta, D.: Counting equilibria of the kuramoto model using birationally invariant intersection index. SIAM J. Appl. Algebra Geometry **2**(4), 489–507 (2018)
9. Chiang, H.D., Chu, C.C.: Theoretical foundation of the BCU method for direct stability analysis of network-reduction power system. Models with small transfer conductances. IEEE Trans. Circuits Syst. I. Fundam. Theory Appl. **42**(5), 252–265 (1995)
10. Chiang, H.D., Wu, F.F., Varaiya, P.P.: Foundations of the potential energy boundary surface method for power system transient stability analysis. IEEE Trans. Circuits Syst. **35**(6), 712–728 (1988)
11. Chiang, H.D., Hirsch, M.W., Wu, F.F.: Stability regions of nonlinear autonomous dynamical systems. IEEE Trans. Autom. Control **33**(1), 16–27 (1988)
12. Coletta, T., Jacquod, P.: Linear stability and the Braess paradox in coupled-oscillator networks and electric power grids. Phys. Rev. E **93**(3), 032222 (2016)
13. Delabays, R., Coletta, T., Jacquod, P.: Multistability of phase-locking and topological winding numbers in locally coupled kuramoto models on single-loop networks. J. Math. Phys. **57**(3) (2016)
14. Dörfler, F., Bullo, F.: On the critical coupling for kuramoto oscillators. SIAM J. Appl. Dynam. Syst. **10**(3), 1070–1099 (2011)
15. Dörfler, F., Bullo, F.: Synchronization in complex networks of phase oscillators: a survey. Automatica **50**(6), 1539–1564 (2014)
16. Dörfler, F., Simpson-Porco, J.W., Bullo, F.: Breaking the hierarchy: distributed control and economic optimality in microgrids. IEEE Trans. Control Netw. Syst. **3**(3), 241–253 (2016)
17. Hasler, M., Wang, C., Ilic, M., Zobian, A.: Computation of static stability margins in power systems using monotonicity. In: 1993 IEEE International Symposium on Circuits and Systems, vol. 4, pp. 2196–2199, May 1993
18. Ilić, M.D., Zaborszky, J.: Dynamics and Control of Large Electric Power Systems. Wiley (2000)
19. Khalil, H.K.: Nonlinear Systems, 3rd edn. Prentice Hall, Upper Saddle River, New Jersey 07458 (2002)
20. Khayat, Y., Shafiee, Q., Heydari, R., Naderi, M., Dragicevic, T., Simpson-Porco, J.W., Dorfler, F., Fathi, M., Blaabjerg, F., Guerrero, J.M., Bevrani, H.: On the secondary control architectures of ac microgrids: an overview. IEEE Trans. Power Electron. 1–1 (2019)
21. Kim, Y., Mesbahi, M.: On maximizing the second smallest eigenvalue of a state-dependent graph laplacian. IEEE Trans. Autom. Control **51**(1), 116–120 (2006)
22. Kundur, P.: Power System Stability and Control. McGraw-Hill (1994)
23. Lee, J., Chiang, H.D.: A singular fixed-point homotopy method to locate the closest unstable equilibrium point for transient stability region estimate. IEEE Trans. Circuits Syst. II, Exp. Briefs **51**(4), 185–189 (2004)
24. Liu, C.W., Thorp, J.S.: A novel method to compute the closest unstable equilibrium point for transient stability region estimate in power systems. IEEE Trans. Circuits Syst. I, Fundam. Theory Appl. **44**(7), 630–635 (1997)
25. Lozano, S., Buzna, L., Díaz-Guilera, A.: Role of network topology in the synchronization of power systems. Eur. Phys. J. B **85**(7), 231 (2012)
26. Luxemburg, L.A., Huang, G.: On the number of unstable equilibria of a class of nonlinear systems. In: 26th IEEE Conference Decision Control, vol. 20, pp. 889–894. IEEE (1987)
27. Manik, D., Timme, M., Witthaut, D.: Cycle flows and multistability in oscillatory networks. Chaos **27**(8), 083123 (2017)
28. Marris, E.: Energy: upgrading the grid. Nature **454**, 570–573 (2008)

29. Mehta, D., Daleo, N.S., Dörfler, F., Hauenstein, J.D.: Algebraic geometrization of the Kuramoto model: equilibria and stability analysis. Chaos **25**(5), 053103 (2015)
30. Mehta, D., Nguyen, H.D., Turitsyn, K.: Numerical polynomial homotopy continuation method to locate all the power flow solutions. IET Gener. Transm. Distrib. **10**(12), 2972–2980 (2016)
31. Menck, P.J., Heitzig, J., Kurths, J., Schellnhuber, H.J.: How dead ends undermine power grid stability. Nat. Commun. **5**, 3969 (2014)
32. Menck, P.J., Heitzig, J., Marwan, N., Kurths, J.: How basin stability complements the linear-stability paradigm. Nat. Phys. **9**(2), 89–92 (2013)
33. Milano, F.: Power Systems Analysis Toolbox. University of Castilla, Castilla-La Mancha, Spain (2008)
34. Motter, A.E., Myers, S.A., Anghel, M., Nishikawa, T.: Spontaneous synchrony in power-grid networks. Nat. Phys. **9**(3), 191–197 (2013)
35. Nishikawa, T., Molnar, F., Motter, A.E.: Stability landscape of power-grid synchronization. IFAC-PapersOnLine **48**(18), 1–6 (2015). 4th IFAC Conference on Analysis and Control of Chaotic Systems CHAOS 2015
36. Nishikawa, T., Motter, A.E.: Comparative analysis of existing models for power-grid synchronization. New J. Phys. **17**(1), 015012 (2015)
37. Nusse, H.E., Yorke, J.A.: Basins of attraction. Science **271**(5254), 1376–1380 (1996)
38. Ochab, J., Góra, P.F.: Synchronization of coupled oscillators in a local one-dimensional Kuramoto model. Acta. Phys. Pol. B Proc. Suppl. **3**, 453–462 (2010)
39. Pecora, L.M., Carroll, T.L.: Master stability functions for synchronized coupled systems. Phys. Rev. Lett. **80**(10), 2109–2112 (1998)
40. Rogge, J.A., Aeyels, D.: Stability of phase locking in a ring of unidirectionally coupled oscillators. J. Phys. A Math. Gen. **37**(46), 11135–11148 (2004)
41. Rohden, M., Sorge, A., Witthaut, D., Timme, M.: Impact of network topology on synchrony of oscillatory power grids. Chaos **24**(1), 013123 (2014)
42. Schavemaker, P., van der Sluis, L.: Electrical Power System Essentials. Wiley (2008)
43. Schiffer, J., Goldin, D., Raisch, J., Sezi, T.: Synchronization of droop-controlled microgrids with distributed rotational and electronic generation. In: 52nd IEEE Conference Decision and Control, pp. 2334–2339, Dec 2013
44. Schiffer, J., Ortega, R., Astolfi, A., Raisch, J., Sezi, T.: Conditions for stability of droop-controlled inverter-based microgrids. Automatica **50**(10), 2457–2469 (2014)
45. Simpson-Porco, J.W., Dörfler, F., Bullo, F.: Voltage collapse in complex power grids. Nat. Commun. **7**, 10790 (2016)
46. Skar, S.J.: Stability of multi-machine power systems with nontrivial transfer conductances. SIAM J. Appl. Math. **39**(3), 475–491 (1980)
47. Skardal, P.S., Taylor, D., Sun, J.: Optimal synchronization of complex networks. Phys. Rev. Lett. **113**(14), 144101 (2014)
48. Tchuisseu, E.B.T., Gomila, D., Colet, P., Witthaut, D., Timme, M., Schäfer, B.: Curing braess' paradox by secondary control in power grids. New J. Phys. **20**(8), 083005 (2018)
49. Treinen, R.T., Vittal, V., Kliemann, W.: An improved technique to determine the controlling unstable equilibrium point in a power system. IEEE Trans. Circuits Syst. I, Fundam. Theory Appl. **43**(4), 313–323 (1996)
50. Van Mieghem, P.: Graph Spectra of Complex Networks. Cambridge University Press (2008)
51. Varaiya, P.P., Wu, F.F., Chen, R.L.: Direct methods for transient stability analysis of power systems: recent results. Proc. IEEE **73**(12), 1703–1715 (1985)
52. Witthaut, D., Timme, Marc: Braess's paradox in oscillator networks, desynchronization and power outage. New J. Phys. **14**(8), 083036 (2012)
53. Wood, A.J., Wollenberg, B.F., Sheble, G.B.: Power Generation, Operation, and Control, 3rd edn. Wiley-IEEE, Hoboken, New Jersey (2013)
54. Xi, K., Lin, H.X., Shen, C., van Schuppen, J.H.: Multi-level power-imbalance allocation control for secondary frequency control of power systems. IEEE Trans. Autom. Control, pp 1 (2019)
55. Xi, K., Dubbeldam, J.L.A., Lin, H.X.: Synchronization of cyclic power grids: equilibria and stability of the synchronous state. Chaos **27**(1), 013109 (2017)

56. Xi, K., Dubbeldam, J.L.A., Lin, H.X., van Schuppen, J.H.: Power imbalance allocation control of power systems-secondary frequency control. Automatica **92**, 72–85 (2018)
57. Zaborsky, J., Huang, G., Leung, T.C., Zheng, B.: Stability monitoring on the large electric power system. In: 24th IEEE Conference Decision Control, vol. 24, pp. 787–798. IEEE (1985)
58. Zaborszky, J., Huang, G., Zheng, B., Leung, T.C.: On the phase portrait of a class of large nonlinear dynamic systems such as the power system. IEEE Trans. Autom. Control **33**(1), 4–15 (1988)

Chapter 15
Geometric Series Method and Exact Solutions of Differential-Difference Equations

Aleksandr I. Zemlyanukhin, Andrey V. Bochkarev, Anna A. Orlova, and Aleksandr V. Ratushny

Abstract A modification of the geometric series method is considered, which is suitable for obtaining exact solutions of nonlinear differential-difference equations. The features of the method are shown in examples of solving three-point and five-point equations, the right-hand sides of which can contain polynomials, rational fractions, explicitly given elementary functions and implicitly defined functions that are solutions of some differential equations. The advantages and disadvantages of the approach are noted in comparison with other methods for constructing exact solutions.

15.1 Introduction

The successes of the theory of integrable and close to integrable systems have led to a revival of interest in the development of asymptotic methods of nonlinear dynamics. Mathematical models in the form of systems of nonlinear difference and differential-difference equations (DDE) are often much more complicated than their continuum analogues. The study of the analytical structure of such systems is fraught with serious difficulties, and the construction of exact solutions is possible in the simplest cases.

A. I. Zemlyanukhin (✉) · A. V. Bochkarev
Yuri Gagarin State Technical University of Saratov, Politekhnicheskaya str. 77, Saratov 410054, Russia
e-mail: azemlyanukhin@mail.ru

A. A. Orlova
Saratov Social and Economic Institute, Plekhanov Russian University of Economics, Radischeva str. 89, Saratov 410003, Russia
e-mail: kostylevaaa@mail.ru

A. V. Ratushny
Saratov State University, Astrakhanskaya str. 83, Saratov 410012, Russia
e-mail: sania.ratushnyy@gmail.com

A. K. Abramian et al. (eds.), *Nonlinear Dynamics of Discrete and Continuous Systems*, Advanced Structured Materials 139,
https://doi.org/10.1007/978-3-030-53006-8_15

In [1–4], a modified version of the asymptotic method of multiscale expansions for linear and weakly nonlinear difference equations of the second order is developed, for which it is possible to construct approximations of exact solutions and first integrals.

Periodic atomic spatial structures in physics are naturally modeled by DDE and systems of DDEs. When considering large atomic volumes, an asymptotic transition to well-studied evolutionary and quasi-hyperbolic partial differential equations (PDE) is carried out. In recent decades, interest has arisen in the study of nanostructures, such as carbon nanotubes [5], in which this asymptotic transition requires special justification. An analysis of publications on DDEs shows the following. As an integrability criterion, the existence of Lax pairs or the presence of an infinite system of symmetries is used [6]. Unlike PDE, exact DDE solutions are provided only for some integrable cases. The latter is due to the complexity of the Darboux transformation procedure for Lax pairs, with the help of which exact solutions are obtained [7]. Meanwhile, the presence of an exact solution in a closed form is important in terms of understanding the properties of the equation, and for verification of numerical methods of solution.

The algorithmic foundations of direct methods for constructing exact solutions of DDE and DDE systems using symbolic mathematics packages are presented in [8]. Solitary wave solutions are found in the form of truncated expansions in powers of hyperbolic functions.

In this paper, we develop a modification of the geometric series method [9, 10], suitable for constructing exact solutions of single DDE and DDE systems. The proposed approach, like the Hirota method [11], which is asymptotic in nature, is a direct method. According to the geometric series method, the solution is sought in the form of a series in powers of the exponential function. After substituting a series in DDE, a sequence of linear equations is solved to determine the coefficients of the series. Then, for the resulting series, a sequence of diagonal Pade approximants is constructed. If all approximants, starting from a certain order, coincide, then the series is geometric and the coincident approximants determine exact sum of the series, and, therefore, the exact solution of DDE. In some cases, the requirement of coincidence of the Pade approximants allows us to identify conditions for the coefficients of the equation/solution under which the series becomes geometric. Our studies show that the vast majority of integrable DDEs with polynomial, rational, or more complex dependence on an the desired function have an exact solution detected by the geometric series method. Therefore, the presence of such a solution can serve as a simple empirical criterion for integrability.

15.2 Volterra Chain

We demonstrate the features of the geometric series method using a simple example. We find a traveling wave solution for the well-known Volterra chain equation [12]

$$\frac{d}{dt}u_n(t) = u_n(t)\left(u_{n+1}(t) - u_{n-1}(t)\right). \tag{15.1}$$

After the transition in Eq. (15.1) to the traveling wave variable

$$z = dn + \omega t, \tag{15.2}$$

where $n \in \mathbb{Z}$, we obtain

$$-\omega \frac{d}{dz} u_n(z) + u_n(z)\left(u_{n+1}(z) - u_{n-1}(z)\right) = 0. \tag{15.3}$$

We will seek a solution in the form of an exponential functions series with unknown coefficients

$$u_n = \sum_{k=0}^{\infty} M_k \mathrm{e}^{kz}. \tag{15.4}$$

If index variable n in (15.3) increases by one, then traveling wave variable z increases by d, therefore

$$u_{n+1} = \sum_{k=0}^{\infty} M_k \delta^k \mathrm{e}^{kz}, \quad u_{n-1} = \sum_{k=0}^{\infty} M_k \delta^{-k} \mathrm{e}^{kz}, \tag{15.5}$$

where $\delta = \mathrm{e}^d$. We substitute (15.4) and (15.5) into (15.3) and collect the result by powers of the exponential function. Equating to zero the coefficients at e^z, e^{2z}, $\mathrm{e}^{3z}, \ldots$, we have a system of equations

$$\left(\delta^2 - 1\right) M_0 M_1 - \delta\omega M_1 = 0, \tag{15.6}$$

$$\left(\delta^4 - 1\right) M_0 M_2 + \delta\left(\delta^2 - 1\right) M_1^2 - 2\delta^2\omega M_2 = 0, \tag{15.7}$$

$$\left(\delta^6 - 1\right) M_0 M_3 + \delta\left(\delta^4 + \delta^3 - \delta - 1\right) M_1 M_2 - 3\delta^3\omega M_3 = 0, \tag{15.8}$$

$$\left(\delta^8 - 1\right) M_0 M_4 + \delta\left(\delta^6 + \delta^4 - \delta^2 - 1\right) M_1 M_3 + \delta^2\left(\delta^4 - 1\right) M_2^2 - 4\delta^4\omega M_4 = 0,$$

$$\ldots$$

Note that the k-th equation of the system is linear with respect to the coefficient M_k. To find a nontrivial solution, the frequency ω should be determined from the Eq. (15.6) to obtain an analog of the dispersion relation:

$$\omega = M_0 \delta^{-1}\left(\delta^2 - 1\right). \tag{15.9}$$

From the Eq. (15.7) we find the coefficient M_2, from the next Eq. (15.8) we find M_3 and so on. Substituting expressions for the coefficients $M_2, M_3, \ldots$ into Eq. (15.4) and replacing $\mathrm{e}^z = Z$, we obtain a power series with respect to the variable Z:

$$\begin{aligned} u_n &= M_0 + M_1 Z - \frac{\delta M_1^2 Z^2}{(\delta-1)^2 M_0} + \frac{\delta^2\left(\delta^2+\delta+1\right) M_1^3 Z^3}{(\delta+1)^2(\delta-1)^4 M_0^2} \\ &- \frac{\delta^3\left(\delta^2+1\right) M_1^4 Z^4}{(\delta+1)^2(\delta-1)^6 M_0^3} + \frac{\delta^4\left(\delta^4+\delta^3+\delta^2+\delta+1\right) M_1^5 Z^5}{(\delta+1)^4(\delta-1)^8 M_0^4} \\ &- \frac{\delta^5\left(\delta^4+\delta^2+1\right) M_1^6 Z^6}{(\delta+1)^4(\delta-1)^{10} M_0^5} + \frac{\delta^6\left(\delta^6+\delta^5+\delta^4+\delta^3+\delta^2+\delta+1\right) M_1^7 Z^7}{(\delta+1)^6(\delta-1)^{12} M_0^6} - \cdots \end{aligned} \tag{15.10}$$

We verify that the series (15.10) is geometric by calculating for it the first few diagonal Pade approximants [13]:

$$\begin{aligned} [1/1] &= M_0 \frac{(\delta-1)^2 M_0 + \left(\delta^2-\delta+1\right) M_1 Z}{(\delta-1)^2 M_0 + \delta M_1 Z}, \\ [2/2] &= M_0 \frac{\left((\delta+1)(\delta-1)^2 M_0 + M_1 Z\right)\left((\delta+1)(\delta-1)^2 M_0 + \delta^3 M_1 Z\right)}{\left((\delta+1)(\delta-1)^2 M_0 + \delta M_1 Z\right)\left((\delta+1)(\delta-1)^2 M_0 + \delta^2 M_1 Z\right)}, \\ [3/3] &= [2/2]. \end{aligned}$$

The coincidence of two successive approximants is a claim that the series (15.10) is geometric. It can be rigorously proved that the series is geometric if we derive the general formula for the Pade approximants of arbitrary order $[N/N]$, which is not an easy task. Fortunately, there is an easier way. After the reverse substitution $Z = e^z$, we substitute the approximant [2/2] into Eq. (15.3) instead of function $u_n(z)$:

$$u_n = M_0 + \frac{M_0^2 M_1 (\delta+1)^2 (\delta-1)^4 e^z}{\left((\delta+1)(\delta-1)^2 M_0 + \delta M_1 e^z\right)\left((\delta+1)(\delta-1)^2 M_0 + \delta^2 M_1 e^z\right)}. \tag{15.11}$$

We replace function $u_{n+1}(z)$ by approximant [2/2], using the substitution $Z = \delta e^z$, and replace function $u_{n-1}(z)$ by [2/2], using $Z = \delta^{-1} e^z$. After simplification, taking into account (15.9) Eq. (15.3) turns into an identity, therefore, the expression on the right-hand side of (15.11) is an exact solution to Eq. (15.3), and series (15.10) is geometric and its sum coincides with the approximant [2/2]. Solution (15.11) contains three arbitrary constants M_0, M_1, δ, which are subject to the following conditions: M_0 and M_1 do not vanish simultaneously; $\delta > 0,\ \delta \neq 1$. The solution is bounded and has a soliton-like form when $M_0 M_1 > 0$.

15.3 Modified Discrete Sawada-Kotera Equation

Volterra equation (15.1) is a classic example of an integrable three-point chain. The approach proposed in Sect. 15.2 can easily be extended to 5-point chains and chains of higher orders. Let us consider a modification of the discrete Sawada-Kotera equation [7] (hereinafter, for brevity, we will not explicitly indicate the arguments of the functions)

$$\frac{d}{dt}u_n = u_{n+1}u_n^3u_{n-1}\left(u_{n+2}u_{n+1} - u_{n-1}u_{n-2}\right) - u_n^2\left(u_{n+1} - u_{n-1}\right). \tag{15.12}$$

After passing in (15.12) to the running variable (15.2) we have

$$-\omega\frac{d}{dz}u_n + u_{n+1}u_n^3u_{n-1}\left(u_{n+2}u_{n+1} - u_{n-1}u_{n-2}\right) - u_n^2\left(u_{n+1} - u_{n-1}\right) = 0. \tag{15.13}$$

Supplementing the substitutions (15.4) and (15.5) with the following equalities

$$u_{n+2} = \sum_{k=0}^{\infty} M_k\,\delta^{2k}\,\mathrm{e}^{kz}, \quad u_{n-2} = \sum_{k=0}^{\infty} M_k\,\delta^{-2k}\,\mathrm{e}^{kz}, \tag{15.14}$$

we apply (15.4), (15.5) and (15.14) to the Eq. (15.13). Collecting by the powers of the exponential function and equating to zero a coefficient at e^z, we obtain

$$\omega = M_0^2\left(1 - \delta^{-2}\right)\left(M_0^4\left(\delta^2 + \delta + 1\right) - \delta\right). \tag{15.15}$$

As before in Sect. 15.2, successively equating the factors at $\mathrm{e}^{2z},\ \mathrm{e}^{3z}, \ldots$ to zero, we find the coefficients $M_2, M_3, \ldots$ Series (15.4) after the replacement $\mathrm{e}^z = Z$ takes the form

$$\begin{aligned} u_n &= M_0 + M_1 Z - \frac{2\delta M_1^2 Z^2}{M_0\left(\delta - 1\right)^2} + \frac{3\delta^2 M_1^3 Z^3}{M_0^2\left(\delta - 1\right)^4} - \frac{4\delta^3 M_1^4 Z^4}{M_0^3\left(\delta - 1\right)^6} + \cdots \\ &= M_0 + M_1 Z + \sum_{n=2}\left(-\frac{\delta}{M_0}\right)^{n-1}\frac{n M_1^n Z^n}{\left(\delta - 1\right)^{2n}}. \end{aligned} \tag{15.16}$$

The calculation of the Pade approximants for (15.16) shows that $[1/1] \neq [2/2],\ \ [2/2] = [3/3]$. After reverse substitution $Z = \mathrm{e}^z$, the approximant

$$[2/2] = M_0 + \frac{M_0^2 M_1\left(\delta - 1\right)^4\mathrm{e}^z}{\left(M_0\left(\delta - 1\right)^2 + M_1\delta\,\mathrm{e}^z\right)^2} \tag{15.17}$$

becomes an exact solution of Eq. (15.13) under the condition (15.15). Solution (15.17) is bounded when $M_0M_1 > 0,\ \ \delta > 0$ and has bell-shaped form. Phase velocity

$$\begin{aligned} \frac{\omega}{d} &= \frac{M_0^2}{d}\left(1 - \mathrm{e}^{-2d}\right)\left(M_0^4\left(\mathrm{e}^{2d} + \mathrm{e}^{d} + 1\right) - \mathrm{e}^{d}\right) \\ &= \frac{2M_0^2}{d}\left(\sinh\left(2d\right) + \left(1 - M_0^{-4}\right)\sinh\left(d\right)\right) \end{aligned}$$

is an even positive function that increases with increasing of $|d|$. A solitary wave corresponding to solution (15.17), propagates to the left along the spatial axis On.

15.4 Fraction Term DDE

The right-hand sides of DDEs (15.1) and (15.12) have a simple polynomial form with respect to u_n, $u_{n\pm1}$ and $u_{n\pm2}$. A more difficult situation for analysis arises when the right side of the equation contains rational fractions. Let's consider a DDE

$$\begin{aligned}\frac{d}{dt}u_n &= (u_n+1)\left(\frac{u_{n+2}u_n\,(u_{n+1}+1)^2}{u_{n+1}}\right.\\&\left.-\frac{u_{n-2}u_n\,(u_{n-1}+1)^2}{u_{n-1}}+(2u_n+1)\,(u_{n+1}-u_{n-1})\right),\end{aligned}\tag{15.18}$$

which first appeared in the work [14]. After passing to the traveling wave variable (15.2), we reduce the terms (15.18) to the common denominator and assuming $u_{n+1}u_{n-1}\neq 0$, consider the numerator of the resulting expression:

$$\begin{aligned}&-\omega u_{n+1}u_{n-1}\frac{d}{dz}u_n+(u_n+1)\left(u_{n-1}u_nu_{n+2}\,(u_{n+1}+1)^2\right.\\&\left.-u_{n-2}u_nu_{n+1}\,(u_{n-1}+1)^2+u_{n-1}u_{n+1}\,(2u_n+1)\,(u_{n+1}-u_{n-1})\right)=0.\end{aligned}\tag{15.19}$$

Repeating with Eq. (15.19) the steps described above in Sect. 15.3, we have

$$\omega=\left(1-\delta^{-2}\right)(M_0+1)^3\left(\delta^2+\frac{M_0\,(M_0+2)}{(M_0+1)^2}\delta+1\right),$$

$$\begin{aligned}u_n &= M_0+M_1Z\\&-\frac{\delta M_1^2\left(\left(M_0^2-1\right)\delta^2+\left(M_0^2+4M_0+2\right)\delta+M_0^2-1\right)Z^2}{(\delta-1)^2M_0\,(M_0+1)\left((M_0+1)\,\delta^2+(M_0+2)\,\delta+M_0+1\right)}+\cdots\end{aligned}\tag{15.20}$$

Calculation of the Pade approximants for the series (15.20) shows that $[1/1]\neq[2/2]\neq[3/3]\neq[4/4]$, $[4/4]=[5/5]$. After the reverse substitution $Z=\mathrm{e}^z$ the exact solution of Eq. (15.19) is given by a fourth-order approximant

$$[4/4]=M_0+\frac{\mathrm{e}^z\left(A_2\mathrm{e}^{2z}+A_1\mathrm{e}^z+A_0\right)}{\left(B_2\mathrm{e}^{2z}+B_1\mathrm{e}^z+B_0\right)^2},\tag{15.21}$$

where

$$
\begin{aligned}
A_2 &= \delta^3 (\delta+1)^2 (\delta-1)^4 M_0 (M_0+1)^2 M_1^3 P Q, \\
A_1 &= \delta (\delta+1)^4 (\delta-1)^6 M_0 (M_0+1)^3 M_1^2 P Q, \\
A_0 &= (\delta+1)^4 (\delta-1)^8 M_0 (M_0+1)^4 Q^2, \\
B_2 &= \delta^3 M_1^2 P, \\
B_1 &= \delta (\delta+1)^2 (\delta-1)^2 M_0 (M_0+1) M_1 Q, \\
B_0 &= (\delta+1)^2 (\delta-1)^4 M_0 (M_0+1)^2 \left(\left(\delta^2+\delta+1\right) M_0 + (\delta+1)^2\right), \\
P &= (M_0+1)^2 \delta^2 + \left(M_0^2 - 2\right) \delta + (M_0+1)^2, \\
Q &= (M_0+1) \delta^2 + (M_0+2) \delta + M_0 + 1.
\end{aligned}
$$

Solution (15.21) is real and bounded if all roots of the square trinomial $B_2 x^2 + B_1 x + B_0$ are complex or negative.

15.5 DDE with Arbitrary Coefficients

The geometric series method is suitable for solving DDE, the right-hand side of which includes one or more arbitrary coefficients. Let's try to find the exact solution of the 3-point integrable DDE [12]:

$$
\frac{d}{dt} u_n = \left(\alpha u_n^4 + \beta u_n^3 + \gamma u_n^2 + \lambda u_n + \mu\right) \left(\frac{1}{u_{n+1} - u_n} + \frac{1}{u_n - u_{n-1}}\right).
$$

After passing to the running variable (15.2) we obtain

$$
\begin{aligned}
&- \omega \left(u_n - u_{n-1}\right) \left(u_{n+1} - u_n\right) \frac{d}{dz} u_n \\
&+ \left(\alpha u_n^4 + \beta u_n^3 + \gamma u_n^2 + \lambda u_n + \mu\right) \left(u_{n+1} - u_{n-1}\right) = 0.
\end{aligned}
\tag{15.22}
$$

After substituting (15.4) and (15.5) into (15.22) and collecting by powers of exponential function, for the coefficient at e^z we have

$$
\left(\delta^2 - 1\right) M_1 \left(\alpha M_0^4 + \beta M_0^3 + \gamma M_0^2 + \lambda M_0 + \mu\right) = 0.
$$

Under condition $M_1 = 0$ we obtain the trivial solution $u_n = M_0$, if we take $\delta = 1$ then the chain degenerates, therefore, we must require the third factor to vanish:

$$
\mu = -M_0 \left(\alpha M_0^3 + \beta M_0^2 + \gamma M_0 + \lambda\right). \tag{15.23}
$$

For the coefficient at e^{2z} we obtain

$$
\left(\delta^2 - 1\right) M_1^2 \left(4\alpha M_0^3 + 3\beta M_0^2 + 2\gamma M_0 + \lambda\right) = 0,
$$

which implies

$$\lambda = -M_0 \left(4\alpha M_0^2 + 3\beta M_0 + 2\gamma\right). \tag{15.24}$$

Equating to zero the coefficient at e^{3z}, you can determine the frequency ω:

$$\omega = \frac{\left(6\alpha M_0^2 + 3\beta M_0 + \gamma\right)(\delta + 1)}{\delta - 1}. \tag{15.25}$$

Equating the factors at e^{4z}, e^{5z}, e^{6z}, ... to zero, we determine the first few coefficients $M_2, M_3, M_4, \ldots$ of the series (15.4), in order to calculate the Pade approximants for it: $[1/1] \neq [2/2]$, $[2/2] = [3/3]$. After substitution $Z = e^z$, the exact solution of Eq. (15.22) is given by a second-order approximant

$$[2/2] = M_0 + \frac{A_1 e^z}{B_2 e^{2z} + B_1 e^z + B_0}, \tag{15.26}$$

where

$$\begin{aligned}
A_1 &= 4(\delta + 1)^2 M_1 \left(6\alpha M_0^2 + 3\beta M_0 + \gamma\right)^2, \\
B_2 &= -\delta M_1^2 \left(8\alpha^2 M_0^2 + 4\alpha\beta M_0 + 4\alpha\gamma - \beta^2\right), \\
B_1 &= -6(\delta + 1)^2 M_1 (4\alpha M_0 + \beta)\left(2\alpha M_0^2 + \beta M_0 + \frac{1}{3}\gamma\right), \\
B_0 &= 36(\delta + 1)^2 \left(2\alpha M_0^2 + \beta M_0 + \frac{1}{3}\gamma\right)^2,
\end{aligned}$$

under conditions (15.23), (15.24) and (15.25). The special structure (15.22), when ω appears only in the third equation, forces one to set two coefficients λ and μ through the remaining coefficients α, β, γ and M_0. From the last set, only M_0 is not predefined, therefore, only one of the two coefficients λ and μ can be chosen arbitrarily and the solution (15.26) is not general.

15.6 Toda-Type DDE

Let us demonstrate the possibility of solving chain equations containing transcendental functions by the example of an Toda-type equation [12], which in the traveling wave variable has the form

$$-\omega^2 \frac{d^2}{dz^2} v_n + e^{v_{n+1} - 2v_n + v_{n-1}} + \lambda = 0. \tag{15.27}$$

Using the well-known expansion

$$e^x = 1 + x + \frac{1}{2}x^2 + \frac{1}{6}x^3 + \frac{1}{24}x^4 + \cdots ,$$

we write (15.27) as follows

$$\begin{aligned} &-\omega^2 \frac{d^2}{dz^2} v_n + \lambda + 1 + (v_{n+1} - 2v_n + v_{n-1}) + \frac{1}{2}(v_{n+1} - 2v_n + v_{n-1})^2 \\ &+ \frac{1}{6}(v_{n+1} - 2v_n + v_{n-1})^3 + \frac{1}{24}(v_{n+1} - 2v_n + v_{n-1})^4 + \cdots = 0. \end{aligned} \tag{15.28}$$

After the substitution $v_n = \sum_{k=0}^{\infty} M_k\, e^{kz}$, we find that the constant term on the left-hand side of (15.28) is $(1 + \lambda)$. Therefore, continuing the steps of the method, we can only hope for a particular solution when $\lambda = -1$. To find a more general solution, note that for function

$$w_n = A + Bz + Cz^2 \tag{15.29}$$

the following expressions have constant values:

$$\begin{aligned} \frac{d^2}{dz^2} w_n &= 2C, \\ w_{n+1} - 2w_n + w_{n-1} &= \left(A + B(z + d) + C(z + d)^2\right) - 2\left(A + Bz + Cz^2\right) \\ &\quad + \left(A + B(z - d) + C(z - d)^2\right) = 2Cd^2 = 2C \ln^2\delta. \end{aligned}$$

We will seek a solution to the problem in a modified form

$$v_n = w_n + u_n, \tag{15.30}$$

where u_n is determined by the equality (15.4). Substituting (15.30) into (15.27), we have

$$-\Omega^2 \frac{d^2}{dz^2} u_n + e^{u_{n+1} - 2u_n + u_{n-1}} + \Lambda = 0, \tag{15.31}$$

where

$$\Omega^2 = \frac{\omega^2}{e^{2C \ln^2\delta}}, \quad \Lambda = \frac{\lambda - 2\omega^2 C}{e^{2C \ln^2\delta}}. \tag{15.32}$$

We select the value of the constant C from condition $\Lambda = -1$, after which we substitute (15.4), (15.5) into (15.31). Equating to zero the coefficient at e^z, we obtain

$$\Omega = \pm \frac{\delta - 1}{\sqrt{\delta}}. \tag{15.33}$$

Successively equating to zero the coefficients at $e^{2z}, e^{3z}, \ldots$ and determining $M_2, M_3, \ldots$ we obtain a series

$$u_n = M_0 + M_1 Z - \frac{1}{2} M_1^2 Z^2 + \frac{1}{3} M_1^3 Z^3 - \frac{1}{4} M_1^4 Z^4 + \frac{1}{5} M_1^5 Z^5 - \cdots, \quad (15.34)$$

which, despite its simplicity, is not geometric. However, the derivative of (15.34) with respect to Z

$$\frac{d}{dZ} u_n = M_1 - M_1^2 Z + M_1^3 Z^2 - M_1^4 Z^3 + M_1^5 Z^4 - \cdots,$$

forms a geometric series with sum

$$\frac{d}{dZ} u_n = \frac{M_1}{M_1 Z + 1}. \quad (15.35)$$

After integrating (15.35) over Z and returning to the variable z, we obtain

$$u_n = M_0 + \ln\left(M_1 \, e^z + 1\right), \quad (15.36)$$

where M_0 plays the role of integration constant. Thus, an exact solution to Eq. (15.27) is given by equality (15.30) under the conditions (15.29), (15.32), (15.33), (15.36) and $\Lambda = -1$. Constants A, B, M_0, M_1, δ can be assigned arbitrarily.

15.7 Volterra-Type DDE

The geometric series method can be used to solve DDEs that contain functions that do not have an explicit analytical representation, but are specified in the form of a solution to some differential equation. Consider, for example, the Volterra-type equation [12]

$$\frac{d}{dt} u_n = y\left(u_{n+1} - u\right) + y\left(u - u_{n-1}\right), \quad (15.37)$$

where function $y(x)$ satisfies the Riccati equation

$$\frac{dy}{dx} = \alpha y^2 + \beta y + \gamma. \quad (15.38)$$

Let's find the expansion of $y(x)$ in the Maclaurin series. To do this, we successively differentiate both sides of (15.38) with respect to x, replacing the derivative dy/dx in the right-hand sides in accordance with (15.38):

$$
\begin{aligned}
\frac{d^2y}{dx^2} &= \left(\alpha y^2 + \beta y + \gamma\right)(2\alpha y + \beta), \\
\frac{d^3y}{dx^3} &= \left(\alpha y^2 + \beta y + \gamma\right)\left(6\alpha^2 y^2 + 6\alpha\beta y + \beta^2 + 2\alpha\gamma\right), \\
\frac{d^4y}{dx^4} &= \left(\alpha y^2 + \beta y + \gamma\right)(2\alpha y + \beta)\left(12\alpha^2 y^2 + 12\alpha\beta y + \beta^2 + 8\alpha\gamma\right), \ldots
\end{aligned}
$$

Taking into account initial condition $y(0) = 0$, we have

$$
\begin{aligned}
\left.\frac{d^2y}{dx^2}\right|_{x=0} &= \beta\gamma, \\
\left.\frac{d^3y}{dx^3}\right|_{x=0} &= \left(2\alpha\gamma + \beta^2\right)\gamma, \\
\left.\frac{d^4y}{dx^4}\right|_{x=0} &= \left(8\alpha\gamma + \beta^2\right)\beta\gamma, \ldots
\end{aligned}
$$

and desired expansion in the Maclaurin series takes the form:

$$
\begin{aligned}
y(x) &= y(0) + \sum_{k=1}^{\infty}\left(\left.\frac{d^k y}{dx^k}\right|_{x=0}\right)\frac{x^k}{k!} = \gamma x + \frac{1}{2}\beta\gamma x^2 + \frac{1}{6}\left(2\alpha\gamma + \beta^2\right)\gamma x^3 \\
&+ \frac{1}{24}\left(8\alpha\gamma + \beta^2\right)\beta\gamma x^4 + \frac{1}{120}\left(16\alpha^2\gamma^2 + 22\alpha\beta^2\gamma + \beta^4\right)\gamma x^5 \\
&+ \frac{1}{720}\left(136\alpha^2\gamma^2 + 52\alpha\beta^2\gamma + \beta^4\right)\beta\gamma x^6 + \cdots
\end{aligned}
\tag{15.39}
$$

Passing to variable (15.2) in (15.37) and using (15.39), we have

$$
\begin{aligned}
&-\omega\frac{d}{dz}u_n + \gamma\left(u_{n+1} - u_{n-1}\right) + \frac{1}{2}\beta\gamma\left(\left(u_{n+1} - u_n\right)^2 + \left(u_n - u_{n-1}\right)^2\right) \\
&+ \frac{1}{6}\left(2\alpha\gamma + \beta^2\right)\gamma\left(\left(u_{n+1} - u_n\right)^3 + \left(u_n - u_{n-1}\right)^3\right) \\
&+ \frac{1}{24}\left(8\alpha\gamma + \beta^2\right)\gamma\left(\left(u_{n+1} - u_n\right)^4 + \left(u_n - u_{n-1}\right)^4\right) + \cdots = 0
\end{aligned}
\tag{15.40}
$$

Further, using the substitution (15.4), (15.5) and repeating the steps of the method, we obtain the dispersion relation

$$
\omega = \frac{\gamma}{\delta}\left(\delta^2 - 1\right)
$$

and the solution in the form of a series

$$\begin{aligned} u_n &= M_0 + M_1 Z - \frac{\beta\left(\delta^2+1\right) M_1^2 Z^2}{2\left(\delta^2-1\right)} \\ &+ \frac{\left(\delta^2-\delta+1\right)\left(\left(\beta^2-\alpha\gamma\right)\left(\delta^2+1\right)+\left(2\alpha\gamma+\beta^2\right)\delta\right) M_1^3 Z^3}{3\left(\delta^2-1\right)^2} \\ &- \frac{\beta\left(\delta^4+1\right)\left(\left(\beta^2-2\alpha\gamma\right)\left(\delta^2+1\right)+4\alpha\delta\gamma\right) M_1^4 Z^4}{4\left(\delta^2-1\right)^3} + \cdots \end{aligned} \quad (15.41)$$

The series (15.41) is not geometric; the diagonal Pade approximants $[N/N]$ for it turn out to be different from each other. However, if we factorize each expression from the sequence $[2/2]-[1/1]$, $[3/3]-[2/2]$, $[4/4]-[3/3]$, ..., we can find that all such expressions, starting from the second one, contain a common factor $\left(\beta^2-4\alpha\gamma\right)$. Equating this factor to zero, i.e., requiring

$$\alpha = \frac{\beta^2}{4\gamma}, \quad (15.42)$$

we turn the series (15.41) into a geometric one, since in this case we have $[2/2] = [3/3] = [4/4] = \cdots$ After the inverse transition from Z to z the approximant $[2/2]$ gives an exact solution to the problem (15.40):

$$u_n = M_0 + \frac{4M_1\left(\delta-1\right)\left(\beta\delta M_1\,\mathrm{e}^z+\delta^2-1\right)\mathrm{e}^z}{\left(\delta+1\right)\left(\beta\delta M_1\,\mathrm{e}^z+2\delta-2\right)\left(\beta M_1\,\mathrm{e}^z+2\delta-2\right)}.$$

Note that under condition (15.42) the solution of the Riccati equation (15.38) degenerates into a rational fraction

$$y\left(x\right) = -\frac{2\gamma\left(\beta\left(x+x_0\right)+2\right)}{\beta^2\left(x+x_0\right)}.$$

15.8 Ablowitz-Ladik Lattice

We show the features of the method when solving DDE systems using the example of the well-known Ablowitz-Ladik lattice [15], which in the traveling wave variable form looks like

$$-\omega\frac{d}{dz}u_n + \alpha\left(u_{n+1}-2u_n+u_{n-1}\right) - u_n v_n\left(u_{n+1}+u_{n-1}\right) = 0, \quad (15.43)$$

$$-\omega\frac{d}{dz}v_n - \alpha\left(v_{n+1}-2v_n+v_{n-1}\right) + u_n v_n\left(v_{n+1}+v_{n-1}\right) = 0. \quad (15.44)$$

Substituting expansions

$$u_n = \sum_{k=0}^{\infty} M_k e^{kz}, \quad u_{n+1} = \sum_{k=0}^{\infty} M_k \delta^k e^{kz}, \quad u_{n-1} = \sum_{k=0}^{\infty} M_k \delta^{-k} e^{kz},$$
$$v_n = \sum_{k=0}^{\infty} N_k e^{kz}, \quad v_{n+1} = \sum_{k=0}^{\infty} N_k \delta^k e^{kz}, \quad v_{n-1} = \sum_{k=0}^{\infty} N_k \delta^{-k} e^{kz},$$

into system (15.43), (15.44), we equate to zero the factors at e^z in both equations:

$$(\alpha - M_0 N_0)\,\delta^2 - (2M_0N_0 + 2\alpha + \omega)\,\delta + \alpha - M_0N_0 - 2\delta \frac{M_0^2 N_1}{M_1} = 0, \quad (15.45)$$
$$(\alpha - M_0 N_0)\,\delta^2 - (2M_0N_0 + 2\alpha - \omega)\,\delta + \alpha - M_0N_0 - 2\delta \frac{N_0^2 M_1}{N_1} = 0. \quad (15.46)$$

Assuming $N_0 = 0$, from (15.45), (15.46) we find

$$\omega = -\frac{\alpha\,(\delta-1)^2}{\delta}, \quad M_0 = \pm(\delta - 1)\sqrt{\frac{\alpha M_1}{\delta N_1}}.$$

Equating to zero the factors at e^{2z} in both equations, we find

$$M_2 = \pm\frac{\delta M_1 N_1}{\alpha\,(\delta-1)}\sqrt{\frac{\alpha M_1}{\delta N_1}}, \quad N_2 = \pm\frac{\delta N_1^2}{\alpha\,(\delta-1)}\sqrt{\frac{\alpha M_1}{\delta N_1}},$$

and so on. As a result, after replacing $e^z = Z$ we obtain the expressions for u_n and v_n as follows

$$\begin{aligned} u_n &= \pm(\delta-1)\sqrt{\frac{\alpha M_1}{\delta N_1}} + M_1 Z \pm \frac{\delta M_1 N_1}{\alpha\,(\delta-1)}\sqrt{\frac{\alpha M_1}{\delta N_1}} Z^2 + \frac{\delta M_1^2 N_1}{\alpha\,(\delta-1)^2} Z^3 \\ &\pm \frac{\delta^2 M_1^2 N_1^2}{\alpha^2\,(\delta-1)^3}\sqrt{\frac{\alpha M_1}{\delta N_1}} Z^4 + \frac{\delta^2 M_1^3 N_1^2}{\alpha^2\,(\delta-1)^4} Z^5 \pm \frac{\delta^3 M_1^3 N_1^3}{\alpha^3\,(\delta-1)^5}\sqrt{\frac{\alpha M_1}{\delta N_1}} Z^6 + \cdots, \end{aligned} \quad (15.47)$$

$$\begin{aligned} v_n &= N_1 Z \pm \frac{\delta N_1^2}{\alpha\,(\delta-1)}\sqrt{\frac{\alpha M_1}{\delta N_1}} Z^2 + \frac{\delta M_1 N_1^2}{\alpha\,(\delta-1)^2} Z^3 \\ &\pm \frac{\delta^2 M_1 N_1^3}{\alpha^2\,(\delta-1)^3}\sqrt{\frac{\alpha M_1}{\delta N_1}} Z^4 + \frac{\delta^2 M_1^2 N_1^3}{\alpha^2\,(\delta-1)^4} Z^5 \pm \frac{\delta^3 M_1^2 N_1^4}{\alpha^3\,(\delta-1)^5}\sqrt{\frac{\alpha M_1}{\delta N_1}} Z^6 + \cdots \end{aligned} \quad (15.48)$$

Calculations show that [2/2] and [3/3] Pade approximants for u_n—series and v_n—series (15.47), (15.48) coincide with each other and are equal, respectively

$$u_n = \pm(\delta-1)\sqrt{\frac{\alpha M_1}{\delta N_1}} + \frac{\alpha(\delta-1)M_1 e^z}{\alpha(\delta-1)^2 - \delta M_1 N_1 e^{2z}}\left((\delta-1) \pm N_1\sqrt{\frac{\delta M_1}{\alpha N_1}}\, e^z\right) \qquad (15.49)$$

$$v_n = \frac{(\delta-1)N_1 e^z}{\alpha(\delta-1)^2 - \delta M_1 N_1 e^{2z}}\left(\alpha(\delta-1) \pm \delta N_1\sqrt{\frac{\alpha M_1}{\delta N_1}}\, e^z\right). \qquad (15.50)$$

Substitution confirms that (15.49), (15.50) is an exact solution to system (15.43), (15.44).

15.9 Conclusion

The geometric series method presented in this paper is sufficiently versatile and is capable of solving DDEs with a polynomial and rational right-hand side containing both explicitly defined elementary functions of the dependent variable and functions that are solutions of the given ODEs. The solution in all cases is represented by a rational fraction containing the degrees of the exponential function. In the considered examples, the indicated degrees varied from the first to the fourth. Theoretically, solutions of this type can be obtained directly using a rational fraction in exponential functions with arbitrary coefficients as an ansatz. However, in this case, it is necessary to solve a system of nonlinear algebraic equations with respect to these coefficients, and our experiments show that modern computer mathematics systems such as Maple or Mathematica, with the exception of the simplest cases, are not able to solve them. In the geometric series method, unknown coefficients are determined from solving systems of linear equations.

Like any other method, the geometric series method has several limitations. The construction of a solution in the form of a series in this method always starts with equating the constant term to zero, and then zeroing the coefficient at e^z. Therefore, the sum of the geometric series always contains in the numerator the term with e^z. If the exact solution of the equation contains in the numerator the highest degrees e^{2z}, e^{3z} etc., but does not contain e^z, then such a solution cannot be obtained by the geometric series method without preliminary transformations of the equation. Every rational fraction has a pole of natural order. Differentiation increases the order of the fraction pole by one, and integration decreases. If the analysis of the leading terms of the equation shows that its general solution has a fractional or negative pole order, then the solution to such an equation cannot be obtained without transforming the equation itself or its solution, written in the form of a series. So, in Sect. 15.6, we obtained a solution in the form of a fraction (15.35) with a simple pole only after differentiating the series (15.34). The fractional order of the evolutionary equation solution pole can be reduced to a natural number by the corresponding change of variable. However, in the case of DDE, determining the solution pole in general is difficult, since the form of the addition theorem $u_{n\pm k} = F(u_n)$, which is required to express the right-hand side of the equation in terms of a single function u_n, is specific for each class of functions.

Funding: The reported study was funded by RFBR, project number 20-01-00123.

References

1. Rafei, M., van Horssen, W.T.: On asymptotic approximations of first integrals for second order difference equations. Nonlinear Dyn. **61**, 535–551 (2010)
2. Rafei, M., van Horssen, W.T.: Solving systems of nonlinear difference equations by the multiple scales perturbation method. Nonlinear Dyn. **69**, 1509–1516 (2012)
3. Van Horssen, W.T., ter Brake, M.C.: On the multiple scales perturbation method for difference equations. Nonlinear Dyn. **55**, 401–418 (2009)
4. Andrianov, I.V., van Horssen, W.T.: Analytical approximations of the period of a generalized nonlinear van der Pol oscillator. J. Sound Vib. **295**(3), 1099–1104 (2006)
5. Smirnov, V.V., Manevitch, L.I., Strozzi, M., Pellicano, F.: Nonlinear optical vibrations of single-walled carbon nanotubes. 1. Energy exchange and localization of low-frequency oscillations. Physica D Nonlinear Phenom. **325**, 113–125 (2016)
6. Garifullin, R.N., Yamilov, R.I., Levi, D.: Classification of five-point differential–difference equations II. J. Phys. A Math. Theor. **51**, 065204 (2018)
7. Gubbiotti, G.: Algebraic entropy of a class of five-point differential-difference equations. Symmetry **11**(3), 432 (2019)
8. Baldwin, D., Goktas, U., Hereman, W.: Symbolic computation of hyperbolic tangent solutions for nonlinear differential–difference equations. Comput. Phys. Comm. **162**(3), 203–217 (2004)
9. Bochkarev, A.V., Zemlyanukhin, A.I.: The geometric series method for constructing exact solutions to nonlinear evolution equations. Comp. Math. Math. Phys. **57**(7), 1111–1123 (2017)
10. Zemlyanukhin, A.I., Bochkarev, A.V.: Perturbation method, Pade approximants and exact solutions of nonlinear mechanics equations. Materials Phys. Mech. **35**(1), 181–189 (2018)
11. Hirota, R.: Exact solution of the Korteweg – de Vries equation for multiple collisions of solitons. Phys. Rev. Lett. **27**(18), 1192–1194 (1971)
12. Yamilov, R.: Symmetries as integrability criteria for differential-difference equations. J. Phys. A Math. Gen. **39**, R541–R623 (2006)
13. Baker Jr., G.A., Graves-Morris, P.: Pade Approximants. Cambridge University Press, Cambridge (1996)
14. Adler, V.E.: Integrable Möbius invariant evolutionary lattices of second order. arXiv:1605.00018 (2016)
15. Ablowitz, M.J., Prinari, B., Trubatch, A.D.: Discrete and Continuous Nonlinear Schrödinger Systems. Cambridge University Press, Cambridge (2004)

Chapter 16
Harmonic Balance Method for the Stationary Response of Finite and Semi-infinite Nonlinear Dissipative Continua: Three Canonical Problems

Jiangyi Zhang, Enxhi Sulollari, Andrei B. Fărăgău, Federico Pisanò, Pim van der Male, Mario Martinelli, Andrei V. Metrikine, and Karel N. van Dalen

Abstract The Harmonic Balance Method (HBM) is often used to determine the stationary response of nonlinear discrete systems to harmonic loading. The HBM has also been applied to nonlinear continuous systems, but in many cases the nonlinearity consists of *discrete* nonlinear elements. This chapter demonstrates the application of the HBM to dissipative continua with *distributed* nonlinearity by analysing three canonical problems: (a) 1-D layer with a free surface and rigid base (interfering upward and downward propagating shear waves), (b) 1-D half-space with a rigid base (vertically propagating shear waves), and (c) 2-D axially symmetric semi-infinite medium with a circular cavity (radially propagating compressional waves), all of them subject to harmonic excitation at a boundary. Results show that systems (a) and (c) exhibit softening behaviour and super-harmonic resonances, while only the former displays multiple response amplitudes for certain excitation frequencies; the unique frequency-amplitude relationship of system (c) is due to the strong damping (i.e., radiation damping and internal dissipation). Furthermore, although system (b) essentially does not resonate, the third-harmonic component exhibits a maximum caused by the interplay between the dissipative and nonlinear effects, a phenomenon that also occurs in system (c). Finally, the considered systems have applications in earthquake and geotechnical engineering, among others, but the presented methodology is generic.

J. Zhang · E. Sulollari · A. B. Fărăgău · F. Pisanò · P. van der Male · M. Martinelli · A. V. Metrikine · K. N. van Dalen (✉)
Faculty Civil Engineering & Geosciences, Tu Delft, Delft, The Netherlands
e-mail: K.N.vanDalen@tudelft.nl

A. K. Abramian et al. (eds.), *Nonlinear Dynamics of Discrete and Continuous Systems*, Advanced Structured Materials 139,
https://doi.org/10.1007/978-3-030-53006-8_16

16.1 Introduction

The Harmonic Balance Method (HBM) is often applied to compute the stationary response of nonlinear discrete systems to harmonic loading. It is known to be very efficient as it does not require the simulation of the transient response before reaching the stationary/steady-state regime, and it directly yields frequency-response curves (i.e., vibration amplitudes versus excitation frequency). As for the latter, nonlinear systems can vibrate with different amplitudes for certain excitation frequencies (i.e., the frequency-amplitude relationship is not unique), and the different amplitude branches are naturally obtained by the HBM.

The HBM has been applied to nonlinear continuous systems too, but in many cases the nonlinearity consists of discrete nonlinear elements and is thus localized at one or multiple points; both finite [1] and infinite systems have been considered [2, 3]. To the best of the authors' knowledge, only few studies exist in which the HBM is applied to systems having distributed nonlinearity. One example is the paper by Chronopoulos [4], which deals with the response of an infinite composite structure having distributed but still localized nonlinearity (i.e., limited to a finite domain). The paper presents a general framework regarding the application of the HBM to the considered system, but the numerical example still only deals with a discrete nonlinear stiffening spring connecting two linearly behaving parts of the structure.

To fill the niche, the current chapter is devoted to the application of the HBM to continuous systems with distributed nonlinearity. Three canonical problems are considered, one with a finite-size continuum and two with semi-infinite continua, all subject to harmonic excitation at a boundary. The continua are assumed to be viscoelastic with the elastic part of the stress-strain relation being a nonlinear function of the strain. The considered systems and their stationary responses can be described as follows (see also Fig. 16.1):

(a) a 1-D layer with a free surface and rigid base: interfering upward and downward propagating shear waves
(b) a 1-D half-space with a rigid base: vertically propagating shear waves
(c) a 2-D axially symmetric medium of semi-infinite extent with a cylindrical cavity: radially propagating compressional waves.

Both semi-infinite systems ((b) and (c)) can still be considered to have localized nonlinearity in the sense that, due to amplitude decay of the waves propagating away from the source, it is possible to identify a fictitious surface beyond which the behaviour is essentially linear (infinitesimally small amplitude). The region beyond that surface is therefore replaced by an exact frequency-dependent non-reflective boundary condition that is applied at the fictitious surface, so that only a finite domain needs to be discretised when applying the HBM.

The study presented in this chapter aims to (1) demonstrate the application of the HBM to finite and semi-infinite continua, where in all cases the nonlinearity, due to the presence of dissipation, is activated only in a finite domain adjacent to the source, to describe the stationary wave propagation, and (2) use the HBM to

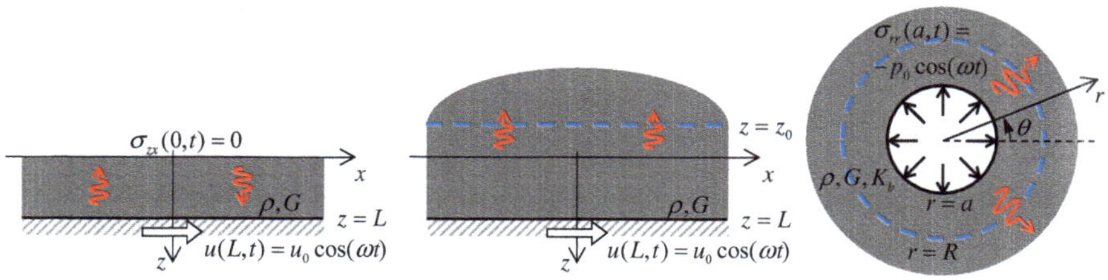

Fig. 16.1 The three canonical problems: the layer problem (panel (a)), the half-space problem (panel (b)), and the cavity problem (panel (c)). The dashed blue lines in panels (b) and (c) represent the location of the non-reflective boundary

reveal fundamental characteristics of the responses. As for the latter part, system (a) serves as a reference because its response resembles that of classical nonlinear discrete systems (e.g., the Duffing oscillator); the responses of systems (b) and (c) are analysed in comparison with that of system (a).

Apart from the use of a non-reflective boundary condition for the semi-infinite systems and the incorporation of distributed nonlinearity in the numerical examples, there are two pronounced differences in the current work compared to that of Chronopoulos [4]. First, the type of nonlinearity is different. The so-called hyperbolic model, well-known in soil mechanics [5], is used to introduce the strain dependence of the shear modulus, which obviously leads to nonlinearity of non-polynomial type. Second, the current study, being focused on frequency-response curves related to fixed observation points in the systems, nicely reveals typical features that have analogues in discrete systems such as the multiple branches of the curves, softening behaviour, as well as super-harmonic resonances.

The three problems considered in this chapter cannot be solved straightforwardly using perturbation methods such as the Method of Multiple Scales [6], because the considered nonlinearity is of non-polynomial type. In earthquake and geo-technical engineering, where the considered problems have applications, they are typically solved numerically by direct time integration or by resorting to a simplified *equivalent linear* representation of the material behaviour (i.e., by iteratively identifying equivalent linear visco-elastic properties corresponding to the strain levels in the soil) [5, 7]. However, the latter approach may overlook fundamental characteristics of the nonlinear system (e.g., multiple branches of the frequency-response curves), while it can be very cumbersome to reveal those using the former method. This chapter demonstrates that the HBM is an effective tool to reveal fundamental characteristics for systems of finite and semi-infinite dimension (with nonlinear behaviour limited to a finite domain). The methodology is generic and also applicable to problems of different nature (e.g., acoustics, electromagnetics).

16.2 Governing Equations for the 1-D Problems

We consider the 1-D nonlinear elastic systems with dissipation shown in Fig. 16.1(a), (b). In this section, the equation of motion as well as the boundary conditions are presented. To this end, the following stress-strain relation is considered:

$$\sigma_{zx} = 2G(\gamma)\varepsilon_{zx} + 2\eta\varepsilon_{zx,t} = G(\gamma)u_{,z} + \eta u_{,zt}, \tag{16.1}$$

where u denotes the horizontal displacement, σ_{zx} denotes the shear stress and $\varepsilon_{zx} = \frac{1}{2}u_{,z}$ the shear strain, $G(\gamma)$ the strain-dependent shear modulus, and η the viscosity; the subscript "$, t$" denotes a partial time derivative (and likewise "$, z$" a partial space derivative). The shear modulus is assumed to depend on the shear strain, which introduces nonlinearity, according to the hyperbolic model [5],

$$G(\gamma) = G_0 \frac{1}{1 + (\gamma/\gamma_{\text{ref}})^{\beta}}, \tag{16.2}$$

where G_0 is the small-strain shear modulus (used in linear continuum mechanics). Equation (16.2) was originally derived for pure shear conditions; however, γ in Eq. (16.2) can be generalised by using a representative deviatoric strain. In this work, we use the deviatoric component ε_{q} of the so-called octahedral strain [8], which for a 3-D continuum reads (ε_{ij}, with $i, j = \{x, y, z\}$, denote the different strain components)

$$\varepsilon_{\text{q}} = \tfrac{1}{3}\sqrt{2}\sqrt{(\varepsilon_{xx} - \varepsilon_{yy})^2 + (\varepsilon_{yy} - \varepsilon_{zz})^2 + (\varepsilon_{xx} - \varepsilon_{zz})^2 + 6\varepsilon_{xy}^2 + 6\varepsilon_{yz}^2 + 6\varepsilon_{zx}^2}. \tag{16.3}$$

For the 1-D case, γ is given as follows:

$$\gamma = \tfrac{3}{2}\varepsilon_{\text{q}} = \sqrt{3}|\varepsilon_{zx}| = \tfrac{1}{2}\sqrt{3}|u_{,z}|. \tag{16.4}$$

The quantity γ_{ref} is a reference shear strain whose value depends on the soil characteristics, and β is another material constant ($0 < \beta < 1$). Finally, a linear damping term is chosen in Eq. (16.1) for simplicity; the goal of this chapter is to demonstrate the application of the HBM, and incorporating a strain-dependent η would make the equation of motion unnecessarily complicated. Particularly, η is chosen to be inversely proportional to the dominant frequency ω_{d} [5]:

$$\eta = 2G_0\xi/\omega_{\text{d}}. \tag{16.5}$$

Here, ξ is the so-called material damping ratio. Note that the damping term would be fully frequency-independent (i.e., hysteretic damping) for the stationary response to harmonic excitation (with frequency $\omega = \omega_{\text{d}}$) if the system were linear. For the nonlinear response, which generally consists of multiple harmonics, this is not exactly the case.

Inserting Eq. (16.1) into Newton's second law, we obtain the equation of motion:

$$\rho u_{,tt} = \sigma_{zx,z} = \mathcal{M} u_{,zz}, \quad \mathcal{M} = G(\gamma)\left[1 - \beta \frac{(\gamma/\gamma_{\text{ref}})^\beta}{1 + (\gamma/\gamma_{\text{ref}})^\beta}\right] + \eta \partial_t, \tag{16.6}$$

where $\mathcal{M}$ is strain-dependent and contains a time-derivative operator $\partial_t = \partial/\partial t$; ρ is the mass density of the material. Clearly, in the small-strain limit where $\gamma \ll \gamma_{\text{ref}}$ (i.e., $G \to G_0$), Eq. (16.6) reduces to the classical linear wave equation (with loss term).

Boundary conditions for the layer

The layer is subject to continuous harmonic excitation at the base ($z = L$) and has a stress-free surface at $z = 0$, which leads to the following boundary conditions:

$$\sigma_{zx}(0, t) = 0 \quad \to \quad u_{,z}(0, t) = 0, \tag{16.7}$$

$$u(L, t) = u_0 \cos(\omega t). \tag{16.8}$$

Boundary conditions for the half-space

For the half-space the boundary condition at $z = L$ is the same (i.e., prescribed displacement, Eq. (16.8)). The boundary condition at $z = 0$, however, is replaced by the zero-displacement condition at infinite distance from the source. To comply with the latter, a non-reflective boundary condition is applied at $z = z_0$, so that the size of the domain that needs to be discretized in the HBM (see Sect. 16.6) can be kept minimal. For this to be correct, however, the behaviour at $z = z_0$ needs to be linear (i.e., $\gamma \ll \gamma_{\text{ref}}$), which implies that L needs to be large enough to ensure that the excited waves decay sufficiently before reaching the upper boundary. Assuming linearity and considering the semi-infinite medium above $z = z_0$, the non-reflective boundary condition can be derived based on the linear wave equation (Eq. (16.6) linearized). Using the analytical solution in the frequency ($\bar{\omega}$) domain, the following relation between applied stress and observed velocity v at $z = z_0$ can be found:

$$\hat{v}(z_0, \bar{\omega}) = \frac{1}{\sqrt{\rho(G_0 + \mathrm{i}\bar{\omega}\eta)}} \hat{\sigma}_{zx}(z_0, \bar{\omega}), \tag{16.9}$$

where the hat signifies that the quantity is considered in the frequency domain, i denotes the imaginary unit, and $\operatorname{Im} \sqrt{\rho(G_0 + \mathrm{i}\bar{\omega}\eta)} < 0$. Equation (16.9) expresses the dynamic reaction of the semi-infinite medium to excitation at the boundary ($z = z_0$), and the square root factor has the meaning of impedance. Using the linearized version of the stress-strain relation (Eq. (16.1)), Eq. (16.9) can be simplified to

$$\hat{u}_{,z}(z_0, \bar{\omega}) = \hat{C}(\bar{\omega}) \hat{v}(z_0, \bar{\omega}), \quad \hat{C}(\bar{\omega}) = \frac{\rho}{\sqrt{\rho(G_0 + \mathrm{i}\bar{\omega}\eta)}}. \tag{16.10}$$

When Eq. (16.10) is prescribed as boundary condition at $z = z_0$, the boundary will be non-reflective for harmonic waves of frequency $\bar{\omega}$ propagating towards $z \to -\infty$.

Transforming to the time domain using the inverse Fourier transform, Eq. (16.10) can be written as follows:

$$u_{,z}(z_0,t)=\int_{-\infty}^{\infty} C(t-\tau)v(z_0,\tau)\mathrm{d}\tau, \quad C(t)=\frac{\exp(-G_0 t/\eta)}{\sqrt{\pi\eta t/\rho}}H(t), \tag{16.11}$$

where $H(\ldots)$ denotes the Heaviside function. Clearly, when Eq. (16.11) is prescribed as boundary condition at $z=z_0$, the boundary will be non-reflective for any type of disturbance propagating towards $z\to-\infty$. To conclude, Eqs. (16.8) and (16.11) are the boundary conditions for the half-space problem.

16.3 HBM Applied to the 1-D Problems

According to the HBM, the following stationary solution, which is truncated at the third harmonic of the excitation frequency, is assumed that is periodic in time with period $T=2\pi/\omega$ [9–11]:

$$u(z,t)=U_{\mathrm{c}1}(z)\cos(\omega t)+U_{\mathrm{s}1}(z)\sin(\omega t)+U_{\mathrm{c}3}(z)\cos(3\omega t)+U_{\mathrm{s}3}(z)\sin(3\omega t). \tag{16.12}$$

Harmonics with higher-order integer multiples of ω (super-harmonics) as well as harmonics with integer fractions of ω (sub-harmonics) have been excluded; numerical examples showed that the importance of those is very small compared to the terms in Eq. (16.12), and results shown in Sect. 16.6 obtained with this solution are converged.

In accordance with the HBM, the adopted solution is substituted into the equation of motion (Eq. (16.6)), which is subsequently projected onto the different harmonics. This converts the partial differential equation into a set of four coupled nonlinear ordinary differential equations:

$$\begin{bmatrix} -\rho\omega^2 U_{\mathrm{c}1} \\ -\rho\omega^2 U_{\mathrm{s}1} \\ -9\rho\omega^2 U_{\mathrm{c}3} \\ -9\rho\omega^2 U_{\mathrm{s}3} \end{bmatrix} = \begin{bmatrix} m_{11} & m_{12} & m_{13} & m_{14} \\ m_{21} & m_{22} & m_{23} & m_{24} \\ m_{31} & m_{32} & m_{33} & m_{34} \\ m_{41} & m_{42} & m_{43} & m_{44} \end{bmatrix} \begin{bmatrix} U_{\mathrm{c}1,zz} \\ U_{\mathrm{s}1,zz} \\ U_{\mathrm{c}3,zz} \\ U_{\mathrm{s}3,zz} \end{bmatrix}, \tag{16.13}$$

with

$$m_{pq}=\frac{\omega}{\pi}\int_0^T \left[\mathcal{M}h_q(t)\right]h_p(t)\mathrm{d}t, \quad \mathbf{h}=[\cos(\omega t)\ \sin(\omega t)\ \cos(3\omega t)\ \sin(3\omega t)]^{\mathrm{T}}. \tag{16.14}$$

Projections of boundary conditions for the layer
Next to the equation of motion, the solution is also substituted into the boundary conditions (Eqs. (16.7) and (16.8)), which are subsequently projected onto the different harmonics as well. This leads to the eight required conditions:

$$U_{c1,z}(0) = 0, \qquad U_{s1,z}(0) = 0, \qquad U_{c3,z}(0) = 0, \qquad U_{s3,z}(0) = 0, \tag{16.15}$$
$$U_{c1}(L) = u_0, \qquad U_{s1}(L) = 0, \qquad U_{c3}(L) = 0, \qquad U_{s3}(L) = 0. \tag{16.16}$$

Projections of boundary conditions for the half-space
For the half-space, the projections of the boundary condition at $z = L$ (Eq. (16.16)) hold as well. Finding the projections of the non-reflective boundary condition is slightly less straightforward. First, the cosine terms of the HBM solution are discussed (which can be done separately because of the superposition principle; the behaviour is linear at z_0). Substituting the complex representation of $u(z_0, t) \propto \cos(\Omega_p t)$, with $\Omega_p = p\omega$ and $p = \{1, 3\}$, into the integral of Eq. (16.11), we obtain

$$\begin{aligned}
&\frac{1}{2}\int\limits_{-\infty}^{\infty} C(t-\tau)\mathrm{i}\Omega_p\left(\mathrm{e}^{\mathrm{i}\Omega_p\tau} - \mathrm{e}^{-\mathrm{i}\Omega_p\tau}\right)\mathrm{d}\tau \\
&= \frac{\mathrm{i}\Omega_p}{4\pi}\int\limits_{-\infty}^{\infty}\int\limits_{-\infty}^{\infty} \hat{C}(\bar{\omega})\mathrm{e}^{\mathrm{i}\bar{\omega}(t-\tau)}\left(\mathrm{e}^{\mathrm{i}\Omega_p\tau} - \mathrm{e}^{-\mathrm{i}\Omega_p\tau}\right)\mathrm{d}\bar{\omega}\mathrm{d}\tau \\
&= \frac{\mathrm{i}\Omega_p}{2}\int\limits_{-\infty}^{\infty} \hat{C}(\bar{\omega})\mathrm{e}^{\mathrm{i}\bar{\omega}t}\left(\delta(\bar{\omega} - \Omega_p) - \delta(\bar{\omega} + \Omega_p)\right)\mathrm{d}\bar{\omega} \\
&= \frac{\mathrm{i}\Omega_p}{2}\left(\hat{C}(\Omega_p)\mathrm{e}^{\mathrm{i}\Omega_p t} - \hat{C}(-\Omega_p)\mathrm{e}^{-\mathrm{i}\Omega_p t}\right) = -\Omega_p|\hat{C}(\Omega_p)|\sin\left(\Omega_p t + \varphi(\Omega_p)\right),
\end{aligned} \tag{16.17}$$

where the well-known integral representation of the Dirac function has been used, as well as $\hat{C}(-\Omega_p) = \hat{C}^*(\Omega_p)$ and $\hat{C}(\Omega_p) = |\hat{C}(\Omega_p)|\exp(\mathrm{i}\varphi(\Omega_p))$; furthermore, $t \to \infty$ is implicit in Eq. (16.17) as the response is assumed to have reached the stationary regime. Next, substituting the complex representation of $u(z_0, t) \propto \sin(\Omega_p t)$ into the integral of Eq. (16.11), we obtain

$$\frac{\Omega_p}{2}\int\limits_{-\infty}^{\infty} C(t-\tau)\left(\mathrm{e}^{\mathrm{i}\Omega_p\tau} + \mathrm{e}^{-\mathrm{i}\Omega_p\tau}\right)\mathrm{d}\tau = \Omega_p|\hat{C}(\Omega_p)|\cos\left(\Omega_p t + \varphi(\Omega_p)\right). \tag{16.18}$$

Now, collecting all terms, the non-reflective boundary condition (Eq. (16.11)) with the stationary solution (Eq. (16.12)) substituted can be expressed as follows:

$$\begin{aligned}U_{\mathrm{c1},z}(z_0)\cos(\omega t)+U_{\mathrm{s1},z}(z_0)\sin(\omega t)+U_{\mathrm{c3},z}(z_0)\cos(3\omega t)+U_{\mathrm{s3},z}(z_0)\sin(3\omega t)=\\+\omega|\hat{C}(\omega)|\left[-U_{\mathrm{c1}}(z_0)\sin\left(\omega t+\varphi(\omega)\right)+U_{\mathrm{s1}}(z_0)\cos\left(\omega t+\varphi(\omega)\right)\right]\\+3\omega|\hat{C}(3\omega)|\left[-U_{\mathrm{c3}}(z_0)\sin\left(3\omega t+\varphi(3\omega)\right)+U_{\mathrm{s3}}(z_0)\cos\left(3\omega t+\varphi(3\omega)\right)\right].\end{aligned} \tag{16.19}$$

Subsequently, we obtain the following four projections (next to the four in Eq. (16.16)):

$$\begin{aligned}U_{\mathrm{c1},z}(z_0)&=\omega|\hat{C}(\omega)|\left[-U_{\mathrm{c1}}(z_0)\sin\left(\varphi(\omega)\right)+U_{\mathrm{s1}}(z_0)\cos\left(\varphi(\omega)\right)\right],\\U_{\mathrm{s1},z}(z_0)&=\omega|\hat{C}(\omega)|\left[-U_{\mathrm{c1}}(z_0)\cos\left(\varphi(\omega)\right)-U_{\mathrm{s1}}(z_0)\sin\left(\varphi(\omega)\right)\right],\\U_{\mathrm{c3},z}(z_0)&=3\omega|\hat{C}(3\omega)|\left[-U_{\mathrm{c3}}(z_0)\sin\left(\varphi(3\omega)\right)+U_{\mathrm{s3}}(z_0)\cos\left(\varphi(3\omega)\right)\right],\\U_{\mathrm{s3},z}(z_0)&=3\omega|\hat{C}(3\omega)|\left[-U_{\mathrm{c3}}(z_0)\cos\left(\varphi(3\omega)\right)-U_{\mathrm{s3}}(z_0)\sin\left(\varphi(3\omega)\right)\right].\end{aligned} \tag{16.20}$$

16.4 Governing Equations for the Cavity Problem

We now consider the problem of a cylindrical cavity (radius a) in an infinite elastic body, with harmonic stress applied at the cavity surface (Fig. 16.2c). It is assumed that the applied stress does not vary with the out-of-plane coordinate, which renders the problem two-dimensional, and is independent of the circumferential coordinate θ as well, which renders the problem axi-symmetric. Waves will thus propagate only in radial (r) direction, and the response will be independent of θ.

The relevant stresses in the current problem are the radial and tangential normal stresses σ_{rr} and $\sigma_{\theta\theta}$. The associated stress-strain relations are

$$\sigma_{rr}=K_{\mathrm{b}}\varepsilon_{\mathrm{vol}}+2G(\gamma)e_{rr}+2\eta e_{rr,t},\quad \sigma_{\theta\theta}=K_{\mathrm{b}}\varepsilon_{\mathrm{vol}}+2G(\gamma)e_{\theta\theta}+2\eta e_{\theta\theta,t}, \tag{16.21}$$

where K_{b} denotes the bulk modulus (constant), η is again the viscosity (Eq. (16.5)), and $\varepsilon_{\mathrm{vol}}=\varepsilon_{rr}+\varepsilon_{\theta\theta}$ is the volume strain, with the radial and tangential strains defined as $\varepsilon_{rr}=u_{,r}$ and $\varepsilon_{\theta\theta}=u/r$, respectively; the quantities $e_{rr}=\varepsilon_{rr}-\frac{1}{3}\varepsilon_{\mathrm{vol}}$ and $e_{\theta\theta}=\varepsilon_{\theta\theta}-\frac{1}{3}\varepsilon_{\mathrm{vol}}$ are the deviatoric strains in radial and tangential directions, respectively; u denotes the radial displacement. The shear modulus $G(\gamma)$ is again strain-dependent (Eq. (16.2)), and considering the deviatoric strain defined in Eq. (16.3) (in cylindrical coordinates instead of Cartesian ones), γ for the current problem becomes

$$\gamma=\tfrac{1}{2}\sqrt{2}\sqrt{\varepsilon_{rr}^2+(\varepsilon_{rr}-\varepsilon_{\theta\theta})^2+\varepsilon_{\theta\theta}^2}. \tag{16.22}$$

Substituting the stresses into Newton's second law [12],

$$\rho u_{,tt}=\sigma_{rr,r}+\tfrac{1}{r}(\sigma_{rr}-\sigma_{\theta\theta}), \tag{16.23}$$

where ρ is again the mass density, and using the definitions above, the following nonlinear equation of motion is obtained:

$$\rho u_{,tt} = \left(K_b + \tfrac{4}{3}G(\gamma) + \tfrac{4}{3}\eta\partial_t\right)\left(u_{,rr} + \tfrac{1}{r}u_{,r} - \tfrac{u}{r^2}\right) + \tfrac{2}{3}G_{,\gamma}(\gamma)\gamma_{,r}\left(2u_{,r} - \tfrac{u}{r}\right). \tag{16.24}$$

In the small-strain limit, the last term of this equation vanishes and $G(\gamma) \to G_0$ in the first term of the right-hand side. Equation (16.24) can be written with $u_{,rr}$ explicitly as

$$\begin{aligned} \rho u_{,tt} &= \mathcal{M} u_{,rr} + \mathcal{Q}, \\ \mathcal{M} &= K_\mathrm{b} + \tfrac{4}{3}G(\gamma) + \tfrac{4}{3}\eta\partial_t + \tfrac{1}{3}G_{,\gamma}(\gamma)\tfrac{1}{\gamma}(2u_{,r} - \tfrac{u}{r})^2, \\ \mathcal{Q} &= \left[K_b + \tfrac{4}{3}G(\gamma) + \tfrac{4}{3}\eta\partial_t + \tfrac{1}{3}G_{,\gamma}(\gamma)\tfrac{1}{\gamma}(2\tfrac{u}{r} - u_{,r})(2u_{,r} - \tfrac{u}{r})\right]\tfrac{1}{r}\left(u_{,r} - \tfrac{u}{r}\right). \end{aligned} \tag{16.25}$$

As boundary condition, there is the continuous excitation at the cavity surface:

$$\sigma_{rr}(r = a, t) = -p_0 \cos(\omega t). \tag{16.26}$$

Next to that, the zero-displacement condition at infinite distance from the source holds. To comply with this, a non-reflective boundary condition is applied like for the half-space problem. Assuming linear behaviour for $r \geq R$, the relation between applied stress and the response at $r = R$ can be derived based on the linearized equation of motion (Eq. (16.24)) in the frequency domain (similar approach as in Sect. 16.2). Rearrangement of terms leads to the following expression (cf. Eq. (16.10)):

$$\hat{u}_{,r}(R, \bar{\omega}) = \hat{C}(\bar{\omega})\hat{v}(R, \bar{\omega}), \quad \hat{C}(\bar{\omega}) = \frac{1}{\mathrm{i}\bar{\omega}}\frac{kRH_0^{(2)}(kR) - H_1^{(2)}(kR)}{RH_1^{(2)}(kR)}, \tag{16.27}$$

where $H_n^{(2)}(\dots)$ denotes the Hankel function of the second kind and order n, and $k = \bar{\omega}/((K_\mathrm{b} + \tfrac{4}{3}(G_0 + \mathrm{i}\bar{\omega}\eta))/\rho)^{1/2}$ is the complex-valued wavenumber of the compressional wave whose imaginary part is taken negative. Transforming Eq. (16.27) to the time domain, the following expression is obtained:

$$u_{,r}(R, t) = \int_{-\infty}^{\infty} C(t - \tau)v(R, \tau)\mathrm{d}\tau. \tag{16.28}$$

The expression for $C(t)$ is not given here, as it is not needed for the application of the HBM; the frequency-domain expression suffices (cf. Eqs. (16.19) and (16.20)).

16.5 HBM Applied to the Cavity Problem

Similar to that in Sect. 16.3, the HBM solution is taken as

$$u(r,t) = U_{c1}(r)\cos(\omega t) + U_{s1}(r)\sin(\omega t) + U_{c3}(r)\cos(3\omega t) + U_{s3}(r)\sin(3\omega t). \tag{16.29}$$

Substituting this into Eq. (16.25) and projecting each term onto the different harmonics, we get a system of four coupled nonlinear ordinary differential equations:

$$\begin{bmatrix} -\rho\omega^2 U_{c1} - q_1 \\ -\rho\omega^2 U_{s1} - q_2 \\ -9\rho\omega^2 U_{c3} - q_3 \\ -9\rho\omega^2 U_{s3} - q_4 \end{bmatrix} = \begin{bmatrix} m_{11} & m_{12} & m_{13} & m_{14} \\ m_{21} & m_{22} & m_{23} & m_{24} \\ m_{31} & m_{32} & m_{33} & m_{34} \\ m_{41} & m_{42} & m_{43} & m_{44} \end{bmatrix} \begin{bmatrix} U_{c1,rr} \\ U_{s1,rr} \\ U_{c3,rr} \\ U_{s3,rr} \end{bmatrix}, \tag{16.30}$$

where $q_p = \frac{\omega}{\pi}\int_0^T Qh_p(t)\mathrm{d}t$ (with $\mathbf{h}$ defined in Eq. (16.14)), and the expressions for m_{pq} are the same as those in Eq. (16.14), although $\mathcal{M}$ is defined differently.

Substituting the HBM solution (Eq. (16.29)) into the boundary condition at the cavity surface (Eq. (16.26)), employing Eqs. (16.21), (16.22) and the strain-displacement relations, and projecting the result onto the different harmonics, the following four conditions are obtained (for the cosine and sine terms, respectively):

$$K_\mathrm{b}\left(U_{cp,r}(a) + \tfrac{1}{a}U_{cp}(a)\right) - \tfrac{2}{3}\eta\Omega_p\left(2U_{sp,r}(a) - \tfrac{1}{a}U_{sp}(a)\right) + 2\tfrac{\omega}{\pi}g_{cp} = F_p, \tag{16.31}$$

$$K_\mathrm{b}\left(U_{sp,r}(a) + \tfrac{1}{a}U_{sp}(a)\right) + \tfrac{2}{3}\eta\Omega_p\left(2U_{cp,r}(a) - \tfrac{1}{a}U_{cp}(a)\right) + 2\tfrac{\omega}{\pi}g_{sp} = 0, \tag{16.32}$$

where $\Omega_p = p\omega$, $p = \{1,3\}$, $\mathbf{F} = [-p_0\ 0]^\mathrm{T}$, and

$$g_{cp} = \left.\int_0^T G(\gamma)e_{rr}\cos(\Omega_p t)\mathrm{d}t\right|_{r=a}, \quad g_{sp} = \left.\int_0^T G(\gamma)e_{rr}\sin(\Omega_p t)\mathrm{d}t\right|_{r=a}. \tag{16.33}$$

The non-reflective boundary condition (Eq. (16.28)) results in the same expression as in Eq. (16.19), provided replacements $\partial_z \to \partial_r$ and $z = z_0 \to r = R$ are made, and $\hat{C}$ is taken according to Eq. (16.27). When projected onto the different harmonics, the same four conditions are obtained as in Eq. (16.20) (assuming the same replacements).

16.6 Results and Discussion

In this section, the results for the three canonical problems are presented and discussed. The focus of this section is partly on the validation of the HBM and mostly on the investigation of fundamental characteristics of the responses. For all results presented in this section, the boundary-value problems were solved using finite difference approximations (e.g., *bvp4c* algorithm in Matlab) while the numerical evaluation of integrals was performed using the trapezoidal rule. The parameters that were kept constant in the simulations are given in the following: $\gamma_{\text{ref}} = 8.7 \times 10^{-4}$, $\beta = 0.91$, $u_0 = 0.001$ m, $G_0 = 111.86 \times 10^6$ Pa, $\rho = 2009.8\,\text{kg}\,\text{m}^{-3}$, $p_0 = 50$ kPa.

16.6.1 Layer Problem

The response of the harmonically excited 1-D layer (system (a) in Fig. 16.1) obtained with the HBM is compared to the one obtained with the time-integration method (see Appendix) in panel (a) of Fig. 16.2. The results reveal a good agreement between the two methods, thus validating the HBM. To suppress the initial transients in the time-integration method, the prescribed initial conditions are based on the stationary response obtained using HBM. Even though the initial transients have been suppressed, the time-integration method requires significantly more computational effort than the HBM, proving that the latter is a valuable approach for such problems.

Focusing on the response characteristics, we show in Fig. 16.2 the spatial profile of the amplitudes A_1 of the fundamental component (panel (b)) and A_3 of the higher-

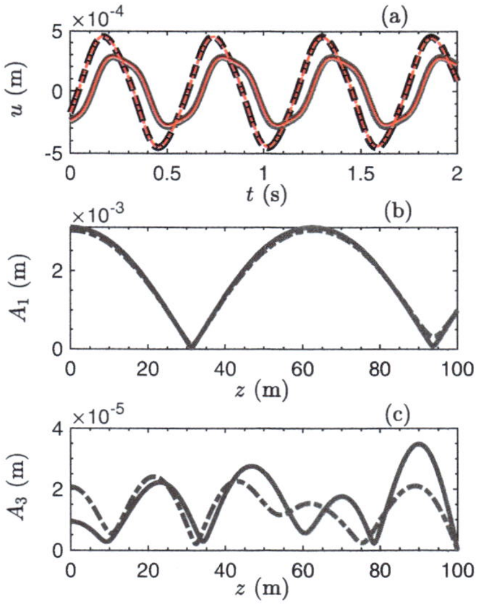

Fig. 16.2 The stationary response of the layer subjected to harmonic excitation ($f = 0.3$ Hz) at $L = 100$ m, obtained with HBM (black lines) and with the time-integration method (red lines). The temporal profile of the response at 93 m is shown in (a). The amplitudes A_1 of the fundamental component and A_3 of the higher-harmonic component are shown in (b) and (c), respectively. $\xi = 0.02$ for the strongly damped case (dashed lines) and $\xi = 0.005$ for the weakly damped case (continuous lines)

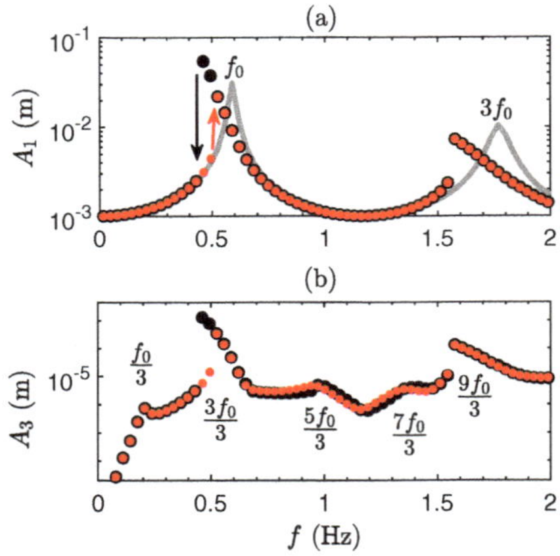

Fig. 16.3 FRCs obtained with HBM evaluated at the free surface ($z = 0$) of the layer subjected to harmonic excitation at $L = 100$ m. The different amplitude branches are obtained by imposing the initial guess as the solution obtained at the previous frequency value: frequency sweep up (red dots) and frequency sweep down (black dots). The grey line shows the FRC for the linear system

harmonic component (panel (c)). These profiles are interference patterns of upward and downward propagating waves. For this system, as for the other ones presented in this chapter, the overall magnitude of the higher-harmonic component is significantly smaller than that of the fundamental component. However, for specific locations (e.g., $z = 93$ m), A_1 and A_3 have comparable magnitudes, especially for the weakly damped case. The significant contribution of the higher harmonic to the response at this specific location causes the time history (weakly damped system in panel (a)) to exhibit a considerable distortion from the harmonic pattern of the excitation. However, the increased dissipation seems to slightly smooth the spatial profiles of A_1 and A_3 (i.e., A_1 slightly increases at $z = 93$ m while A_3 decreases), and consequently, leads to a less pronounced distortion of the response's time history. Furthermore, it is interesting to note that for $z = 93$ m, the overall magnitude of u in the weakly damped system is smaller than that of the strongly damped system. This is caused by the strong destructive interference of the fundamental and super-harmonic components for the weakly damped system, which is less pronounced for the strongly damped system.

The frequency-response curves (FRCs) of the layer problem obtained with the HBM are presented in Fig. 16.3. The observed features of this continuous system (the response of which is composed of propagating shear waves) are very similar to those observed for discrete systems (e.g., the Duffing oscillator [9, 10]). The resonance peaks observed in the FRCs are due to the interference of upward and downward propagating waves. One important characteristic of the layer problem is the non-uniqueness of the stationary response. More specifically, the system can have different stationary responses (which can be reached using different initial conditions in the time-integration method). This is clearly depicted in Fig. 16.3, both for the fundamental component (panel (a)) and for the super-harmonic component (panel (b)); for an excitation frequency of $f = 0.5$ Hz, Fig. 16.3 shows that the system has two solutions (i.e., *branches*). This means that for increasing excitation frequency, the response amplitude follows a smooth increase until a certain frequency value at which it abruptly *jumps* to a different value (depicted in panel (a) by the red arrow);

the opposite occurs for decreasing excitation frequency (black arrow in panel (a)). In actual fact, there are most likely at least three solution branches close to the resonance peak; however, the third branch (which connects the two observed ones) is unstable and cannot be obtained directly. To obtain the unstable branch, additional numerical techniques can be applied (e.g., path-continuation techniques [13]), but this is outside the scope of the present work.

Another important aspect of the system is that super-harmonic resonances can take place (at approximately $N\frac{f_0}{3}$, where f_0 is the first resonance frequency of the linear system and $N = 1, 3, 5, 7, \dots$). This can be observed in the FRC related to A_3. To exemplify this, we take the first peak at approximately $\frac{f_0}{3}$, where the harmonic excitation is $u_0 \cos(\frac{\omega_0}{3}t)$ with $\omega_0 \approx 2\pi f_0$. In this case, the displacement at the free surface is $u = A_1(0)\cos(\frac{\omega_0}{3}t + \Phi_1) + A_3(0)\cos(\omega_0 t + \Phi_3)$. The fundamental component $A_1(0)\cos(\frac{\omega_0}{3}t + \Phi_1)$ is far from the resonance, and thus there is no peak at approximately $\frac{f_0}{3}$ in the FRC related to A_1. However, the generated super-harmonic $A_3(0)\cos(\omega_0 t + \Phi_3)$ reaches the first resonance of the system, and we can therefore observe the peak in the FRC related to A_3.

Finally, as a consequence of the nonlinear model we employed, the resonance peaks of the nonlinear system are shifted and tilted to the left (towards lower frequencies) compared to the linear system. This is caused by the softening behaviour of the hyperbolic model (i.e., larger strains cause a decrease in the shear modulus).

16.6.2 Half-Space Problem

The 1-D half-space subjected to harmonic excitation at the lower boundary is analysed in the following (system (b) in Fig. 16.1). For the corresponding linear half-space, the stationary response contains solely upward propagating shear waves, exhibiting an amplitude decay caused by the system's dissipation. Taking advantage of this amplitude decay for the nonlinear system, the non-reflective boundary condition (NRBC) can be prescribed at a distance from the source (as discussed in Sect. 16.3) where the small strains cause the system to essentially behave linearly (i.e., $G(\gamma_{\max})/G_0 \approx 1$). The normalised minimum shear modulus is presented in panel (a) of Fig. 16.4 for two locations of the NRBC, namely at $z_0 = 0$ m and at $z_0 = -600$ m. The good agreement of the spatial profiles of A_1 (panel (b)) and A_3 (panel (c)) for the two positions of the NRBC suggests that $G(\gamma_{\max})/G_0 > 0.95$ is sufficient to yield correct results.

Figure 16.5 presents the FRCs of the nonlinear half-space obtained with the HBM and with the time-integration method (see Appendix). Again, a good agreement between the results obtained with the two solution methods can be observed. Also for this system, the HBM takes a fraction of the computational effort required by the time-integration method, which enables the computation of complete FRCs for an extensive range of frequencies.

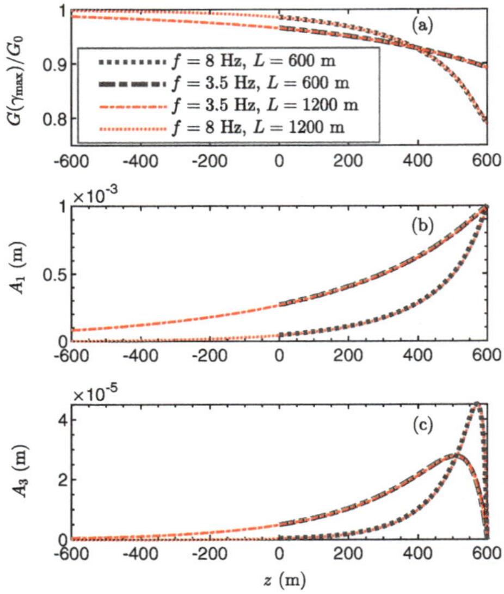

Fig. 16.4 The spatial profile of the response of the half-space with a fictitious surface at $z_0 = 0$ (black lines) and at $z_0 = -600$ m (red lines), and subjected to harmonic excitation at $L = 600$ m. The spatial profile of the normalised minimum (i.e., for the maximum strain) shear modulus is shown in (a). The amplitudes A_1 of the fundamental component and A_3 of the higher harmonic are shown in (b) and (c), respectively

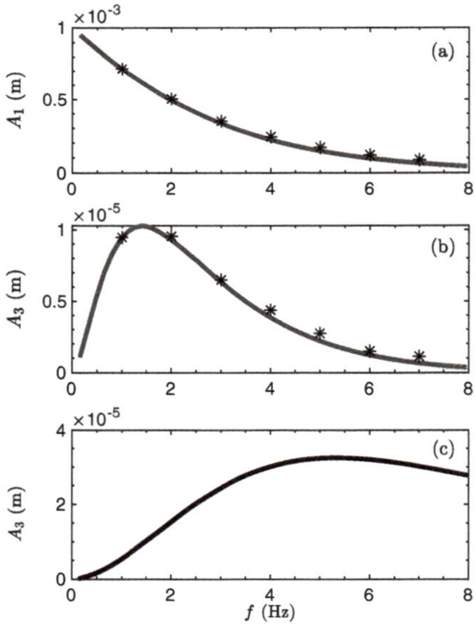

Fig. 16.5 FRCs evaluated at the location of the fictitious surface $z_0 = 0$ (panels (a) and (b)) and at $z = 510$ m (panel (c)) of the half-space subjected to harmonic excitation at $L = 600$ m. The HBM results are represented by solid lines while time-integration method results by asterisks

As for the response characteristics, the half-space differs significantly from the layer (Sect. 16.6.1). Unlike the layer, the spatial profile of the fundamental component (Fig. 16.4(b)) does not exhibit any peaks (i.e., it continuously decreases with increasing distance from the source). As the shear wave propagates away from the source ($z = 600$ m), A_1 decreases because of the nonlinearity (i.e., energy is transferred to the higher harmonic) and the dissipative effect. We observe that A_1 decreases faster in space for $f = 8$ Hz than for $f = 3.5$ Hz, which is due to the shorter wave length of the former; given that a certain number of wavelengths is required for a complete decay, the high-frequency wave clearly needs a smaller distance to be attenuated.

Unlike the fundamental component, the amplitude profile A_3 of the higher-harmonic component (panel (c) in Fig. 16.4) does not decay monotonically, but has a maximum at a certain distance from the source. This is not a resonance, but it can be explained based on the interplay of dissipative and nonlinear effects, as follows. In similar non-dissipative nonlinear systems, as the wave propagates in the so-called pre-shock region, the continuously generated higher harmonics may accumulate leading to the formation of shock waves (i.e., the waveform develops a vertical tangent causing a discontinuity) [14]. On the one hand, the larger the excitation frequency, the shorter the shock distance, that is, a higher excitation frequency is accompanied by a faster generation of higher harmonics (see $f = 3.5$ Hz compared to $f = 8$ Hz in Fig. 16.4(b)). On the other hand, a higher frequency is accompanied by a faster decrease of amplitude, as explained previously for A_1. In the end, the interplay between the dissipative and the nonlinear effects causes the observed peak in the spatial profile of the higher-harmonic component; furthermore, the higher the frequency, the closer the peak lies to the source.

A similar reasoning can be used to explain the FRCs of the response (Fig. 16.5). The half-space essentially does not resonate, and therefore, the FRC of A_1 (panel (a) in Fig. 16.5) does not exhibit peaks. However, like for the spatial profile, also the FRC of A_3 (panels (b) and (c) in Fig. 16.5) exhibits a maximum, meaning that the generation of the third harmonic is relatively powerful at the corresponding frequency. This is caused by the same interplay between the dissipative and the nonlinear forces as explained for the spatial profile. More specifically, at a given point of observation, one may be in the ascending part of the A_3 profile or in the descending part, depending on the frequency, which leads to an extreme at a specific frequency in the FRC of A_3. The frequency at which the peak is observed is expected to vary with distance from the source, which is confirmed by the comparison of panels (b) and (c) of Fig. 16.5.

16.6.3 Cavity Problem

Here, we analyse the cavity problem with harmonic stress applied at the cavity surface (system (c) in Fig. 16.1). In this subsection, we focus only on the characteristics of the response, because the validation of the HBM yields results similar to the ones for the layer and the half-space.

Unlike the half-space, the system with the cavity has an additional mechanism that makes the response decrease with distance from the source (i.e., decay mechanism), namely the geometric energy spreading. For the linear system, the strain decay caused by the geometric spreading alone is proportional to $1/r$ for large values of r [12]. The decay is beneficial from the computational point of view because it causes the nonlinear system to have an almost linear behaviour ($G(\gamma_{\max})/G_0 \approx 1$) just a short distance away from the cavity, and thus enables the NRBC to be placed

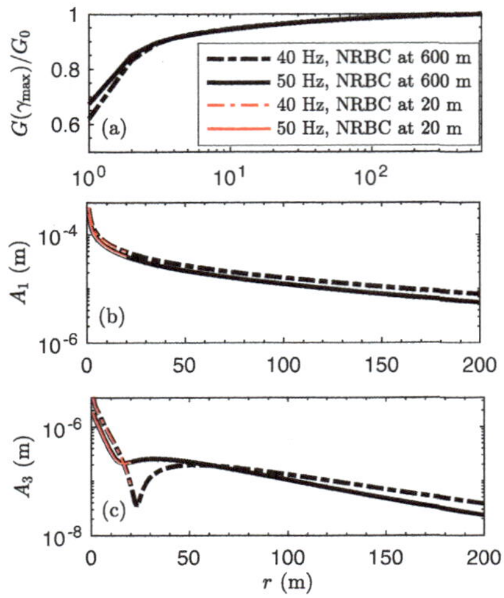

Fig. 16.6 Spatial profile of the response in the 2-D semi-infinite medium with a fictitious surface at $R = 600$ m (thick black lines) and $R = 20$ m (thin red lines) and a circular cavity with $a = 1$ m subjected to harmonic excitation. The spatial profile of the normalised minimum (i.e., for the maximum strain) shear modulus is shown in (a). The amplitudes A_1 of the fundamental component and A_3 of the higher-harmonic component are shown in (b) and (c), respectively

close to the source. This can be seen in Fig. 16.6 where the response for the fictitious surface located at $R = 20$ m and at $R = 600$ m yield the same results (in the computational domain) while the computational effort is significantly smaller for the former. However, by investigating only a short region of the system, we might miss some interesting phenomena as can be seen in Fig. 16.6 beyond $r = 20$ m.

Similar to the half-space, the amplitude A_1 of the fundamental component (panel (b) in Fig. 16.6) exhibits a continuous decay with increasing r and the amplitude A_3 of the higher-harmonic component (panel (c) in Fig. 16.6) exhibits a small peak (at $r \approx 50$ m for $f = 40$ Hz and at $r \approx 35$ m for $f = 50$ Hz). The peak and its shift is caused by the same interplay between nonlinear and dissipative forces encountered in the half-space problem; a higher excitation frequency leads to a faster higher-harmonic generation (nonlinear part) while also leading to a stronger decrease of amplitude during wave propagation (dissipation part). However, we can observe in Fig. 16.6(c) that in the cavity problem the higher harmonic is already generated at location of the source, unlike the half-space problem, caused by the different type of excitation (imposed stress).

The FRCs of the response in the cavity problem are presented in Fig. 16.7. Unlike the half-space, the semi-infinite 2-D system with radially propagating waves does have one resonance peak. At the cavity surface, the resonance is observed close to f_0 (i.e., the resonance frequency of the linear system) for the fundamental component (panel (a)) and, corresponding to that, there is a super-harmonic resonance close to $\frac{f_0}{3}$ in the higher-harmonic component (panel (b)). The resonance can occur because of the constrained rigid body motion of the system due to the symmetry in the loading, which provides a static stiffness/restoring force even in the limit of $\omega \to 0$. The resonance peak of the fundamental component is shifted slightly to the left (towards lower frequencies) compared to the linear system due to the softening behaviour of the material, similarly to the layer problem (Fig. 16.3). Moreover, the softening

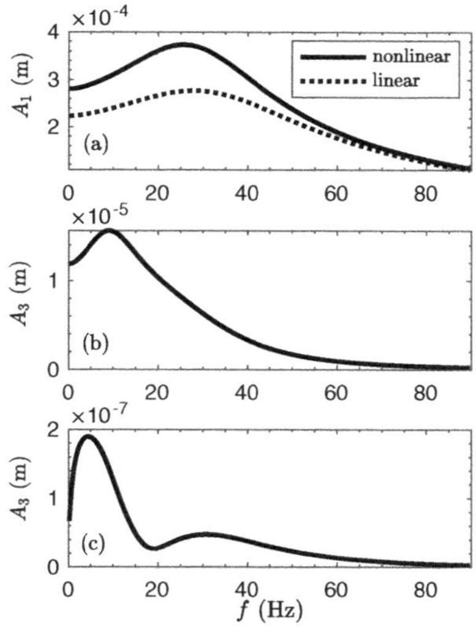

Fig. 16.7 FRCs evaluated at $r = 1$ m (panels (a) and (b)) and at $r = 200$ m (panel (c)) of the 2-D semi-infinite medium with a fictitious surface at $r = 600$ m and a circular cavity with $a = 1$ m subjected to harmonic stress excitation

behaviour causes an increase in the magnitude of A_1 compared to the linear system. However, unlike for the layer problem, in the cavity problem, the FRCs do not exhibit multiple branches, meaning that the system can only vibrate with one amplitude for a single excitation frequency. The uniqueness of the response is probably due to the relatively strong damping mechanisms present in this system (radiation damping and internal dissipation); the multiple branches can be observed to vanish also in the layer problem if the damping is increased.

Similar to the half-space, the FRC of the higher harmonic exhibits an additional peak compared to the fundamental component (in the half-space problem, A_3 has one peak while A_1 has none). In panel (c) of Fig. 16.7, the FRC of the higher-harmonic component is presented for $r = 200$ m. Here, the FRC exhibits two peaks. The one at the lower frequencies is related to the super-harmonic resonance encountered at the cavity surface (panel (b)), although shifted slightly, while the one at higher frequencies ($f \approx 30$ Hz) is the additional peak. The shift of the peak related to the super-harmonic resonance and the additional peak are again caused by the interplay between the dissipation effect and the nonlinear effect, as was explained for the half-space.

16.7 Conclusion

In this chapter, the application of the Harmonic Balance Method (HBM) to dissipative continua with distributed nonlinearity has been demonstrated. Three canonical problems have been analysed: (a) 1-D layer with a free surface and rigid base (interfering upward and downward propagating shear waves), (b) 1-D half-space with a rigid base (vertically propagating shear waves), and (c) 2-D axially symmetric semi-infinite medium with a circular cavity (radially propagating compressional waves), and all of them subject to harmonic excitation at a boundary. The HBM has been validated against a more conventional time-domain method, and it proved to be more efficient as it does not require the simulation of the transient response before reaching the stationary regime, while also directly yielding frequency-response curves (FRCs).

The HBM was used to reveal fundamental response characteristics of the three systems. Results show that for system (a), the FRCs related to the different harmonics display softening behaviour and multiple branches as well as super-harmonic resonances at approximately integer fractions of the resonance frequencies of the corresponding linear system, respectively. A super-harmonic resonance is also encountered for the semi-infinite cavity system (system (c)), even though the response consists of propagating waves. Its FRCs exhibit softening behaviour too, although they are not found to have multiple branches. The latter is probably due to the strong damping mechanisms, namely the radiation damping and internal dissipation; increased dissipation tends to smoothen resonance peaks. System (b) essentially does not resonate, but the FRC as well as the spatial profile related to the third-harmonic amplitude does exhibit a maximum caused by the interplay between the dissipative and nonlinear effects. Moreover, this interplay also causes an additional peak to emerge in the third-harmonic component of system (c).

To conclude, the HBM is an effective tool for revealing fundamental characteristics of nonlinear dissipative continua of finite and semi-infinite dimension. The considered systems have applications in earthquake and geotechnical engineering, among others, but the presented methodology is generic.

Appendix: Numerical Solution for 1-D Problems

As reference for the HBM solution, the governing equations for the 1-D systems (Eq. (16.6) with Eqs. (16.7) and (16.8), or with Eqs. (16.8) and (16.11)) are solved numerically. Here, we pay special attention to the incorporation of the non-reflective boundary condition (Eq. (16.11)) in the solution scheme (for the half-space problem). First of all, the range of integration is limited to $\tau = [0, t]$ accounting for the Heaviside function in $C(t)$ and the fact that the excitation starts only at $t = 0$ in the numerical solution. Furthermore, we apply a finite-difference discretization in time to the response quantities at the boundary:

$$u_{,z}(z_0, t_j) = \sum_{i=1}^{i=j-1} \frac{u(z_0, t_{i+1}) - u(z_0, t_i)}{\Delta t} \int_{t_i}^{t_{i+1}} C(t_j - \tau)\mathrm{d}\tau, \tag{16.34}$$

where Δt is the time step, $t_1 = 0$, and the integral of the weakly singular memory function is known analytically:

$$\int_{t_i}^{t_{i+1}} C(t_j - \tau)\mathrm{d}\tau = \sqrt{\frac{\rho}{G_0}}\left[\mathrm{erf}\Big(\sqrt{\tfrac{G_0}{\eta}(t_j - t_i)}\Big) - \mathrm{erf}\Big(\sqrt{\tfrac{G_0}{\eta}(t_j - t_{i+1})}\Big)\right], \tag{16.35}$$

where $t_{i+1} \leq t_j$. Now, the non-reflective boundary condition is no longer an integral equation, but an ordinary differential equation in z, in which the response at t_j is unknown and can be solved for together with the other boundary condition and the equation of motion.

For both 1-D problems, the system of equations is solved using a standard finite-difference discretization of the spatial domain, while the time integration is done using a fourth-order Runge-Kutta scheme. To save computational time, we employ the stationary responses computed using the HBM as initial conditions, which makes that the simulated behaviours reach the stationary states immediately (provided the HBM solution is converged).

References

1. Chouvion, B.: Vibration analysis of beam structures with localized nonlinearities by a wave approach. J. Sound Vib. **439**, 344–361 (2019)
2. Fang, X., Wen, J., Yu, D., Huang, G., Yin, J.: Wave propagation in a nonlinear acoustic metamaterial beam considering the third harmonic generation. New J. Phys. **20**, 123028 (2018)
3. Lombard, B., Piraux, J.: Propagation of compressional elastic waves through a 1-D medium with contact nonlinearities. Ultrasonic Wave Propagation in Non Homogeneous Media, vol. 128, pp. 183–194. Springer (2009)
4. Chronopoulos, D.: Calculation of guided wave interaction with nonlinearities and generation of harmonics in composite structures through a wave finite element method. Compos. Struct. **186**, 375–384 (2018)
5. Kramer, S.L.: Geotechnical Earthquake Engineering. Pearson Education (1996)
6. Jeffrey, A., Engelbrecht, J.: Nonlinear Waves in Solids. Springer (1994)
7. Régnier, J., Bonilla, L., Bard, P.Y., Bertrand, E., Hollender, F., Kawase, H., Sicilia, D., Arduino, P., Amorosi, A., Asimaki, D., Boldini, D., Chen, L., Chiaradonna, A., De Martin, F., Ebrille, M., Elgamal, A., Falcone, G., Foerster, E., Foti, S., Garini, E., Gazetas, G., Gélis, C., Ghofrani, A., Giannakou, A., Gingery, J.R., Glinsky, N., Harmon, J., Hashash, Y., Iai, S., Jeremić, B., Kramer, S., Kontoe, S., Kristek, J., Lanzo, G., Lernia, A., Caballero, F.L., Marot, M., McAllister, G., Mercerat, E.D., Moczo, P., Noguera, S.M., Musgrove, M., Ferro, A.N., Pagliaroli, A., Pisanò, F., Richterova, A., Sajana, S., d'Avila, M.P.S., Shi, J., Silvestri, F., Taiebat, M., Tropeano, G., Verrucci, L., Watanabe, K.: International benchmark on numerical simulations for 1D, nonlinear site response (PRENOLIN): verification phase based on canonical cases. Bull. Seismol. Soc. Am. **106**, 2112–2135 (2016)
8. Polakowski, N., Rifling, E.J.: Strength and Structure of Engineering Materials. Prentice-Hall, New York (1966)
9. Genesio, R., Tesi, A.: Harmonic balance methods for the analysis of chaotic dynamics in nonlinear systems. Automatica **28**(3), 531–548 (1992)

10. Liu, L., Thomas, J.P., Dowell, E.H., Attar, P., Hall, K.C.: A comparison of classical and high dimensional harmonic balance approaches for a duffing oscillator. J. Comp. Phys. **215**(1), 298–320 (2006)
11. Krack, M., Gross, J.: Harmonic Balance for Nonlinear Vibration Problems. Springer (2019)
12. Verruijt, A.: An Introduction to Soil Dynamics. Springer (2010)
13. Lacarbonara, W.: Nonlinear Structural Mechanics. Springer (2013)
14. Hamilton, M.F., Blackstock, D.T.: Nonlinear Acoustics. Academic Press (1997)

Appendix
Biographical Sketch of Dr. Wim van Horssen

Wim van Horssen was born in December 9th 1960, in Delft, The Netherlands. He graduated in 1984 (cum laude) from Delft University of Technology, where he also obtained the Ph.D. degree (thesis: An asymptotic analysis of a class of nonlinear hyperbolic equations) in 1988.

The main research interests of Wim van Horssen are in the field of asymptotic analysis and perturbation theory for nonlinear ODEs, PDEs, difference and delay equations.

Dynamical systems theory and bifurcation theory also belong to the core of Wim's research studies. He has developed perturbation methods to obtain asymptotic approximations for a great variety of linear and nonlinear problems. His work is application-driven and has a direct impact on many engineering problems at the TU Delft positioned on the applied side of the mathematical analysis of PDEs.

Wim van Horssen is in a perfect situation to bridge the gap between the more theoretical research at DIAM of TU Delft and challenges in engineering. The combination of world leading mathematical research with good interaction with engineering create a unique and internationally very strong position of his department in the field of PDE.

Wim van Horssen has a very good scientific track record. He published more than 86 journal papers, almost all in Q1 journals that are cited very frequently. He has a wide scientific network, both in applied mathematics as in engineering, and collaborated with many well-known researchers all over the world. He is invited often as keynote lecturer at international conferences. He is a member of the Editorial Advisory Board of Journal of Sound and Vibration. He also has realized many successful scientific visits at universities all over the world. Among them, University of Washington, Seattle, USA; the Beijing Institute of Technology; the Shanghai Maritime University; Peter the Great Polytechnic University, St. Petersburg, Russia; Universitas Yogyakarta, Indonesia.

Valorisation of the work of Wim van Horssen is accomplished through his many joint research activities with the engineering disciplines. This is visible from his papers focussing on concrete practical problems that are written jointly with

A. K. Abramian et al. (eds.), *Nonlinear Dynamics of Discrete and Continuous Systems*,
Advanced Structured Materials 139,
https://doi.org/10.1007/978-3-030-53006-8

colleagues from other disciplines. He is also often invited as lecturer of post-graduate courses on differential equations. An example of this is the J. M. Burgerscentrum course "Fundamentals and Applications of Perturbation Methods in Fluid Mechanics", organised in 2018.

Wim van Horssen is also an excellent teacher. Evaluations are always very good, both with respect to the courses of the mathematical program as with service teaching for the engineering programmes. Students are also enthusiastic about his supervision of Bachelor and Master Projects. His role as supervisor of Ph.D. students is exceptional: he supervised already 16 Ph.D. students successfully. Wim is not only a brilliant scientist, but also a splendid scientific adviser who is able to create an atmosphere of scientific creativity by generating bright and productive scientific ideas. His enthusiasm, true passion, and uncompromising attitude toward science are transferred to all who have the good fortune to communicate with him. We are amazed by his human and scientific generosity which he shares his time, forces and ideas with us.

His colleagues, collaborators and students are highly appreciated him as an eminent teacher, science manager and outstanding researcher personality.

Printed by Printforce, the Netherlands